High-Performance Liquid Chromatography of Proteins and Peptides

PROCEEDINGS OF THE FIRST INTERNATIONAL SYMPOSIUM

High-Performance Liquid Chromatography of Proteins and Peptides

PROCEEDINGS OF THE FIRST INTERNATIONAL SYMPOSIUM

Edited by

Milton T. W. Hearn
St. Vincent's School of Medical Research
Fitzroy, Victoria
Australia

Fred E. Regnier
Department of Biochemistry
Purdue University
Lafayette, Indiana

C. Timothy Wehr
Varian Instrument
Walnut Creek, California

ACADEMIC PRESS, INC.

(Harcourt Brace Jovanovich, Publishers)

Orlando San Diego San Francisco New York
London Toronto Montreal Sydney Tokyo

ACADEMIC PRESS, INC.
Orlando, Florida 32887

United Kingdom Edition published by
ACADEMIC PRESS, INC. (LONDON) LTD.
24/28 Oval Road, London NW1 7DX

Library of Congress Cataloging in Publication Data

Main entry under title:

High performance liquid chromatography of proteins
    and peptides.

    Includes bibliographical references and index.
    1. Proteins--Analysis--Congresses. 2. Peptides
--Analysis--Congresses. 3. Liquid chromatography--
Congresses. I. Hearn, Milton T. W.
II. Regnier, Fred E. III. Wehr, C. Timothy.
[DNLM: 1. Chromatography, High pressure liquid--
Congresses. 2. Proteins--Analysis--Congresses.
3. Peptides--Analysis--Congresses. QU 55 H6383
1982]
QP551.H49 1983   574.19'245   83-3867
ISBN 0-12-335780-2

PRINTED IN THE UNITED STATES OF AMERICA

84 85 86 87   9 8 7 6 5 4 3 2

Contents

Contributors   ix

Preface   xv

High-Performance Ion-Exchange Chromatography of Proteins: The Current Status   1

FRED E. REGNIER

Analysis of Peptic Fragmentation of Human Immunoglobulin G Using High-Performance Liquid Chromatography   9

TSUGIKAZU TOMONO, TOHRU SUZUKI, AND EIICHI TOKUNAGA

Cord Blood Screening for Hemoglobin Disorders by High-Performance Liquid Chromatography   17

ADOOR AMANULLAH, SAMIR HANASH, KIRSTEN BUNNELL, JOHN STRAHLER, DONALD L. RUCKNAGEL, AND STEVEN J. FERRUCI

A Simple and Rapid Purification of Commercial Trypsin and Chymotrypsin by Reverse-Phase High-Performance Liquid Chromatography   23

KOITI TITANI, TATSURU SASAGAWA, KATHERYN RESING, AND KENNETH A. WALSH

Adaptation of Reverse-Phase High-Performance Liquid Chromatography for the Isolation and Sequence Analysis of Peptides from Plasma Amyloid P-Component   29

JACQUELINE K. ANDERSON AND JOHN E. MOLE

Characterization of Human Alcohol Dehydrogenase Isoenzymes by High-Performance Liquid Chromatographic Peptide Mapping   39

DANIEL J. STRYDOM AND BERT L. VALLEE

Purification of Glycopeptides of Human Ceruloplasmin and Immunoglobulin D by High-Pressure Liquid Chromatography   47

DANIEL TETAERT, NOBUHIRO TAKAHASHI, AND FRANK W. PUTNAM

High-Performance Size-Exclusion Chromatography of Hydrolyzed Plant Proteins   55

HOWARD G. BARTH

The Isolation of Peptides by High-Performance Liquid Chromatography Using Predicted Elution Positions   65
C. A. BROWNE, H. P. J. BENNETT, AND S. SOLOMON

The Application of High-Performance Liquid Chromatography for the Resolution of Proteins Encoded by the Human Adenovirus Type 2 Cell Transformation Region   73
MAURICE GREEN AND KARL H. BRACKMANN

The Importance of Silica Type for Reverse-Phase Protein Separations   81
JAMES D. PEARSON, NAN T. LIN, AND FRED E. REGNIER

Variables in the High-Pressure Cation-Exchange Chromatography of Proteins   95
CHARLES A. FROLIK, LINDA L. DART, AND MICHAEL B. SPORN

Application of High-Performance Liquid Chromatography to Competitive Labeling Studies: The Chemical Properties of Functional Groups of Glucagon   103
STEPHEN A. COCKLE, HARVEY KAPLAN, MARY A. HEFFORD, AND N. MARTIN YOUNG

Purification by Reverse-Phase High-Performance Liquid Chromatography of an Epidermal Growth Factor-Dependent Transforming Growth Factor   111
MARIO A. ANZANO, ANITA B. ROBERTS, JOSEPH M. SMITH, LOIS C. LAMB, AND MICHAEL B. SPORN

Preparative Size-Exclusion Chromatography of Human Serum Apolipoproteins Using an Analytical Liquid Chromatograph   119
C. TIMOTHY WEHR, ROBERT L. CUNICO, GARY S. OTT, AND VIRGIE G. SHORE

Use of Reverse-Phase High-Performance Liquid Chromatography in Structural Studies of Neurophysins, Photolabeled Derivatives, and Biosynthetic Precursors   129
DAVID M. ABERCROMBIE, CHRISTOPHER J. HOUGH, JONATHAN R. SEEMAN, MICHAEL J. BROWNSTEIN, HAROLD GAINER, JAMES T. RUSSELL, AND IRWIN M. CHAIKEN

Purification of Radiolabeled and Native Polypeptides by Gel Permeation High-Performance Liquid Chromatography   141
C. LAZURE, M. DENNIS, J. ROCHEMONT, N. G. SEIDAH, AND M. CHRÉTIEN

Nonideal Size-Exclusion Chromatography of Proteins: Effects of pH at
Low Ionic Strength   151
W. KOPACIEWICZ AND FRED E. REGNIER

Factors Influencing Chromatography of Proteins on Short
Alkylsilane-Bonded Large Pore-Size Silicas   161
M. J. O'HARE, M. W. CAPP, E. C. NICE, N. H. C. COOKE, AND B. G. ARCHER

High-Performance Liquid Chromatography Analysis in the Synthesis,
Characterization, and Reactions of Neoglycopeptides   173
HARVARD MOREHEAD, PATRICK MC KAY, AND RONALD WETZEL

Use of Size-Exclusion High-Performance Liquid Chromatography for the
Analysis of the Activation of Prothrombin   181
DAVID P. KOSOW, SAM MORRIS, CAROLYN L. ORTHNER, AND
MOO-JHONG RHEE

High-Performance Liquid Chromatography of Proteins: Purification of the
Acidic Isozyme of Adenylosuccinate Synthetase from Rat Liver   189
FREDERICK B. RUDOLPH AND SANDRA W. CLARK

Application of High-Performance Liquid Chromatographic Peptide
Purification to Protein Microsequencing by Solid-Phase Edman
Degradation   195
JAMES J. L'ITALIEN AND JAMES E. STRICKLER

Measurement of Endogenous Leucine Enkephalin in Canine Caudate
Nuclei and Hypothalami with High-Performance Liquid Chromatography
and Field-Desorption Mass Spectrometry   211
SHIGETO YAMADA AND DOMINIC M. DESIDERIO

Fluorescent Techniques for the Selective Detection of Chromatographically
Separated Peptides   221
TIMOTHY D. SCHLABACH AND C. TIMOTHY WEHR

High-Performance Liquid Chromatographic Purification of
Peptide Hormones: Ovine Hypothalamic Amunine (Corticotropin Releasing
Factor)   233
J. RIVIER, C. RIVIER, J. SPIESS, AND W. VALE

Isolation and Purification of Escherichia coli Heat-Stable Enterotoxin of
Porcine Origin   243
R. LALLIER, F. BERNARD, M. GENDREAU, C. LAZURE, N. G. SEIDAH,
M. CHRÉTIEN, AND S. A. ST-PIERRE

viii   CONTENTS

α-N-*Acetyl*-β-*Endorphin*$_{1-26}$ *from the Neurointermediary Lobe of the Rat Pituitary: Isolation, Purification, and Characterization by High-Performance Liquid Chromatography*   253
H. P. J. BENNETT, C. A. BROWNE, AND S. SOLOMON

Index   *263*

Contributors

Numbers in parentheses indicate the pages on which the authors' contributions begin.

David M. Abercrombie (129), Laboratory of Chemical Biology, National Institute of Arthritis, Diabetes, and Digestive and Kidney Diseases, and Department of Chemistry, University of Maryland, College Park, Maryland 20742

Adoor Amanullah (17), Department of Pediatrics and Human Genetics, University of Michigan School of Medicine, Ann Arbor, Michigan 48109

Jacqueline K. Anderson (29), Department of Biochemistry, University of Massachusetts Medical School, Worcester, Massachusetts 01605

Mario A. Anzano (111), Laboratory of Chemoprevention, National Cancer Institute, Bethesda, Maryland 20205

B. G. Archer (161), Altex Scientific, Subsidiary of Beckman Instruments, Berkeley, California 94710

Howard G. Barth (55), Hercules Incorporated, Research Center, Wilmington, Delaware 19899

H. P. J. Bennett (65, 253), Endocrine Laboratory, Royal Victoria Hospital, and Departments of Biochemistry and Experimental Medicine, McGill University, Montreal, Quebec, Canada

F. Bernard (243), Département de Pathologie et Microbiologie, Faculté de Médecine Vétérinaire, Université de Montréal, St-Hyacinthe, Montreal, Canada

Karl H. Brackmann (73), Institute for Molecular Virology, St. Louis University Medical Center, St. Louis, Missouri 63110

C. A. Browne[1] (65, 253), Endocrine Laboratory, Royal Victoria Hospital, and Departments of Biochemistry and Experimental Medicine, McGill University, Montreal, Quebec, Canada

Michael J. Brownstein (129), Laboratory of Clinical Science, National Institute of Mental Health, Bethesda, Maryland 20205

Kirsten Bunnell (17), Department of Pediatrics and Human Genetics, University of Michigan School of Medicine, Ann Arbor, Michigan 48109

M. W. Capp (161), Ludwig Institute for Cancer Research (London Branch), Royal Marsden Hospital, Sutton, Surrey SM2 5PX, Great Britain

[1]Present address: Department of Physiology, Monash University, Clayton, Victoria 3168, Australia.

Irwin M. Chaiken (129), Laboratory of Chemical Biology, National Institute of Arthritis, Diabetes, and Digestive and Kidney Diseases, College Park, Maryland 20742

M. Chrétien (141, 243), Institute de Recherche Clinique, Montreal, Canada

Sandra W. Clark (189), Department of Biochemistry, Rice University, Houston, Texas 77001

Stephen A. Cockle[2] (103), Division of Biological Sciences, National Research Council of Canada, Ottawa K1A 0R6, Canada

N. H. C. Cooke (161), Altex Scientific, Subsidiary of Beckman Instruments, Berkeley, California 94710

Robert L. Cunico (119), Varian Instrument Group, Walnut Creek Instrument Division, Walnut Creek, California 94598

Linda L. Dart (95), Laboratory of Chemoprevention, National Cancer Institute, Bethesda, Maryland 20205

M. Dennis (141), Clinical Research Institute of Montreal, Montreal H2W 1R7, Canada

Dominic M. Desiderio (211), Department of Neurology, University of Tennessee, Center for Health Sciences, Memphis, Tennessee 38163

Steven J. Ferruci (17), Department of Pediatrics and Human Genetics, University of Michigan School of Medicine, Ann Arbor, Michigan 48109

Charles A. Frolik (95), Laboratory of Chemoprevention, National Cancer Institute, Bethesda, Maryland 20205

Harold Gainer (129), Laboratory of Developmental Neurobiology, National Institute of Child Health and Human Development, National Institutes of Health, Bethesda, Maryland 20205

M. Gendreau (243), Département de Physiologie et Pharmacologie, Faculté de Médecine, Université de Sherbrooke, Sherbrooke, Canada

Maurice Green (73), Institute for Molecular Virology, St. Louis University Medical Center, St. Louis, Missouri 63110

Samir Hanash (17), Department of Pediatrics and Human Genetics, University of Michigan School of Medicine, Ann Arbor, Michigan 48109

Mary A. Hefford (103), Department of Biochemistry, University of Ottawa, Ottawa K1N 6N5, Canada

Christopher J. Hough (129), Laboratory of Chemical Biology, National Institute of Arthritis, Diabetes, and Digestive and Kidney Diseases, College Park, Maryland 20742

Harvey Kaplan (103), Department of Biochemistry, University of Ottawa, Ottawa K1N 6N5, Canada

W. Kopaciewicz (151), Department of Biochemistry, Purdue University, West Lafayette, Indiana 47907

[2]Present address: Department of Chemistry, University of Winnipeg, Winnipeg, MB, R3B 2E9, Canada.

David P. Kosow (181), Plasma Derivatives Laboratory, American Red Cross Blood Services Laboratories, Bethesda, Maryland 20814

James J. L'Italien[3] (195), Department of Internal Medicine, Yale University School of Medicine, New Haven, Connecticut 06510

R. Lallier (243), Département de Pathologie et Microbiologie, Faculté de Médecine Vétérinaire, Université de Montréal, St-Hyacinthe, Montreal, Canada

Lois C. Lamb (111), Laboratory of Chemoprevention, National Cancer Institute, Bethesda, Maryland 20205

C. Lazure (141, 243), Institute de Recherche Clinique, Montreal, and Chercheur Boursier, Conseil de la Recherche en Santé du Quebec, Quebec, Canada

Nan T. Lin (81), Department of Biochemistry, Purdue University, West Lafayette, Indiana 47907

Patrick McKay (173), Protein Biochemistry Department, Genentech, Inc., South San Francisco, California 94080

John E. Mole (29), Department of Biochemistry, University of Massachusetts Medical School, Worcester, Massachusetts 01605

Harvard Morehead (173), Protein Biochemistry Department, Genentech, Inc., South San Francisco, California 94080

Sam Morris (181), Plasma Derivatives Laboratory, American Red Cross Blood Services Laboratories, Bethesda, Maryland 20814

E. C. Nice (161), Ludwig Institute for Cancer Research (London Branch), Royal Marsden Hospital, Sutton, Surrey SM2 5PX, Great Britain

M. J. O'Hare (161), Ludwig Institute for Cancer Research (London Branch), Royal Marsden Hospital, Sutton, Surrey SM2 5PX, Great Britain

Carolyn L. Orthner (181), Plasma Derivatives Laboratory, American Red Cross Blood Services Laboratories, Bethesda, Maryland 20814

Gary S. Ott[4] (119), Lawrence Livermore National Laboratory, Biomedical Sciences Division, University of California, Livermore, California 94550

James D. Pearson (81), Department of Biochemistry, Purdue University, West Lafayette, Indiana 47907

Frank W. Putnam (47), Department of Biology, Indiana University, Bloomington, Indiana 47405

Fred E. Regnier (1, 81, 151), Department of Biochemistry, Purdue University, West Lafayette, Indiana 47907

Katheryn Resing (23), Howard Hughes Medical Institute Laboratory and Department of Biochemistry, University of Washington, Seattle, Washington 98195

[3]Present address: Molecular Genetics, Inc., Minnetonka, Minnesota 55343.
[4]Present address: Bio-Rad Laboratories, Richmond, California 94804.

Moo-Jhong Rhee (181), Plasma Derivatives Laboratory, American Red Cross Blood Services Laboratories, Bethesda, Maryland 20814

J. Rivier (233), The Salk Institute, San Diego, California 92038

C. Rivier (233), The Salk Institute, San Diego, California 92038

Anita B. Roberts (111), Laboratory of Chemoprevention, National Cancer Institute, Bethesda, Maryland 20205

J. Rochemont (141), Clinical Research Institute of Montreal, Montreal H2W 1R7, Canada

Donald L. Rucknagel (17), Department of Pediatrics and Human Genetics, University of Michigan School of Medicine, Ann Arbor, Michigan 48109

Frederick B. Rudolph (189), Department of Biochemistry, Rice University, Houston, Texas 77001

James T. Russell (129), Laboratory of Developmental Neurobiology, National Institute of Child Health and Human Development, National Institutes of Health, Bethesda, Maryland 20205

Tatsuru Sasagawa (23), Howard Hughes Medical Institute Laboratory and Department of Biochemistry, University of Washington, Seattle, Washington 98195

Timothy D. Schlabach (221), Varian Instrument Group, Walnut Creek Instrument Division, Walnut Creek, California 94598

Jonathan R. Seeman (129), Laboratory of Chemical Biology, National Institute of Arthritis, Diabetes, and Digestive and Kidney Diseases, College Park, Maryland 20742

N. G. Seidah (141, 243), Institute de Recherche Clinique, Montreal, Canada

Virgie G. Shore (119), Lawrence Livermore National Laboratory, Biomedical Sciences Division, University of California, Livermore, California 94550

Joseph M. Smith (111), Laboratory of Chemoprevention, National Cancer Institute, Bethesda, Maryland 20205

S. Solomon (65, 253), Endocrine Laboratory, Royal Victoria Hospital, and Departments of Biochemistry and Experimental Medicine, McGill University, Montreal, Quebec, Canada

J. Spiess (233), The Salk Institute, San Diego, California 92038

Michael B. Sporn (95, 111), Laboratory of Chemoprevention, National Cancer Institute, Bethesda, Maryland 20205

S. A. St-Pierre (243), Département de Physiologie et Pharmacologie, Faculté de Médecine, Université de Sherbrooke, Sherbrooke, Canada, and Chercheur Boursier, Conseil de la Recherche en Santé du Québec, Quebec, Canada

John Strahler (17), Department of Pediatrics and Human Genetics, University of Michigan School of Medicine, Ann Arbor, Michigan 48109

James E. Strickler (195), Department of Internal Medicine, Yale University School of Medicine, New Haven, Connecticut 06510

Daniel J. Strydom (39), Center for Biochemical and Biophysical Sciences and Medicine, Harvard Medical School, Boston, Massachusetts 02115

Tohru Suzuki (9), Plasma Fractionation Department, The Japanese Red Cross Central Blood Center, 1-31, Hiroo 4, Shibuya-ku, Tokyo 150, Japan

Nobuhiro Takahashi (47), Department of Biology, Indiana University, Bloomington, Indiana 47405

Daniel Tetaert[5] (47), Department of Biology, Indiana University, Bloomington, Indiana 47405

Koiti Titani (23), Howard Hughes Medical Institute Laboratory and Department of Biochemistry, University of Washington, Seattle, Washington 98195

Eiichi Tokunaga (9), Plasma Fractionation Department, The Japanese Red Cross Central Blood Center, 1-31, Hiroo 4, Shibuya-ku, Tokyo 150, Japan

Tsugikazu Tomono (9), Plasma Fractionation Department, The Japanese Red Cross Central Blood Center, 1-31, Hiroo 4, Shibuya-ku, Tokyo 150, Japan

W. Vale (233), The Salk Institute, San Diego, California 92038

Bert L. Vallee (39), Center for Biochemical and Biophysical Sciences and Medicine, Harvard Medical School, Boston, Massachusetts 02115

Kenneth A. Walsh (23), Howard Hughes Medical Institute Laboratory and Department of Biochemistry, University of Washington, Seattle, Washington 98195

C. Timothy Wehr (119, 221), Varian Instrument Group, Walnut Creek Instrument Division, Walnut Creek, California 94598

Ronald Wetzel (173), Protein Biochemistry Department, Genentech, Inc., South San Francisco, California 94080

Shigeto Yamada (211), Stout Neuroscience Mass Spectrometry Laboratory, University of Tennessee, Center for Health Sciences, 800 Madison Avenue, Memphis, Tennessee 38163

N. Martin Young (103), Division of Biological Sciences, National Research Council of Canada, Ottawa K1A 0R6, Canada

[5]Present address: Unite INSERM 16, Place de Verdun, F59045 Lille Cedex, France.

Preface

The first International Symposium on HPLC of Proteins and Peptides, held in Washington, D.C., from November 16 to 17, 1981, was attended by over 400 participants representing North American, European, Asian, and Australian research centers in universities, government institutes, and industry. Topics ranged from the physicochemical basis of peptide retention with chemically bonded hydrocarbonaceous silicas to the isolation of biologically active peptides from tissue extracts. The symposium encompassed six sessions covering size exclusion, ion exchange, and reversed phase chromatography, as well as the use of HPLC techniques in protein structural studies and peptide isolation. A final session was dedicated to detector technology and specialized separation methods.

Two themes became apparent during the course of this first symposium. First, the application of HPLC technologies is revolutionizing the analysis, characterization, and isolation of peptides and proteins. During the past decade, methods of biochemical analysis have undergone major development and refinement. The ready acceptance by the scientific community of HPLC techniques in particular reflects in no small measure the excellent resolution and recovery that can be achieved with these sophisticated separation methods, which can now be applied at the microanalytical level to the fully preparative capacity scale. Second, recent emphasis placed by a number of research groups on the theoretical aspects of the technique has allowed rational approaches to be developed for the evaluation of mobile phase and stationary phase parameters, thus giving wider meaning to the more numerous, but largely empirical, applications of HPLC involved with biological studies. In this context, developments in our understanding of column technology and separation mechanisms have proved most important.

In this book, we have assembled 28 selected papers from the lecture and poster communications presented at this first symposium. Not unexpectedly, contributions involving reversed phase HPLC studies dominate this collection. However, each of the included papers should help the reader become more familiar with the concepts, methodologies, and instrumentation now in vogue in this field. This volume should consequently be of wide interest to all researchers contemplating, or currently engaged in, the use of HPLC techniques for the isolation of less abundant peptide, polypeptide, and protein components from biological extracts, for the biochemical analysis and structure determination of polypeptides and proteins, or for the purification of synthetic peptides. High-performance liquid chromatography in its various elution modes utilizing chemically bonded microparticulate silicas has unquestionably emerged, following the first systematic reports in 1976, as the technique par excellence for the rapid, sensitive, and selective separation of amino acids, peptides, and many types of protein substances. The papers collected here serve to highlight the profound impact these new techniques are having on biochemistry and the attendant biomedical disciplines. From the studies described at this first symposium, and from the fruitful dialogs that ensued during the round-table sessions, it can be confidently concluded that HPLC of peptides and proteins will remain a fertile field of investigation for many years.

REVIEW

High-Performance Ion-Exchange Chromatography of Proteins: The Current Status[1]

FRED E. REGNIER

Department of Biochemistry, Purdue University, West Lafayette, Indiana 47907

Ion-exchange chromatography of proteins and peptides has been most successfully achieved historically on hydrophilic gel matrices. The poor mechanical strength of these organic gels has necessitated the development of new supports for high-performance separations. High-performance supports are of three types: totally inorganic, totally organic, and composite inorganic–organic materials. Several ionic species such as diethylaminoethyl ethanol and polyethylene imine have been used as stationary phases with similar results. Pore-diameter selection has been shown to be important in both resolution and loading capacity. Capacity is maximum for proteins of 50 to 100 kilodaltons on 300-Å-pore-diameter supports. Maximum resolution of high-molecular-weight species also requires macroporous supports. Interestingly, column length is of minor importance in the resolution of proteins. Columns of 5-cm length have approximately the same resolution as those of 30-cm length. Application of high-performance ion-exchange chromatography to a variety of protein mixtures has now been reported. These supports generally give recoveries of enzyme activity equivalent to the classical supports.

A major limitation of conventional gel-type ion-exchange media is their lack of mechanical strength. It is obvious that the high-mobile-phase velocities used in high-performance liquid chromatography (HPLC) require packing materials that are rigid. Interestingly, the first rigid ion-exchange supports for proteins were prepared before diethylaminoethyl (DEAE)- and carboxymethyl (CM)[2]-cellulose. In 1954, Boardman (1,2) prepared a cation exchanger in a two-step process consisting of styrene polymerization on the surface of diatomaceous earth followed by sulfonation of the polystyrene matrix. This strong cation exchanger was used in the chromatography of chymotrypsinogen and cytochrome *c*. Unfortunately, the very hydrophobic nature of polystyrene causes an irreversible adsorption of many proteins and limits the utility of this packing in protein separations. More recently, Eltekov (3) prepared rigid anion-exchange packings by derivatizing the surface of controlled porosity glass with aminopropyltrimethoxy silane (APS). The resulting weak anion exchanger separated proteins but showed a high degree of nonspecific adsorption. This undesirable property is probably the result of residual silanols on the support surface. When an inorganic surface is exposed to this silylating agent, two types of bonding may occur simultaneously: (i) siloxane bond formation at the surface and (ii) ion pairing of the APS amino groups with surface silanols. Siloxane bond formation and the concomitant sequestering of surface silanol is inhibited by ion pairing. The undesirable properties of packings with a high level of surface silanols have been noted repeatedly (4–7).

The propensity of APS monomer to form

[1] This is Journal Paper No. 8930 from the Purdue University Agricultural Experiment Station.

[2] Abbreviations used: CM, carboxymethyl; APS, aminopropyltrimethoxy silane; HPIEC, high-performance ion-exchange chromatography; PEI, polyethylenimine; SEC, size-exclusion chromatography; BSA, bovine serum albumin.

siloxane polymers in the presence of traces of water is another property of triethoxysilanes that has bearing on the quality of ion-exchange coatings (8). Unfortunately, these siloxane polymers are not completely stable when eluted with thousands of volumes of aqueous buffer. Aminopropyl silane monomer gradually leaks from the surface of inorganic supports. The erosion of ion-exchange stationary phase causes both the ion-exchange capacity and separation characteristics of columns to change with aging. Chang (9,10) found that thin surface layers of organic polymers produce far more stable ion exchangers than organosilane monomer bonded phases. Both epoxy and polyamine polymers produced packings that behaved like the conventional gel-type ion-exchange packings used in the chromatography of proteins but achieved separations in less than 30 min. Based on these studies it was concluded (10) that microparticulate inorganic supports coated with a thin film of ion-exchanging organic polymer had a substantial advantage in protein separations over both gel-type and organosilane monomer ion-exchange packings.

PACKING MATERIALS

The extensive literature on both conventional gel-type ion-exchange resins for proteins and HPLC packings indicate that the ideal HPIEC packing for proteins should be (i) mechanically stable to mobile phase velocities of 1 mm/s, (ii) completely hydrophilic, (iii) of high ion-exchange capacity, (iv) chemically stable over a broad pH range, (v) available in 5- to 10-μm particle size, (vi) available in pore diameters from 300 to 1000 Å, (vii) spherical, (viii) of a pore volume between 0.5 and 1.0 ml/g, (ix) easy to pack, and (x) inexpensive. No packing material available at the present time has all of these properties.

Available HPIEC materials for proteins are of three types: (i) organic resins, (ii) composite inorganic–organic matrices, and (iii) inorganic supports with a thin surface layer of organic bonded phase. More experience with HPIEC is necessary to judge which of these materials is superior.

Organic Resins

The only known organic ion-exchange resin that can be classed as an HPLC packing for proteins is the polymethacrylate-based Spheron (11) material of Lachema. Separation of proteins on both the CM and DEAE versions of Spheron have been reported by Mikes et al. (12,13). Unfortunately, these ion exchangers are not commercially available in less than 20-μm particle size. The weak cation-exchange resin Bio-Rex 70 is also useful in protein separations but is only available in greater than 40-μm particle size (1,14–16). Van der Wal (17) has examined the utility of several nonrigid ion exchangers in the fractionation of cytochrome c and its derivatives by HPIEC. Cation exchangers with a polystyrene–divinylbenzene or cellulose matrix were generally found to be unsatisfactory. Polymethacrylate matrices had much greater utility. In general, retention increased as either pH or ionic strength was decreased. It was reported that temperature had minimal effect on retention of derivatized cytochrome c.

Composite Organic–Inorganic Packings

Vanecek and Regnier (18) have recently reported the preparation of weak anion-exchange packings with polymer layers 100 Å thick in 1000- and 4000-Å pores. Since the whole polymer layer exchanged ions and was permeable to small molecules, the ion-exchange capacity of the packings was much higher than that of a silica support with a surface monolayer of ion-exchange groups. After heavy polymer layers in the silica-support pores were bonded, the functional porosity to macromolecules was still greater than 600 Å with an ion-exchange capacity of 300 meq/g. The function of the inorganic support was to provide a rigid matrix within which gel-type ion-exchangers could survive the pressure of a high-performance system.

The principal advantage of packings with such large pores is said to be in the analysis of proteins exceeding 10^5 daltons. Determination of the ultimate utility of these composite materials will require much more testing.

Surface-Modified Inorganic Packings

The most widely used technique for the preparation of HPIEC packings for proteins is through the application of a 20- to 30-Å polymer layer to the surface of inorganic supports. Ion-exchange groups are either coupled to the surface (i) during the polymerization reaction, (ii) after the polymer layer has been deposited, or (iii) as preformed ionic polymers. An example of the first technique may be found in Chang's preparation of an inorganic DEAE ion exchanger (19). Deposition of diethylaminoethanol and a multifunctional oxirane on the surface of a glycidoxypropylsilane-bonded phase support followed by polymerization caused both DEAE and glyceryl residues to become incorporated into the epoxy polymer formed. The DEAE groups served the function of partitioning proteins electrostatically while the glyceryl residues anchored the coating to the surface at multiple sites.

Gupta (20) has reported a second synthetic route to polymeric ion-exchange coatings. The first step in this process is the bonding of a simple organosilane monomer with a reactive organic functional group to the surface of a silica support. This silylation reaction is followed by the covalent attachment of polyethylenimine (PEI) at multiple sites to the silica support through functional groups on the organosilanes. This PEI packing may either be used directly as a weak anion exchanger, further derivatized with oxiranes to form a tertiary amine ion exchanger, or reacted with acid anhydrides (succinic or diglycolic) to produce weak cation exchangers. A single synthetic route is used in the preparation of both anion- and cation-exchange packings.

A third type of HPIEC packing has been described by Alpert (21). In this process, PEI is first adsorbed onto the surface of inorganic supports as a monolayer and then crosslinked into a thin polyamine skin with multifunctional oxiranes. This crosslinked coating is adsorbed to the surface at many sites and held so tightly that it cannot be desorbed even under severe conditions. The high surface density of amino groups completely titrates all surface silanols and prevents them from interacting with proteins.

In 1978, SynChrom introduced the first commercial HPIEC supports for proteins. These weak anion exchangers had a polyamine-bonded phase that produced separations similar to DEAE-cellulose. Although the materials were available in 100-, 300-, 500-, and 1000-Å-pore-diameter sizes, the 300-Å-pore-diameter packing was the most widely used because of its greater loading capacity. This weak anion exchanger was followed by a weak cation-exchanging CM support in 1982. Pharmacia and Toya Soda Corporation also introduced complete lines of HPIEC columns for proteins in 1982. Although these columns are so new that few people have examined them, they are reported to give separations equivalent to the conventional CM and DEAE columns. It should also be noted that the strong cation-exchange material from Whatman, Partisil SCX, has been used effectively in the separation of low-molecular-weight proteins.

PORE-DIAMETER EFFECTS

The ratio of support pore diameter to molecular size can contribute to both the resolution and loading capacity of a column.

Resolution

Retention in ion-exchange chromatography is controlled by two independent phenomena: (i) the inherent size-exclusion contribution from differential penetration by solutes of macroporous matrices and (ii) electrostatic partitioning at the surface of the ion exchanger. It has been observed in size-exclusion chromatography (SEC) that as solute dimensions approach those of the support pores, severe limitations of molecular diffusion occur within the pores (22). The net effect of this phenomenon is resistance to mass transfer between the stationary and mobile phases and a concomitant loss of resolution. The inherent SEC properties of HPIEC packings cause this same bandspreading phenomenon to occur as solute size approaches pore diameter. These pore-diameter limitations of stagnant mobile-phase mass transfer are easily overcome in HPIEC by using larger pore-diameter packings (18). The optimum relationship between pore diameter and solute size has not been determined. Preliminary studies would indicate that 300-Å-pore-diameter packings give less than optimum resolution with proteins greater than 10^5 daltons.

Loading Capacity

Because ion-exchange chromatography is a surface-directed process, it would seem that ion-exchange capacity should be proportional to the surface area of a packing. Actually, not all of the surface of a packing is available to macromolecules. Greater than 95% of the surface area of a porous support is inside the pore network, and a molecule must be able to penetrate the pore matrix to reach the surface. The surface area (A_a) available for ion-exchange partitioning is expressed by the equation

$$A_a = A_0 + K_a A_i$$

where A_0 is the surface area on the outer surface of particles in a bed, A_i is the internal surface area of the pores of particles in the bed, and K_a is a surface accessability coefficient. The term K_a indicates the fraction of the internal surface available for ion-exchange adsorption and thus ranges from 0 to 1. Obviously, K_a will be different for each molecular species and approach zero as molecular dimensions approach the dimensions of pores. For molecules such as hemoglobin and bovine serum albumin (BSA), 250- to 300-Å-pore-diameter packings have the

greater value of $K_a A_i$ and therefore give the highest ion-exchange capacity. A BSA loading capacity of 100 mg/g of packing is common (23). Smaller proteins such as carbonic anhydrase exhibit highest loading capacity on 100-Å-pore-diameter packings as opposed to proteins of several hundred thousand kilodaltons that require 500- to 1000-Å pores for maximum loading.

For both loading capacity and resolution, 250- to 300-Å-pore-diameter ion-exchange packings are probably of broadest utility because they bracket the 50- to 100-kilodalton range of proteins. A 0.41 × 25-cm column will show no signs of overloading with up to 10 mg of ovalbumin (23). Under overloading conditions, the column will still separate most impurities in an ovalbumin sample even at a 50-mg injection.

COLUMN LENGTH

It has been determined that column length plays a minimal role in resolution of proteins (23). Resolution on 5-cm columns is almost equivalent to that of 25-cm columns. These small columns have some definite advantages in analytical work such as (i) the elution of proteins in a smaller elution volume and therefore a more concentrated form, (ii) up to 6 times greater sensitivity because eluants are more concentrated, (iii) fewer problems with underloading columns, (iv) lower operating pressures, (v) columns last longer, and (vi) columns are cheaper. The disadvantages of short (<5 cm) columns are that maximum loading capacity is a few milligrams of protein and the small elution volumes promote extracolumn band spreading in some equipment. Both low dead-volume connecting tubing and flow cells are essential.

MOBILE-PHASE VELOCITY

The influence of mobile-phase velocity on band spreading and resolution has been treated in depth by Giddings (24) and is discussed in a number of books (25–27). The Giddings treatment indicates that as the diffusion coefficient of a molecule decreases, its ability to diffuse into and out of support pores decreases accordingly. Thus, as molecular weight increases, mass transfer problems are expected to escalate because of the decreasing diffusion coefficients. Obviously increasing mobile-phase velocity in a column aggravates the mass transfer problem. It has been calculated that to obtain equivalent column efficiency, mobile-phase velocity must be 10 times slower with a molecule of 70 kilodaltons than one of several hundred daltons (28). Practically speaking, this means

TABLE 1

PROTEINS PURIFIED BY HIGH-PERFORMANCE ION-EXCHANGE CHROMATOGRAPHY

Protein	Column	Type of ion exchanger	References
Lactate dehydrogenase	DEAE-glycophase[d]	WAX[a]	(10,19,30,31,33)
isoenzymes	SynChropak AX300[e]	WAX	(30)
Creatine kinase	DEAE-glcyophase	WAX	(31–34)
isoenzymes	SynChropak AX300	WAX	(30,35)
Alkaline phosphatases	DEAE-glycophase	WAX	(36)
Hexokinase isoenzymes	SynChropak AX300	WAX	(37)
Arylsulfatase isoenzymes	DEAE-glycophase	WAX	(38)
Hemoglobins	SynChropak AX300	WAX	(39,41,43,45)
	DEAE-glycophase	WAX	(10,19)
	IEX 545 DEAE[f]	WAX	(44)
	IEX 535 CM[f]	WCX[b]	(44)
	Bio-Rex 70[g]	WCX	(42)
Cytochrome c	CM-polyamide	WCX	(20)
Lysozyme	CM-polyamide	WCX	(20)
Myoglobin	CM-polyamide	WCX	(20)
	IEX 535 CM	WCX	(46)
Soybean trypsin inhibitor	CM-glycophase[d]	WCX	(19)
Interferon	Partisil SCX[h]	SCX[c]	(48)
Lipoxygenase	SynChropak AX300	WAX	(18)
Trypsin	DEAE-glycophase	WAX	(19)
Chymotrypsinogen	SP-glycophase	SCX	(19)
	IEX 535 CM	WCX	(46)
Immunoglobulin G	SynChropak AX300	WAX	(54)
Ovalbumin	SynChropak AX300	WAX	(23)
	IEX 545 DEAE	WAX	(47)
Albumin	SynChropak AX300	WAX	(23)
	DEAE-Glycophase	WAX	(19)
	IEX 545 DEAE	WAX	(46)
Apolipoproteins	SynChropak AX300	WAX	(50,55)
Adenylsuccinate synthetase	SynChropak AX300	WAX	(40)
Insulin	Partisil SCX	SCX	(49)
	IEX 535 CM	WCX	(46)
β-Lactoglobulin	Partisil SCX	SCX	(49)
Carbonic anhydrase	Partisil SCX	SCX	(49)
Monoamine oxidase	SynChropak AX300	WAX	(51)

[a] WAX designates weak anion exchanger.
[b] WCX designates weak cation exchanger.
[c] SCX indicates strong cation exchanger.
[d] DEAE and CM-glycophase are products of Pierce Chemical Company, Rockford, Illinois.
[e] SynChropak AX300 is produced by SynChrom, Linden, Indiana.
[f] IEX 535 CM and 545 DEAE are the products of Toya Soda Corporation, Yamaguchi, Japan.
[g] Bio-Rex 70 is supplied by Bio-Rad, San Francisco, California.
[h] Partisil SCX is manufactured by Whatman, Clifton, New Jersey.

that good resolution will only be obtained at a linear velocity of 0.1 mm/s or less. Columns of 0.41 × 25 cm will give best resolution at 0.25 ml/min when developed in a gradient of several hours duration (23). Although molecules of 200 kilodaltons may be eluted from columns in less than 10 min, a very large price is paid in resolution. In most cases, the demands on the column for resolution are sufficiently low that 0.5- to 1-ml/min flow rates and 30-min total analysis times are tolerable.

MOBILE PHASES

Relatively little work has been done on mobile-phase selection in HPIEC. Because most commercial columns are silica based, there is an upper limit of pH 8.0 to prevent destruction of the silica support matrix. Other than this limitation, column packing materials will withstand a broad range of organic and aqueous mobile phases.

Ion-exchange chromatography of proteins is based on the use of mobile-phase pH to manipulate the charge on a protein. At any pH above its pI a protein will have a net negative charge as opposed to a positive charge below the pI. Depending on the pI of a protein, one selects either an anion-exchange or cation-exchange column. Proteins with a high pI are best separated on cation exchangers, and those with a low pI separate best on anion exchangers. Both cation- and anion-exchange columns may be eluted with an ionic strength or pH gradient.

A variety of mobile phases have been used in the elution of conventional ion-exchange columns that function well in HPIEC columns. The elution profiles obtained with both the conventional and HPIEC columns are usually very similar. Differences will be in ionic strength required for elution, resolution, and separation time. Because the ionic strength required for protein desorption is a function of a packing's ligand density and it is unlikely that different manufacturers materials will be the same, only relative elution position in chromatograms can be compared.

As a general rule, recoveries of enzyme activity from HPIEC columns are equivalent to those obtained with conventional ion-exchange columns (21,30). The short residence time on HPIEC columns will also be useful in recovery of enzyme activity from very labile proteins. Addition of stabilizing agents to mobile phases will be an additional technique for recovering labile proteins from HPIEC columns.

APPLICATIONS

HPIEC has now been applied to the resolution of a variety of proteins. As noted above, there is considerable similarity among the mobile phases and elution protocols used to develop all ion-exchange columns. Rather than discuss each case of HPIEC individually, see Table 1 for a list of the essential features of the chromatography of a spectrum of proteins.

ACKNOWLEDGMENT

The author gratefully acknowledges the support of U. S. Public Health Grant GM 25431.

REFERENCES

1. Boardman, N. K. (1955) *Biochim. Biophys. Acta* **18**, 290.
2. Boardman, N. K., and Partridge, S. M. (1955) *Biochem. J.* **59**, 543.
3. Eltekov, Y. A., Kiselev, A. V., Khokhlova, T. D., and Nikitin, Y. S. (1973) *Chromatographia* **6**; 187–189.
4. Mizutani, T., and Mizutani, A. (1975) *J. Chromatogr.* **111**, 214–215.
5. Regnier, F. E., and Noel, R. (1976) *J. Chromatogr. Sci.* **14**, 316–320.
6. Engelhardt, H., and Mathers, D. (1977) *J. Chromatogr.* **142**, 311–320.
7. Schmidt, D. E., Jr., Giese, R. W., Conron, D., and Karger, B. L. (1980) *Anal. Chem.* **52**, 177–182.
8. Pluddeman, E., Dow Corning Corporation, Midland, Mich., personal communication.
9. Chang, S. H. (1976) PhD Thesis, Purdue University.
10. Chang, S. H., Gooding, K. M., and Regnier, F. E. (1976) *J. Chromatogr.* **125**, 103–114.
11. Mikes, O., Strop, P., Zbrozek, J., and Coupek, J. J. (1976) *J. Chromatogr.* **119**, 339–354.
12. Mikes, O., Strop, P., and Sedlackova, J. (1978) *J. Chromatogr.* **148**, 237–245.
13. Mikes, O., Sedlackova, J., Rexova-Benkova, L., and Omelkova, J. (1981) *J. Chromatogr.* **207**, 99.

14. Davis, J. E., McDonald, J. M., and Jarett, L. (1978) *Diabetes* 27, 289.
15. Cole, R. A., Soeldener, J. S., Dunn, T. J., and Bunn, H. F. (1978) *Metabolism* 27, 289.
16. Wajcman, H., Dastugue, B., and Labie, D. (1979) *Clin. Chem. Acta* 92, 33.
17. Van Der Wal, S., and Huber, J. F. K. (1980) *Anal. Biochem.* 105, 219–229.
18. Vanecek, G., and Regnier, F. E. (1982) *Anal. Biochem.*, in press.
19. Chang, S. H., Noel, R. N., and Regnier, F. E. (1976) *Anal. Chem.* 48, 1839–1845.
20. Gupta, S., Pfannkoch, E., and Regnier, F. E. (1982) *Anal. Biochem.*, in press.
21. Alpert, A. J., and Regnier, F. E. (1979) *J. Chromatogr.* 185, 375–392.
22. Pfannkoch, E., Lu, K. C., Regnier, F. E., and Barth, H. (1980) *J. Chromatogr. Sci.* 18, 430–441.
23. Vanecek, G., and Regnier, F. E. (1980) *Anal. Biochem.* 109, 345–353.
24. Giddings, J. C. (1965) *in* Dynamics of Chromatography, pp. 13–94, Dekker, New York.
25. Synder, L. R., and Kirkland, J. J. (1979) *in* Introduction to Modern Liquid Chromatography, p. 27, Wiley–Interscience, New York.
26. Karger, B. L., Synder, L. R., and Horvath, C. (1973) *in* An Introduction to Separation Science, p. 135, Wiley–Interscience, New York.
27. Morris, C. J. O. R., and Morris, P. (1976) *in* Separation Methods in Biochemistry, 2nd ed., p. 17, Halsted Press, New York.
28. Chang, S. H., Gooding, K. M., and Regnier, F. E. (1977) *Contemp. Top. Clin. Anal. Chem.* 1, 1–27.
29. Schroeder, R. R., Kudirka, P. J., and Toren, E. C., Jr. (1977) *J. Chromatogr.* 134, 83–90.
30. Schlabach, T. D., Alpert, A. J., and Regnier, F. E. (1978) *Clin. Chem.* 24, 1351–1360.
31. Schlabach, T. D., Fulton, J. A., Mockridge, P. B., and Toren, E. C., Jr. (1979) *Clin. Chem.* 25, 1600–1607.
32. Fulton, J. A., Schlabach, T. D., Kerl, J. E., Toren, E. C., Jr., and Miller, A. R. (1979) *J. Chromatogr.* 175, 269–281.
33. Schlabach, T. D., Fulton, J. A., Mockridge, P. B., and Toren, E. C., Jr. (1980) *Anal. Chem.* 52, 729–733.
34. Denton, M. S., Bostick, W. D., Dinsmore, S. R., and Mrochek, J. E. (1978) *Clin. Chem.* 24, 1408–1413.
35. Bostick, W. D., Denton, M. S., and Dinsmore, S. R. (1980) *Clin. Chem.* 26, 712.
36. Schlabach, T. D., Chang, S. H., Gooding, K. M., and Regnier, F. E. (1977) *J. Chromatogr.* 134, 91–106.
37. Alpert, A. J. (1979) PhD Thesis, Purdue University.
38. Bostick, W. D., Dinsmore, S. R., Mrochek, J. R.,

and Waalkes, T. P. (1978) *Clin. Chem.* 24, 1305–1316.
39. Gooding, K. M., Lu, K. C., and Regnier, F. E. (1979) *J. Chromatogr.* 164, 506–509.
40. Rudolph, F. B., and Clark, S. W. (1981) Paper No. 202, International Symposium on HPLC of Proteins and Peptides, Washington, D. C.
41. Hanash, S. M., and Shapiro, D. N. (1980) *Hemoglobin* 5, 165.
42. Abraham, E. C., Cope, N. D., Braziel, N. N., and Huisman, T. H. J. (1979) *Biochim. Biophys. Acta* 577, 159–169.
43. Hanash, S. M., Kavadella, M., Amanulla, A., Scheller, K., and Bunnell, K. (1981) *in* Advances in Hemoglobin Analysis (Hanash, S. M., and Brewer, G. J., eds.), pp. 53–67.
44. Umino, M., Watanbe, H., Komiya, K., and Mori, N. (1981) Paper No. 208, International Symposium on HPLC of Proteins and Peptides, Washington, D. C.
45. Gardiner, M. B., Wilson, J. B., Carver, J., Abraham, B. L., and Huisman, T. H. J. (1981) Paper No. 203, International Symposium on HPLC of Proteins and Peptides, Washington, D. C.
46. Umino, M., Watanabe, H., and Komiya, K. (1981) Paper No. 204, International Symposium on HPLC of Proteins and Peptides, Washington, D. C.
47. Kato, Y., Komiya, K., and Hashimoto, T. (1981) Paper No. 214, International Symposium on HPLC of Proteins and Peptides, Washington, D. C.
48. Radhakrishnan, A. N., Stein, S., Licht, A., Gruber, K. A., and Udenfriend, S. (1977) *J. Chromatogr.* 132, 552–555.
49. Frolick, C. A., Dart, L. L., and Sporn, M. B. (1981) Paper No. 205, International Symposium on HPLC of Proteins and Peptides, Washington, D. C.
50. Alpert, A. J., and Beaudet, A. L. (1981) Paper No. 210, International Symposium on HPLC of Proteins and Peptides. Washington, D. C.
51. Ansari, G. A. S., Patel, N. T., Fritz, R. R., and Abell, C. W. (1981) Paper No. 211, International Symposium on HPLC of Proteins and Peptides, Washington, D. C.
52. Barford, R. A., Sliwinski, B. J., and Rothbart, H. L. (1979) *J. Chromatogr.* 185, 393–402.
53. Jones, B. N., Lewis, R. V., Paabo, S., Kojima, K., Kimura, S., and Stein, S. (1980) *J. Liq. Chromatogr.* 3, 1373–1383.
54. Lu, K.-C., Gooding, K. M., and Regnier, F. E. (1979) *Clin. Chem.* 25, 1608–1612.
55. Ott, G. S., and Shore, V. G. (1981) Paper No. 201, International Symposium on HPLC of Proteins and Peptides, Washington, D. C.

Analysis of Peptic Fragmentation of Human Immunoglobulin G Using High-Performance Liquid Chromatography*

TSUGIKAZU TOMONO,[1] TOHRU SUZUKI, AND EIICHI TOKUNAGA

Plasma Fraction Department, The Japanese Red Cross Central Blood Center, 1-31, Hiroo 4, Shibuya-ku, Tokyo 150, Japan

High-performance liquid chromatography using a hydrophilic hard-gel TSK-G3000SW column was employed to investigate the peptic fragmentation of a commercial human immunoglobulin (IgG) and its components, i.e., monomeric, dimeric, and aggregated IgG. The monomeric IgG was digested to yield two major fragments, $F(ab')_2$ and pFc', and at least six minor fragments. The dimeric IgG was digested to give three major fragments including $F(ab')_2$, pFc', and a fragment similar to $F(ab')_2$ dimer. The main product from the aggregated IgG was an unknown fragment, and the $F(ab')_2$ fragment was scarcely observed. These observations demonstrated that three IgG components were different from each other in the susceptibilities to peptic digestion and also suggested that the dimer and aggregates were associated mainly through their Fab regions. The TSK-G3000SW column was very useful as a means of fractionating monomer, dimer, and aggregates of IgG as well as a potential means of analyzing the peptic fragment of IgG.

The fragmentation of human immunoglobulin G $(IgG)^2$ with a wide range of proteolytic enzymes has been used to determine immunoglobulin structure as well as to investigate the relationships between structure and function of IgG (1). The peptic fragmentation of IgG has also been studied in detail from the view of the clinical medicine, because a peptic fragment of IgG can be intravenously administered with smaller side effects for passive immunotherapy as compared with commercial IgG (2). One possible explanation of these side effects is that Cohn Fraction II, commonly used as an IgG preparation, contains small amounts of IgG dimer and aggregates which may possess some biological activities as well as immune complex. Little is known, however, of the structural features of the IgG dimer and aggregates which impart these biological activities. The same is true of the behaviors of the dimeric and aggregated IgG in the peptic digestions. Therefore, it is of interest to clarify the reactivities of the dimeric and aggregated IgG toward a proteolytic enzyme such as pepsin. This kind of investigation has been very difficult because there was no quantitative and rapid procedure to analyze the proteolytic degradation of IgG, a high molecular weight protein. It was timely, therefore, that we developed a potential technique to isolate and analyze plasma proteins by high-performance liquid chromatography (HPLC) using a TSK-G3000SW-type column (3). This paper describes the HPLC analysis of the peptic digests of the commercial IgG and the monomeric, dimeric, and aggregated IgG which were isolated from the commercial IgG with the aid of preparative HPLC using TSK-G3000SW columns.

---

* This paper was presented at the International Symposium on HPLC of Proteins and Peptides, November, 16–17, 1981, Washington, D. C.

[1] To whom correspondence should be addressed.

[2] Abbreviations used: IgG, immunoglobulin G; HPLC, high-performance liquid chromatography; SDS, sodium dodecyl sulfate; CD, circular dichroism.

MATERIALS AND METHODS

Materials. Cohn Fraction II, industrially prepared from pooled human plasma by Cohn's method (4) in the Japanese Red Cross Central Blood Center (JRCCBC), was used as commercial IgG in the present experiment. Pepsin (EC 3.4.23.1; 2× crystallized; grade 1:60,000) was purchased from Sigma Chemical Company. The concentrations of proteins used were determined spectrophotometrically using 13.8 (IgG) (5) and 14.9 (pepsin) (6) as $E_{280\,nm}^{1\%}$.

Peptic digestion of IgG. The peptic digestion of the commercial IgG and of the components fractionated from the commercial IgG were carried out at 37°C in 0.1 M acetate buffer (pH 4.0) containing 0.15 M NaCl. The reactions were stopped by adding 0.1 M Tris buffer containing an appropriate amount of NaOH to adjust the pH of the solution to 8.5.

HPLC. Gel-permeation chromatographic analysis of peptic digests of IgG was achieved with a high-performance liquid chromatograph (Model HLC-802, Toyo Soda Manufacturing Co., Ltd.) equipped with two TSK-G3000SW-type columns (7.5 × 600 mm) in tandem. Proteins were detected spectrophotometrically at 280 nm. The mobile phase was 0.05 M sodium acetate buffer, pH 5.0, containing 0.1 M sodium sulfate. The flow rate was 1 ml/min. The sample volume was 100 μl. Preparative fractionation of aggregated, dimeric, and monomeric IgG from commercial IgG was achieved with a preparative high-performance liquid chromatograph (Model HLC-827, Toyo Soda) equipped with two TSK-G3000SW columns (21.5 × 600 mm) in tandem. Proteins were detected by measuring their refractive index. The mobile phase was the same buffer used in the peptic digestion. The flow rate was 8 ml/min. Standard proteins used as the molecular weight markers were human fibrinogen (M_r 340,000), human IgG dimer (M_r 300,000), human IgG (M_r 150,000), and

human transferrin (M_r 77,000), all of which were prepared by JRCCBC, and egg white ovalbumin (M_r 45,000) from Sigma Chemical Company.

SDS–polyacrylamide gel electrophoresis. SDS–polyacrylamide gel electrophoresis was carried out according to Laemmli's method (7), except that 2-mercaptoethanol was not used in the present experiment. The molecular weight markers included in the run were rabbit muscle phosphorylase *b* (M_r 94,000), bovine serum albumin (M_r 67,000), egg white ovalbumin (M_r 45,000), bovine erythrocyte carbonic anhydrase (M_r 30,000), soybean trypsin inhibitor (M_r 20,100), and bovine milk α-lactoalbumin (M_r 14,400), all from Pharmacia Fine Chemical, and human IgG (M_r 150,000) from JRCCBC.

Immunoelectrophoresis. Immunoelectrophoresis, according to the method of Grabar and Williams (8), was performed using Veronal buffer (pH 8.6, μ = 0.06). Antisera used were goat antihuman IgG antiserum (Cappel Laboratory) and horse antihuman serum antiserum (Hoechst).

Circular dichroism (CD) spectra. In order to measure CD spectra of IgG components, IgG components were fractionated using two TSK-G3000SW columns equilibrated with 0.03 M phosphate buffer (pH 7.4, μ = 0.15) containing NaCl. The CD measurements were made using a Jasco Model J-500C spectropolarimeter in cells with a 10-mm path length. Samples generally contained 0.1–1 mg/ml of protein. Ellipticity is expressed as reduced residue ellipticity [θ] which is defined as

$$[\theta] = \frac{M \cdot \theta}{10 \cdot C \cdot l}$$

where M is the mean residue weight, θ is the observed ellipticity in degrees, C is the concentration of optically active solute in grams per cubic centimeter, and l is the path length in centimeters. A mean residue weight of 113 was used for IgG components.

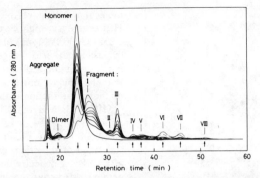

FIG. 1. HPLC analysis of peptic digest of commercial IgG. Digestions of IgG (20 mg/ml) with pepsin (1 wt% against IgG) were carried out at 37°C in 0.1 M acetate buffer (pH 4.0) containing 0.15 M NaCl for 0, 2, 4, 6, 8, 10, 15, 30 and 60 min. One hundred microliters of peptic digest was analyzed on two TSK-G3000SW columns combined in tandem in 0.05 M sodium acetate buffer (pH 5.0) containing 0.1 M sodium sulfate at 1.0 ml/min flow rate. The downward and upward arrows indicate the decrease and increase of each peaks with time, respectively.

RESULTS AND DISCUSSION

Peptic Digestion of Commercial IgG

In our preceding papers (3,9), we reported that the IgG aggregates and IgG dimer could be separated from IgG monomer by HPLC using TSK-G3000SW columns. This finding made it possible to follow the peptic digestion of three states of IgG molecule. The HPLC analysis of peptic digests of commercial IgG is shown in Fig. 1. As the peptic digestion proceeded, the three components of IgG decreased in their concentrations, and at the same time, two major fragments, (I and III) and at least six minor fragments (II, IV, V, VI, VII, and VIII) were produced. The amounts of these fragments, except for fragment II, continued to increase monotonically up to 60 min of digestion, whereas fragment II started to decrease its concentration after 15 min of digestion. The molecular weights of these fragments were estimated by means of HPLC as shown in Fig. 2. The molecular weights estimated for fragments I, II, and III were 102,000, 49,000, and 38,000, respectively.

Earlier studies (10–13) indicated that the major fragments of peptic digestion of human IgG were both $F(ab')_2$ and pFc′ fragments. The $F(ab')_2$ fragment is a dimer associated through interheavy-chain disulfide bonds, with the loss of most of the Fc region from the parent molecule. Its molecular weight is calculated at 96,000 based on Eu sequence of Edelman et al. (14). The pFc′ fragment corresponds closely to the $C_\gamma 3$-do-

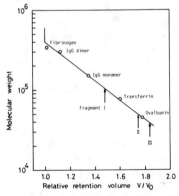

FIG. 2. Calibration graph of molecular weights of proteins by HPLC, where V was the retention volume of solute and V_0 was the void volume.

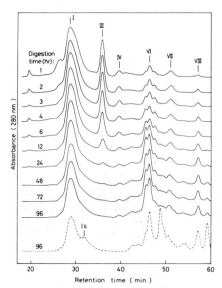

FIG. 3. HPLC analysis of prolonged peptic digests of commercial IgG. Digestions were carried out for the times indicated under the same conditions as shown in Fig. 1, except for the pepsin concentration (—, 1 wt% against IgG; ---, 2 wt% against IgG).

main dimer and so its calculated molecular weight is 25,000. The two major fragments (I and III) observed by HPLC may correspond to $F(ab')_2$ and pFc', respectively. The molecular weight of pFc' estimated by HPLC was larger than the calculated value, whereas the $F(ab')_2$ fragment was in accordance with the calculated molecular weight. Separated SDS–polyacrylamide gel electrophoresis analysis of the digest demonstrated that the molecular weights of fragments I and III were 108,000 and 32,000, respectively, when the latter was calculated from that of SDS-dissociated fragment. Therefore, the molecular weights observed were close to the respective calculated values for $F(ab')_2$ and pFc'. Furthermore, immunoelectrophoretic patterns of the two major fragments I and III were consistent with those of the $F(ab')_2$ and pFc' fragments, respectively, shown by Turner et al. (15). The apparent increase of

molecular weight of the pFc' fragment in HPLC analysis might be due to the fact that TSK-G3000SW-type columns exhibited ion exclusion, cation exchange, and hydrophobic partitioning under certain conditions (16) so that the pFc' fragment was subjected to an unknown interaction with the HPLC gels.

HPLC showed that fragment II was a minor fragment with its molecular weight of 49,000 and was produced only at the initial stage of digestion, disappearing thereafter. SDS-gel electrophoresis analysis also demonstrated that the fragment with such a digestion behavior was observed only in the migration positions corresponding to molecular weights of 33,000 to 21,000, which corresponded closely to the molecular weight of SDS-dissociated fragment II. These results imply that fragment II is a Fc-like fragment (named Fc″ in the present paper) derived with the loss of $F(ab')_2$ from the parent molecule and probably noncovalently associated through the $C_\gamma 3$ region.

Figure 3 shows the chromatograms of prolonged peptic digests of commercial IgG under the same conditions as shown in Fig. 1. In this experiment, the amount of fragment III (pFc') reached a maximum after 1 h of digestion and then decreased. Fragment I $(F(ab')_2)$ also decreased slowly compared with pFc', while at the same time an unknown fragment (Is) appeared as a shoulder

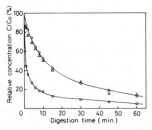

FIG. 4. Decreases of the concentrations of IgG monomer, dimer, and aggregates in the peptic digestion of commercial IgG. Digestion was carried out under the same conditions as indicated in Fig. 1. ○, IgG monomer; △, dimer; and □, aggregates.

peak of fragment I, remarkably after 96 h of digestion with 2 wt% pepsin. The molecular weight of fragment Is was estimated at 68,000 by HPLC and 64,000 by SDS–polyacrylamide gel electrophoresis. The peptic fragmentation of IgG was therefore considered to proceed according to the process

$$IgG \xrightarrow{\text{very fast}} F(ab')_2 \text{ (I)} + Fc'' \text{ (II)}$$

$$Fc'' \text{ (II)} \xrightarrow{\text{fast}} pFc' \text{ (III)}$$

$$+ \text{ oligopeptides}$$

$$pFc' \text{ (III)} \xrightarrow{\text{slow}} \text{oligopeptides}$$

$$F(ab')_2 \text{ (I)} \xrightarrow{\text{very slow}} \text{fragment (Is)}$$

$$+ \text{ oligopeptides}$$

Peptic Digestions of Aggregated, Dimeric, and Monomeric IgG

Figure 4 shows the time dependence of the concentrations of the aggregated, dimeric, and monomeric IgG in the peptic digestion of commercial IgG under the same conditions as indicated in Fig. 1. The relative concentration (C/C_0) of the component was calculated as

$$C/C_0 \text{ (\%)} = (H - H_\infty) \times 100/(H_0 - H_\infty)$$

where C_0 and C were the concentrations of the component at time zero and at a certain digestion time, respectively. H_0, H, and H_∞ were peak heights of the component at time zero, at a certain time, and at an infinite time (24 h in this experiment), respectively. The ratio H_∞/H_0 was 0.0501 for the monomeric IgG and approximately zero for the dimeric and aggregated IgG.

As shown in Fig. 4, the aggregated IgG was digested significantly faster than the dimeric and monomeric IgG. However, it is possible that this susceptibility might not be caused by an intrinsic property of the aggregated IgG, since this was observed in the digestion of Cohn Fraction II, which consists of three components with different concen-

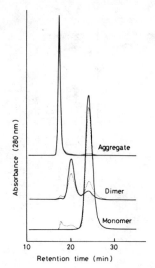

FIG. 5. Thermal stabilities of three IgG components fractionated by means of preparative HPLC. —, Just after fractionation; ---, after heating at 37°C and pH 4.0 for 60 min.

trations. In order to clarify the real susceptibilities of the three components of IgG to the peptic digestion, they were isolated by the use of HPLC. Although the isolation of IgG monomer and aggregates could be easily accomplished, IgG dimer could not be isolated. That is, the dimer fraction contained about 24% of monomer and about 1% of aggregate components. This contamination could be explained by the lesser stability of the dimer component, as shown in Fig. 5. When the three fractions, just after fractionation, were incubated separately for 60 min at 37°C and pH 4.0, 36% of the dimer was converted to monomer or aggregates, whereas only 5% of the aggregates was dissociated and only 9% of the monomer formed aggregates. Therefore, the dimer was considered to be thermally unstable and could be converted into monomer or aggregates by heat.

In order to compare the peptic susceptibilities of thermally stable monomer and

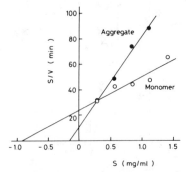

FIG. 6. Hanes–Woolf plots for the peptic digestions of both IgG aggregates and monomer. Digestions with pepsin (0.5 µg/ml) were carried out at 37°C for 10 min in 0.1 M acetate buffer (pH 4.0) containing 0.15 M NaCl. The digestion rate of IgG was estimated from its peak height decreased with digestion on HPLC. Hanes–Woolf equation: $S/V = K_m/V_{max} + S/V_{max}$; where S is IgG concentration (mg/ml); K_m, Michaelis constant (mg/ml); V_{max}, maximum velocity (mg/ml/min).

aggregates of IgG, the kinetic parameters were determined. Figure 6 shows Hanes–Woolf plots (S/V vs S plots) for the aggregated and monomeric IgG. The kinetic parameters obtained are listed in Table 1. The Michaelis constant (K_m) of the aggre-

TABLE 1

KINETIC PARAMETERS FOR THE PEPTIC DIGESTIONS OF BOTH IgG AGGREGATES AND MONOMER

Substrate	Michaelis constant (K_m) (mg/ml)	Maximum velocity (V_{max}) (mg/ml/min)
IgG monomer	0.930	3.85×10^{-2}
IgG aggregates	0.159	1.42×10^{-2}

gated IgG was one-sixth of that of the monomeric IgG, and the maximum velocity of the former was about one-third of the latter. These results suggested that the aggregated IgG could easily form a Michaelis complex with pepsin, although the decomposition rate of IgG in the complex was small. The digestion rate of the aggregated IgG was therefore larger than that of monomeric IgG at low substrate concentration and smaller at high substrate concentration. The K_m value for the aggregated IgG was smaller than that of the monomeric IgG, although it had been expected that its value might have been larger due to the decrease of the diffusion constant with aggregation. The kinetic parameters obtained, therefore, implied that

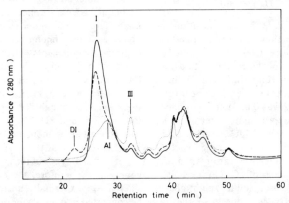

FIG. 7. HPLC analysis of the peptic digests of the monomer, dimer, and aggregate fractions of commercial IgG. Digestions of IgG components (1 mg/ml) with pepsin (0.5 wt% against IgG) were carried out at 37°C in 0.1 M acetate buffer (pH 4.0) containing 0.15 M NaCl for 24 h. —, the digest of monomer fraction; ---, the digest of dimer fraction; ···, the digest of aggregate fraction.

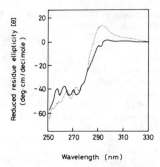

FIG. 8. CD spectra of the monomeric and aggregated IgG fractionated from the commercial IgG by using HPLC. ---, Monomer; —, aggregates.

the aggregated IgG was subjected to conformational change around its cleavage sites or that the cleavage positions of the two components might be different from each other.

Figure 7 shows the chromatograms of peptic digests of the aggregates, dimer, and monomer fractions of the commercial IgG. In the peptic digestion of IgG monomer, fragments I and III were produced similarly to that of commercial IgG. This is consistent with the quantitative predominance of the monomer component in the commercial IgG.

After digestion of the dimeric IgG, the major peptic fragments observed were not only fragments I (F(ab')$_2$) and III (pFc'), but also an unknown fragment (DI). The molecular weight of fragment DI was 202,000; therefore, this fragment presumably corresponds to the dimerized fragment I. Although the structural feature of the dimeric IgG has not been fully elucidated, this result implies that the dimeric IgG was formed from the IgG monomer by binding the Fab regions and that fragment DI, therefore, was formed with the loss of the Fc region from IgG dimer by the attack of pepsin.

In the digestion of the aggregate fraction of IgG, an unknown fragment AI and fragment III (pFc') were formed as major peptic fragments, but fragment I (F(ab')$_2$), a main peptic fragment from IgG monomer, was

hardly produced. The molecular weight of fragment AI was 79,000.

In the peptic digestion of IgG at high temperature, a 3.3 S fragment composed of both C-domain of L-chain and C-domain of H-chain was known to be produced (17,18), whereas in the presence of 8 M urea a fragment with a molecular weight of 57,000 with the loss of V_H and V_L domains, called Fb'$_2$, was produced, giving no F(ab')$_2$ fragment (19). These observations reveal that the denaturation of IgG by urea or heat creates the same susceptibility to peptic digestion in the variable regions of both H- and L-chains. According to Augener et al. (20), heat-aggregated IgG was associated mainly through its variable regions. It was uncertain whether the aggregates in the commercial IgG were formed by heat and/or whether the aggregates were subjected to denaturation. With regard to the CD spectra at the near-uv region, the aggregate component in the commercial IgG differed remarkably from the monomeric IgG (See Fig. 8). Further investigations showed that the CD spectra of the aggregates in the commercial IgG were consistent with one of the aggregates induced by heat. These observations support the fact that the aggregates in the commercial IgG as well as heat aggregates were subjected to conformational changes in their Fab region and, therefore, that the aggregated IgG was associated through its Fab regions. These structural alterations may cause the susceptibility to peptic digestion in their Fab regions. Therefore, one possible explanation on fragment AI is that this fragment was derived with the partial loss of the variable region from F(ab')$_2$ fragment by pepsin. Such an alteration of the cleavage position in the aggregated IgG may explain the difference between the aggregate and monomer components on the digestion rate and the kinetic parameter of peptic digestion.

In the present study, it was found and confirmed for the first time by using HPLC with a TSK-G3000SW column that the major peptic fragments from the dimeric and aggregated IgG were different from those derived from the monomeric IgG and also that the digestion rate of the aggregated IgG differed from the monomeric IgG, which corresponds to the kinetic parameters of digestion. Thus, the TSK-G3000SW column was found to be useful as a means of fractionating monomer, dimer, and aggregates of IgG without dissociation of noncovalent bonds, as well as a potential means of analyzing the peptic fragments of IgG in associating states.

REFERENCES

1. Winkelhake, J. L. (1979) *Immunochemistry* **15**, 695–714.
2. Schultz, H. E., and Schwick, G. (1962) *Deut. Med. Wochenschr.* **87**, 1643–1650.
3. Tomono, T., Yoshida, S., and Tokunaga, E. (1979) *J. Polym. Sci. Polym. Lett. Ed.* **17**, 335–341.
4. Cohn, E. J., Strong, L. E., Hughes, W. L., Mulford, D. J., Ashworth, J. N., Melin, M., and Taylor, H. L. (1946) *J. Amer. Chem. Soc.* **68**, 459–475.
5. Putnam, F. W. (1975) *in* The Plasma Proteins (Putnam, F. W., ed.), Vol. 1, p. 62, Academic Press, New York.
6. Kassell, B., and Meitner, P. A. (1970) *in* Methods in Enzymology (Perlmann, G., ed.), Vol. 19, p. 337, Academic Press, New York.
7. Laemmli, U. K. (1970) *Nature (London)* **227**, 680–685.
8. Grabar, P., and Williams, C. A., Jr. (1953) *Biochim. Biophys. Acta* **10**, 193–194.
9. Tomono, T., Suzuki, T., and Tokunaga, E. (1981) *Biochim. Biophys. Acta* **660**, 186–192.
10. Nisonoff, A., Wissler, F. C., Lipman, L. N., and Woernley, D. L. (1960) *Arch. Biochem. Biophys.* **89**, 230–244.
11. Utsumi, S., and Karush, F. (1965) *Biochemistry* **4**, 1766–1779.
12. Heimer, R., Schnoll, S. S., and Primack, A. (1967) *Biochemistry,* **6**, 127–133.
13. Bennich, H., and Turner, M. W. (1969) *Biochim. Biophys. Acta* **175**, 388–395.
14. Edelman, G. M., Cunningham, B. A., Gall, W. E., Gottlieb, P. D., Putishauser, U., and Waxdal, M. J. (1969) *Proc. Nat. Acad. Sci. USA* **63**, 78–85.
15. Turner, M. W., Bennich, H. H., and Natvig, J. B. (1970) *Clin. Exp. Immunol.* **7**, 603–625.
16. Pfannkoch, E., Lu, K. C., Regnier, F. E., and Barth, H. G. (1980) *J. Chromatogr. Sci.* **18**, 430–441.
17. Seon, B.-K., and Pressman, D. (1974) *J. Immunol.* **113**, 1190–1197.
18. Seon, B.-K., and Pressman, D. (1975) *Immunochemistry* **12**, 333–337.
19. Parr, D. M., Connell, G. E., Kells, D. I. C., and Hofmann, T. (1976) *Biochem. J.* **155**, 31–36.
20. Augener, W., and Grey, H. M. (1970) *J. Immunol.* **105**, 1024–1030.

Cord Blood Screening for Hemoglobin Disorders by High-Performance Liquid Chromatography[*][1]

ADOOR AMANULLAH, SAMIR HANASH, KIRSTEN BUNNELL, JOHN STRAHLER,
DONALD L. RUCKNAGEL, AND STEVEN J. FERRUCI

*Department of Pediatrics and Human Genetics, University of Michigan School of Medicine,
Ann Arbor, Michigan 48109*

Ion-exchange high-performance liquid chromatography was employed as a screening method for abnormal hemoglobins in the newborn period. Samples of cord blood collected in EDTA tubes were used for this analysis. Hemolysates were injected onto 4.1×100-mm Synchropak ion-exchange columns using an automatic injector. Hemoglobin separation was carried out by means of a sodium acetate gradient. A total of 415 samples was analyzed. Hemoglobins A, F, and Bart's, as well as C or S when present, were separately eluted and quantitated using a 35-min gradient program. Four individuals with sickle cell disease, 26 with S or C trait, one with SC disease, and two others with alpha-chain variants were diagnosed with this method. The proportion of Bart's hemoglobin was greater than 1% in 33 individuals. The elution pattern was highly reproducible. The potential for complete automation and the ease with which quality control can be assured make this technique well suited for the detection of abnormal hemoglobins in the newborn period.

Neonatal screening for abnormal hemoglobins provides a practical approach for the detection of sickling syndromes and other hemoglobin disorders such as α-thalassemia. Most screening programs utilize various electrophoretic techniques for the detection of abnormal hemoglobins (1–4). Electrophoresis on cellulose acetate at pH 8.6 identifies increased amounts of Bart's hemoglobin (Hb)[2] associated with alpha thalassemia (1). However, hemoglobins F, A, and S are not always well resolved from each other. Electrophoresis on citrate agar at pH 6.2 provides better separation of Hb F from Hb S; however, Bart's Hb is not easily detectable. Some screening programs utilize both of these techniques because of the limitations

of each when used separately (4). Microcolumn chromatography (5) has also been used for newborn screening. Another recently described technique is thin-layer isoelectric focusing (6). Neither microcolumn chromatography nor isoelectric focusing has been useful for the quantitation of Bart's Hb (7). Recently ion-exchange high-performance liquid chromatography (HPLC) has been utilized for the separation of hemoglobin types (8). We have therefore investigated the use of HPLC in cord blood screening for hemoglobin disorders.

MATERIALS AND METHODS

Blood samples. Cord blood samples from black newborns were collected into EDTA-containing tubes. Red cells were isolated by centrifugation, washed three times with physiological saline, and hemolyzed with equal volumes of distilled water and toluene. The toluene fraction containing cellular de-

---

[*] This paper was presented at the International Symposium on HPLC of Proteins and Peptides, November 16–17, 1981, Washington, D. C.

[1] Supported in part by NIH Grant HL25541.

[2] Abbreviations used: Hb, hemoglobin; HPLC, high-performance liquid chromatography.

17

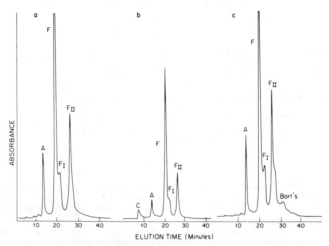

Fig. 1. HPLC separation of hemoglobins in hemolysates from three newborns: (a) normal, (b) HbC trait, and (c) with increased levels of Bart's Hb, using a 35-min gradient-program. Column: Synchropak AX 300, 4.1 × 100 mm. Sample volume: 20 μl. Gradient parameters: flow rate 1.5 ml/min. Mobile phase at time 0: 100% buffer A (0.02 M Tris, 100 mg/liter KCN, pH 8.6) and 100% buffer B at 35 min (0.02 M Tris, 0.2 M Na acetate, pH 7.0). Requilibration was accomplished by pumping buffer A for 10 min. Notice the increased amounts of Hb F II due to aging of the hemolysate. All three samples were separated on the same column.

bris was removed following centrifugation. An aliquot of the hemolysate was immediately utilized for cellulose acetate electrophoresis (9) and the remainder was kept frozen at −4°C for subsequent HPLC analysis. In some cases blood was also collected from other relatives for family studies. Some infants were restudied at 2 to 7 months of age to evaluate changes in the hemoglobin HPLC pattern.

HPLC chromatography. The HPLC system used consisted of a Varian Model 5020 gradient liquid chromatograph (Varian Instruments, Palo Alto, Calif.). The hemolysate was injected onto the column either manually or automatically using an automatic injector (Varian Instruments). Synchropak AX300 (10) ion-exchange columns measuring 4.1 × 100 mm were used (Synchrom, Linden, Ind.). The support consisted of a 10-μm macroporous spherical silica with a bonded polymeric layer of amine. Initially, columns packed by the manufacturer were

utilized. Subsequently, however, because of variability in retention times from batch to batch of ion exchanger, all columns were prepared in the laboratory from the same batch of ion exchanger (11). Elution was accomplished using a linear gradient of buffers A and B. Buffer A consisted of 0.02 M Tris, 100 mg/liter KCN; pH 8.6. Buffer B consisted of 0.02 M Tris, 0.2 M Na acetate, pH 7.0. The pH of all buffers was adjusted with acetic acid. Buffer A was used for equilibration. Initially, several programs with different flow rates and gradient slopes were utilized. Subsequently, however, a 35-min program consisting of 100% buffer A (0% buffer B) at time 0 and 100% buffer B at 35 min with a flow rate of 1.5 ml/min was utilized. Operating column pressure was 500 psi. Column reequilibration was accomplished by pumping buffer A for 10 min at a flow rate of 1.5 ml/min. Separation was carried out at room temperature. Absorbance was monitored at 415 nm with a vari-

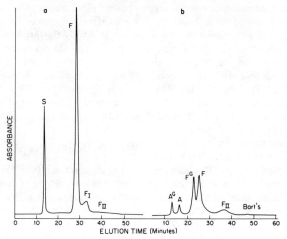

FIG. 2. Chromatographic separation of hemoglobins in hemolysates from (a) a newborn with sickle cell disease and (b) newborn with Hb G Philadelphia. Conditions are the same as in Fig. 1, except a longer 60-min program was used with 100% buffer A at time 0 and 50% buffer B at 60 min with a flow rate of 1.0 ml/min. Notice the small amount of Hb F II in (a).

able wavelength detector. The proportion of the separated hemoglobin peaks was computed using a Beckman Model 126 data system (Beckman, Palo Alto, Calif.). Stored hemolysates were diluted fivefold with buffer A, followed by centrifugation at 10,000g for 10 min prior to chromatography. Ten to twenty microliters of the final hemolysate preparation were injected onto the column.

RESULTS

Elution Pattern

The order in which various hemoglobins were eluted was similar to that observed with conventional anion-exchange chromatography (Figs. 1–3). By means of a 35-min program with a flow rate of 1.5 ml/min, four major peaks were identified in hemolysates from normal newborns. Hb A was eluted separately, followed by the major Hb F peak, a Hb FI shoulder, and a Hb FII peak. The proportion of the Hb FII fraction was variable and gradually increased with the aging of the hemolysates. Freshly prepared samples contained approximately 3–5% of Hb

FII, while hemolysates that had been stored in the refrigerator for 3 to 6 months contained up to 35% of Hb FII. Freezing of the hemolysate reduced the effect of aging on

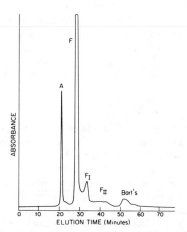

FIG. 3. Chromatographic separation of hemoglobins from a newborn with elevated levels of Bart's Hb. Conditions are the same as in Fig. 2, except a different column was used. A small shoulder is noted following the major Bart's Hb peak.

Hb FII but did not eliminate it completely. Dialysis in the presence of 0.5% 2-mercaptoethanol for 2 to 6 h was asociated with a reduction in the proportion of Hb FII. Storage had little effect on the proportion of Hb FI, which amounted to 8–10% of the total hemoglobin (Fig. 1). To determine the identity of Hb FI, hemoglobins in newborn hemolysates were first separated on Bio-Rex columns (12). Each of the hemoglobin fractions collected was subsequently analyzed by ion-exchange HPLC. Only the acetylated Hb F fraction collected off the Bio-Rex column had an identical elution time as Hb FI.

Abnormal samples encountered among 415 newborns studied included 4 FS, 1 FSC, 22 FAS, and 4 FAC patterns. Two unrelated infants had an alpha-chain variant identified as Hb G Philadelphia by peptide analysis. Hemoglobins A, S, and C when present were eluted separately (Figs. 1 and 2). The proportion of Hb S in samples showing a FAS pattern was quite variable. As little as 2% Hb S was present in some hemolysates. Patterns corresponding to newborns with sickle cell anemia did not include any peaks between S and F when fresh hemolysates were used. However, after storage in the refrigerator for 2 to 4 weeks a small peak amounting to 2–4% of the total hemoglobin was noted in the Hb A position. The appearance of this peak was associated with a corresponding decrease in the size of the Hb S peak.

Hemolysates that showed a striking increase in Bart's Hb by cellulose acetate electrophoresis (6–8% of the total hemolysate) were utilized to determine the elution time of Bart's Hb by HPLC. Newborns from whom these hemolysates were obtained had abnormal red blood cell indices indicative of thalassemia. The HPLC pattern for the hemolysates contained a late-eluting peak not present in patterns obtained from normal newborns (Figs. 1–3). The size of the late-eluting peak corresponded well to the amount of Bart's Hb on cellulose acetate electrophoresis. The major peak corresponding to Bart's Hb was followed by a small shoulder when relatively flat gradients were utilized (Fig. 3). The shoulder was less noticeable using the shorter 35-min gradient program (Fig. 1). Of all the newborn hemolysates studied, 33 contained a Bart's Hb peak that represented 1% or more of the total hemoglobin.

Reproducibility

The elution time for different hemoglobin types was remarkably constant from run to run using the same column and gradient program and did not differ by more than 0.2 min from the expected value. However, the elution time was quite variable from column to column. For example, in the case of the 35-min program, there was as much as a 5-min difference in elution time for a hemoglobin type between columns. The difference was related to the packing material because much less variability was encountered when columns prepared from the same batch of ion exchanger were utilized.

Hb G Philadelphia in hemolysates from two newborns and their relatives consistently eluted in the same position, which was 2 min later than Hb S with the 35-min gradient program. This observation, coupled with the presence of two major peaks in the Hb F region, clearly indicated the presence of an alpha-chain variant (Fig. 2b). The reproducibility of the technique extends to the quantitation of the different hemoglobin types in hemolysates. The proportion of various hemoglobins did not differ by more than 1% in 10 consecutive runs of a hemolysate that gave a FAS pattern. The reproducibility is best illustrated in the case of one of the two newborns with Hb G Philadelphia who was studied at birth and later at 4 and 7 months of age. Both parents of this infant were also studied (Table 1). The presence of equal amounts of normal and variant Hbs A and F, coupled with increased amounts

TABLE 1

QUANTITATION OF HEMOGLOBIN TYPES IN HEMOLYSATES FROM A FAMILY
WITH HEMOGLOBIN G PHILADELPHIA

Subject	G_2 (%)	A_2 (%)	G (%)	A (%)	F/G (%)	F (%)	Bart's (%)	Total G (%)
Infant								
Cord blood	—	—	5.8	7.2	38.2	46.3	2.5	44.0
4 Mos	1.0	1.1	40.4	43.1	5.8	8.6	—	47.2
7 Mos	1.0	1.4	43.6	46.2	3.8	4.0	—	48.4
Father	0.8	1.6	34.5	60.3	—	2.8	—	35.5
Mother	—	2.1	—	95.0	—	2.9	—	—

of Bart's Hb at birth was strongly suggestive of a homozygous state for a deletion of one of the two active alpha genes in the infant (α^G-/α-genotype). The smaller proportion of HB G Philadelphia in the father was suggestive of a deletion of one of the four alpha genes (α^G-/$\alpha\alpha$-genotype). It was therefore concluded that the infant's mother had an alpha gene deletion to account for the presence of only two active alpha genes in the infant. These conclusions, based on the HPLC quantitation of the normal and variant hemoglobins, were confirmed by restriction endonuclease mapping of the alpha globin genes (13).

The proportion of Bart's Hb in hemolysates containing increased amounts of this hemoglobin did not appreciably decrease even after several months of storage of the hemolysates, if they were kept frozen at $-4°C$. However, quantitation of small amounts of Bart's Hb (less than 1%) was not reproducible in multiple runs.

It is noteworthy that in the two cases of Hb G Philadelphia, the presence of an alpha-chain variant was missed by electrophoresis because there was no splitting of the Hb F band and the only abnormal Hb band noted was in the Hb S position. In addition, a Hb A band in four individuals who had a FAS pattern by HPLC was not obvious by electrophoresis because of poor resolution from Hb F.

DISCUSSION

Screening for abnormal hemoglobins in the newborn period differs from screening in other age groups because of the presence of large amounts of Hb F and the need to detect and quantitate Bart's Hb, which is only transiently increased at birth in alpha thalassemia. In contrast, quantitation of the small amounts of Hb A2, which is increased in beta-thalassemia, is an important part of screening in the older group but not in the newborn period. Hb A2 is an early eluting peak, whereas Bart's Hb is the last peak to come off the column. Consequently, HPLC analysis of hemoglobins in the newborn period requires the use of gradient programs different from those used for the older age group. Adequate resolution was achieved using a higher ionic strength for buffer B and a higher flow rate than have been used before (14,15).

Several factors have an important effect on the separation of hemoglobins by ion-exchange HPLC. These have been discussed in detail elsewhere (14). In particular, the use of columns prepared from different batches of ion exchanger requires slight modifications in the gradient program to ensure optimal resolution. Obviously, it is necessary to use hemoglobin standards with every new column to establish the elution time of different hemoglobin types.

The results described above indicate that hemoglobin analysis by ion-exchange HPLC combines the advantages of several other techniques. This includes excellent separation of hemoglobins A, S, and F. Furthermore, the elution time for Hb S was different from that of Hb G Philadelphia. Also, in a previous study, Hb D Punjab differed in its elution time from Hb S.

Bart's Hb eluted as a distinct peak even with the short 35-min gradient program. The importance of quantitating Bart's Hb in the newborn period is emphasized by recent studies utilizing restriction endonuclease mapping of alpha globin genes. It has been demonstrated that more than 25% of American blacks have a heterozygous alpha-thalassemia-2 genotype $(-\alpha/\alpha\alpha)$ with an expected frequency of the homozygous state of 2 to 3% (16). While the heterozygous state might not necessarily be associated with increased amounts of Bart's Hb, the presence of more than 2% Bart's Hb in black newborns has been found to be indicative of a homozygous alpha-thalassemia-2 state $(-\alpha/-\alpha)$ (17,18).

The importance of counseling and follow-up evaluations of newborns with abnormal patterns, as essential components of a screening program, has previously been emphasized (19). For example, all four newborns with a FS pattern by HPLC had subsequent evaluations, including family studies, that confirmed the diagnosis of sickle cell anemia.

Most screening programs involve large numbers of samples. Therefore, the potential for automation of HPLC, coupled with excellent reproducibility, makes this technique ideally suited for the detection of hemoglobin disorders in the newborn period.

REFERENCES

1. Evans, D. I. K., and Carmel, A. (1976) *Arch. Dis. Childhood* 51, 127–130.
2. Pearson, H. A., O'Brien, R. T., McIntosh, S., Aspnes, G. T., and Yang, M. (1974) *J. Amer. Med. Assoc.* 227, 420–421.
3. Sexauer, C. L., Graham, H. L., Starling, K. A., and Fernbach, D. J. (1976) *Amer. J. Dis. Child.* 130, 805–806.
4. Serjeant, B. E., Forbes, M., Williams, L. L., and Serjeant, G. R. (1974) *Clin. Chem.* 20, 666–669.
5. Powars, D., Schroeder, W. A., and White, L. (1975) *Pediatrics* 55, 630–635.
6. Galacteros, F., Kleman, K., Cabari-Martin, J., Beuzard, Y., Rosa, J., and Lubin, B. (1980) *Blood* 56, 1068–1071.
7. Beuzard, Y., Galacteros, F., Braconnier, F., Dubart, A., Chen-Marotel, J., Caburi-Martin, J., Monplaisir, N., Sellaye, M., Martin, C. S., Seytor, S., Bassett, P., and Rosa, J. (1981) *in* Advances in Hemoglobin Analysis (Hanash, S. M., and Brewer, G. J., eds.), pp. 177–195, Alan R. Liss, New York.
8. Hanash, S. M., and Shapiro, D. N. (1980) *Hemoglobin* 5(2), 165–175.
9. Huisman, T. H. J., and Jonxis, J. H. P. (1977) The Hemoglobinopathies. Techniques of Identification, 105, Dekker, New York.
10. Gooding, K. M., Lu, K. C., and Regnier, F. E. (1979) *J. Chromatogr.* 164, 506–509.
11. Broquaire, M. (1979) *J. Chromatogr.* 170, 43–52.
12. Abraham, E. C., Cope, N. D., Braziel, N. N., and Huisman, T. H. J. (1979) *Biochim. Biophys. Acta* 577, 159–169.
13. Jeffereys, A. J., and Flavell, R. A. (1977) *Cell* 12, 429–439.
14. Hanash, S. M., Kavadella, M., Amanulla, A., Scheller, K., and Bunnell, K. (1981) *in* Advances in Hemoglobin Analysis (Hanash, S. M., and Brewer, G. J., eds.), pp. 53–67, Alan R. Liss, New York.
15. Huisman, T. H. J., Gardiner, M. B., and Wilson, J. B. (1981) *in* Advances in Hemoglobin Analysis (Hanash, S. M., and Brewer, G. J., eds.), pp. 69–82, Alan R. Liss, New York.
16. Dozy, A. M., Kan, Y. W., Embury, S. H., Mentzer, W. C., Yang, W. C., Lubin, B., Davis, J. R., Jr., and Koenig, H. M. (1979) *Nature (London)* 280, 605–607.
17. Higgs, D. R., Pressley, L., Clegg, J. B., Weatherall, D. J., and Serjeant, G. R. (1980) *Johns Hopkins Med. J.* 146, 300–310.
18. Ohene-Frempong, K., Rappaport, E., Atwater, J., Schwartz, E., and Surrey, S. (1980) *Blood* 56, 931–933.
19. Miller, D. R. (1979) *Amer. J. Dis. Child.* 133, 1235–1236.

A Simple and Rapid Purification of Commercial Trypsin and Chymotrypsin by Reverse-Phase High-Performance Liquid Chromatography*

Koiti Titani, Tatsuru Sasagawa, Katheryn Resing, and Kenneth A. Walsh

Howard Hughes Medical Institute Laboratory and Department of Biochemistry, University of Washington, Seattle, Washington 98195

Commercial trypsin and chymotrypsin were further purified with respective recoveries of approximately 80 and 50% of the activity in a reverse-phase high-performance liquid chromatography system using acetonitrile in dilute trifluoroacetic acid at pH 2. The purified enzymes showed single enzymatic activities toward synthetic and protein substrates. The enzymes can be rapidly purified in amounts appropriate for structural analysis of proteins.

Trypsin and chymotrypsin have been essential tools in the analysis of protein sequences. It is, however, always difficult to obtain from commercial sources entirely pure enzyme, either trypsin or chymotrypsin, which lacks contamination with the other. Classical techniques to inhibit chymotrypsin-like activity in trypsin included incubation in dilute acid (1), denaturation in 8 M urea (2), and treatment with small quantities of diisopropylphosphorofluoridate (3). Schoellman and Shaw (4) and Mares-Guia and Shaw (5) introduced more specific methods, namely, treatment with TPCK[1] or TLCK to inhibit contaminant activity. In our experience, however, "TPCK–trypsin" prepared in this way still retains some chymotrypsin-like activity which is variable from one batch to another. Although it is controversial whether it is an inherent property of the trypsin molecule, it can be re-

duced by retreatment with TPCK. Chromatographic procedures have also been suggested to achieve this same result: ion-exchange (6) and affinity chromatography (7). The latter procedure has several advantages: It is capable of separating the single-chain β-trypsin from the double-chain α-trypsin and it is also applicable to purification of trypsin from a crude mixture (e.g., activated pancreatic juice). However, these procedures are time consuming.

The present paper describes a simple and rapid procedure allowing purification by reverse-phase HPLC of commercial trypsin and chymotrypsin in a small amount which is sufficient for protein-sequencing studies.

MATERIALS AND METHODS

Crystalline trypsin, TPCK–trypsin, and thrice crystallized α-chymotrypsin were purchased from Worthington. Commercial TPCK–trypsin was retreated with TPCK as described by Carpenter (8). Columns for reverse-phase HPLC were obtained from sources indicated in parentheses; μ-Bondapak C18 and CN (3.9 × 300 mm) (Waters), SynChropak RP-P (4.1 × 250 mm) (SynChrom), ODS-HC/Sil-X-1 (2.6 × 250 mm) (Perkin–Elmer), and Partisil-5

---

* This paper was presented at the International Symposium on HPLC of Proteins and Peptides, November 16–17, 1981, Washington, D. C.

[1] Abbreviations used: HPLC, high-performance liquid chromatography; CN, cyanopropyl; TFA, trifluoroacetic acid; TPCK, L-1-tosylamido-2-phenylethyl chloromethyl ketone; TLCK, L-1-tosylamido-5-aminopentyl chloromethyl ketone; BAEE, N-α-benzoylarginine ethyl ester; BTEE, N-α-benzoyltyrosine ethyl ester.

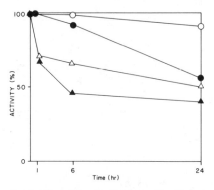

FIG. 1. Effect of organic solvent at acidic pH on stability of trypsin. Trypsin (0.15 mg/ml) was incubated at room temperature for 24 h in 0.1% TFA containing 50% methanol or 50% acetonitrile in the presence or absence of 2 mM CaCl₂. Trypsin-like activity was measured using 10-μl aliquots after different times of incubation. (O) CH₃OH with CaCl₂; (●) CH₃OH; (△) CH₃CN with CaCl₂; (▲) CH₃CN.

ODS-3 (5 × 250 mm) (Whatman). Methanol, 2-propanol, and acetonitrile were purchased from Matheson, Coleman and Bell or Burdick & Jackson. TFA (Pierce) was refluxed twice over solid CrO_3, dried over anhydrous $CaSO_4$, and fractionally distilled. [³H]BAEE, labeled in the ethyl moiety, was a gift of New England Nuclear and diluted with cold BAEE (Vega–Fox). BTEE was a product of Calbiochem–Behring. Bovine oxidized insulin B chain and TPCK were purchased from Boehringer–Mannheim and Sigma, respectively.

HPLC was carried out with a Varian Model 5000 liquid chromatograph using a reverse-phase system at 2 ml/min. In most cases, the mobile phase was 0.1% TFA and the mobile-phase modifier was methanol, 2-propanol, or acetonitrile containing 0.07% TFA (9,10). In some cases, 2 mM $CaCl_2$ was added to the mobile phase.

Trypsin-like activity was assayed by the method of Anderson et al. (11) using radiolabeled BAEE. Chymotrypsin-like activity was assayed using BTEE (12). The substrate specificities of both enzymes were studied using oxidized insulin B chain.

Amino acid analyses were performed with a Dionex amino acid analyzer (Model D-500). Automated sequence analyses were performed with a Beckman sequencer (Model 890) according to Edman and Begg (13) as modified by Brauer et al. (14) and using polybrene (15). Phenylthiohydantoin derivatives of amino acids were identified by two complementary reverse-phase HPLC systems (16,17) and the results were interpreted in a semiquantitative manner (18).

RESULTS AND DISCUSSION

Effect of organic solvents at acidic pH on stability of trypsin. It is well known that pancreatic serine proteases are extremely stable at acidic pH. Their stability in combinations of acid and organic solvent was explored prior to purification by reverse-phase HPLC. Trypsin was found to be less stable in 50% acetonitrile than in 50% methanol at pH 2 but is partially protected by 2 mM $CaCl_2$ in each case (Fig. 1). Thus, 2 mM $CaCl_2$ was added to the mobile phase in order to obtain better yields.

Choice of mobile-phase modifier. Figure 2 shows the elution profiles of trypsin from a Waters CN column using 0.1% TFA and

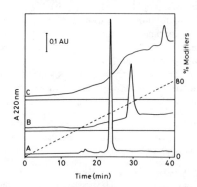

FIG. 2. Comparison of mobile-phase modifiers. Trypsin (100 μg) was chromatographed on a μ-Bondapak CN column using acetonitrile (A), 2-propanol (B), or methanol (C) as the mobile-phase modifier. The linear gradient is shown by the broken line.

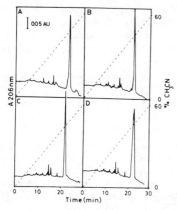

FIG. 3. Chromatography of trypsin on various C18 and CN columns. Trypsin (25 μg) was chromatographed on four different columns using 0.1% TFA and aceto-nitrile as the mobile-phase modifiers. The linear gradient is shown by the broken line. (A) μ-Bondapak C18 (Waters); (B) SynChropak RP-P (SynChrom); (C) ODS-HC/Sil-X-1 (Perkin-Elmer); (D) μ-Bondapak CN (Waters).

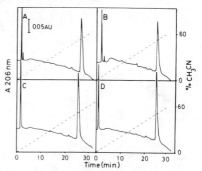

FIG. 4. Chromatography of chymotrypsin (25 μg) on various C18 columns using 0.1% TFA and acetonitrile as the mobile-phase modifiers as in Fig. 3. The linear gradient is shown by the broken line. (A) μ-Bondapak C18 (Waters); (B) Partisil ODS-3 (Whatman); (C) SynChropak RP-P (SynChrom); (D) ODS-HC/Sil-X-1 (Perkin-Elmer).

either methanol, 2-propanol, or acetonitrile containing 0.07% TFA. Although 2-pro-panol has been recommended for large pep-tides or proteins by several investigators (10,19), it is clear that acetonitrile is supe-rior to 2-propanol or methanol for purifi-cation of trypsin. Acetonitrile gives a shorter retention time and a higher recovery of en-zymatic activity. Acetonitrile also gives a lower back pressure and the elution can pro-ceed at a faster flow rate.

Choice of reverse-phase column. Figures 3 and 4 illustrate analytical separations of trypsin and chymotrypsin using various com-mercial C18 and CN columns with identical elution conditions. There are slight differ-ences in the retention time of each enzyme; for example, trypsin was eluted at 21.9, 22.3, 22.4, and 23.8 min from the four tested col-umns, but this may be due to slight differ-ences in the sizes of the columns. A signif-icant difference was reproducibly observed with a Waters CN column as a marginal resolution of trypsin into two overlapping peaks, whereas the other columns each gave

a single symmetrical peak. In fact, trypsin was partially separated into α and β forms on the CN column in the preparative ex-periment described below.

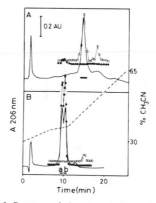

FIG. 5. Recovery of chymotrypsin-like and trypsin-like activities. One hundred micrograms of commercial α-chymotrypsin (A) or trypsin (B) was applied to a Waters μ-Bondapak CN column in 0.1% TFA, pH 2, containing 2 mM CaCl₂, and eluted with an acetonitrile gradient (broken line). The elution was monitored by uv absorption at 206 nm, and by assay of 10-μl aliquots for chymotrypsin-like (O) or trypsin-like (●) activity. Fractions were pooled (horizontal bars) for substrate specificity studies.

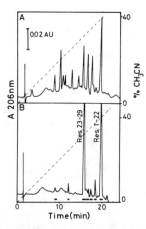

FIG. 6. Comparison of digestion products generated by trypsin before (A) and after (B) purification. The oxidized B chain of insulin (5 nmol) was digested (with 2.3 μg of trypsin) for 6 h in 50 μl of 0.1 M NH₄HCO₃, pH 8, then chromatographed on a column of Waters μ-Bondapak C18. The fractions indicated by the horizontal bars were collected and identified by amino acid analysis. The linear gradient is shown by the broken line.

Recovery of trypsin and chymotrypsin. Commercial trypsin and α-chymotrypsin were eluted from a Waters CN column by a shallow, nonlinear gradient of 0.1% TFA and acetonitrile containing 0.07% TFA in the presence of 2 mM CaCl₂, pH 2, as shown in Fig. 5. Their activities were monitored by using BTEE and radiolabeled BAEE and the recovery of protein was evaluated by amino acid analysis. Total trypsin-like activity was recovered in the two partially separated peaks with yields of 70–104% (the average of four separations was approximately 80%). A trace of chymotrypsin-like activity was clearly separated from the trypsin. The protein was recovered in only 40% yield, presumably as a result of removal of autolysis products and denatured enzyme. In fact, trypsin denatured by reduction and S-alkylation eluted in very poor yield from the column under the same conditions (not shown). A similar result was obtained with α-chymotrypsin (Fig. 5) where the recovery of the

enzymatic activity was approximately 50% (in two peaks). A trace of trypsin-like activity was clearly separated from the chymotrypsin. The minor species of chymotrypsin was not identified among the known intermediates from chymotrypsinogen activation (20).

The two partially separated trypsin peaks were pooled as shown in Fig. 5 and subjected to amino-terminal sequence analysis. Pool b showed a single sequence of Ile–Val–Gly–Gly–, indicating that it is the single-chain form, whereas pool a showed two sequences of Ile and Ser (first cycle), Val and Ser (second cycle), and Gly (third cycle), indicating that it is the double-chain form, α-trypsin, with an internal nick at Lys₁₃₁–Ser₁₃₂ (21).

Substrate specificity of purified trypsin and chymotrypsin toward protein substrates. Each purified enzyme showed the single expected substrate specificity of trypsin or chymotrypsin (Fig. 5). The major peaks of trypsin and α-chymotrypsin were pooled as shown in Fig. 5 and lyophilized. They were tested for substrate specificity toward oxidized insulin B chain. The results of enzymatic digestion (Figs. 6 and 7) are

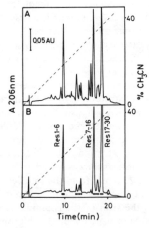

FIG. 7. Comparison of digestion products of insulin B chain generated by chymotrypsin before (A) and after (B) purification. All conditions were as in Fig. 6.

compared before and after purification of the enzymes. Only two peptides were generated by the purified trypsin, and amino acid analysis identified them as residues 1–22 and 23–29. Free alanine released from the carboxyl terminus of insulin B chain eluted under the breakthrough peak. Several very small peaks at retention times 9.2, 12.1, 16.4, 16.9, 17.5, and 18.4 min were analyzed for their amino acid content but could not be clearly identified. They were assumed to be either impurities from the column or minor peptides generated from insulin in less than 2% yields. In contrast to the complex mixture resulting from digestion with commercial thrice crystallized chymotrypsin, the purified enzyme also yielded a simple mixture of peptides as shown in Fig. 7. Peptides identified by amino acid analysis corresponded to residues 1–6, 7–16, and 17–30.

In summary, commercial trypsin and α-chymotrypsin were purified in a TFA/water/acetonitrile system by reverse phase HPLC. The procedure was very simple and rapid (approximately 30 min). The purified enzymes were recovered by lyophilization and showed the expected enzymatic activity in each case toward synthetic and protein substrates. The amount purified in each case was approximately 1 mg, which is appropriate for protein-sequencing studies. We have recently applied this system to separate a mixture of trypsin-like isoenzymes from an invertebrate source (22), and anticipate that the procedure could be useful for purification of any acid-stable protein. We have not yet explored analogous conditions of separation in a more physiological pH range.

ACKNOWLEDGMENTS

The authors are grateful for the assistance of Santosh Kumar. This work was supported in part by Grant GM-15731 from the NIH.

REFERENCES

1. Northrop, J. H., and Kunitz, M. (1936) in Handbuch der biologischen Arbeitsmethoden (Abderhalden, E., ed.), Vol. IV, Pt. 2, pp. 2213–2223, Urban & Schwarzenberg, Berlin.
2. Harris, J. I. (1956) Nature (London) 177, 471–473.
3. Potts, J. T., Berger, A., Cooke, J., and Anfinsen, C. B. (1962) J. Biol. Chem. 237, 1851–1855.
4. Schoellman, G., and Shaw, E. (1962) Biochem. Biophys. Res. Commun. 7, 36–40.
5. Mares-Guia, M., and Shaw, E. (1963) Fed. Proc. 22, 528.
6. Rovery, M. (1967) in Methods in Enzymology (Hirs, C. H. W., ed.), Vol. 11, pp. 231–236, Academic Press, New York.
7. Robinson, N. C., Tye, R. W., Neurath, H., and Walsh, K. A. (1971) Biochemistry 10, 2743–2747.
8. Carpenter, F. H. (1967) in Methods in Enzymology (Hirs, C. H. W., ed.), Vol. 11, p. 237, Academic Press, New York.
9. Dunlap, C. E., III, Gentleman, S., and Lowney, L. I. (1978) J. Chromatogr. 160, 191–198.
10. Mahoney, W. C., and Hermodson, M. A. (1980) J. Biol. Chem. 255, 11199–11203.
11. Anderson, L. E., Walsh, K. A., and Neurath, H. (1977) Biochemistry 16, 3354–3360.
12. Miller, D. D., Horbett, T. A., and Teller, D. C. (1971) Biochemistry 10, 4641–4648.
13. Edman, P., and Begg, G. (1967) Eur. J. Biochem. 1, 80–91.
14. Brauer, A. W., Margolies, M. N., and Haber, E. (1975) Biochemistry 14, 3029–3035.
15. Tarr, G. E., Beecher, J. F., Bell, M., and McKean, D. J. (1978) Anal. Biochem. 84, 622–627.
16. Ericsson, L. H., Wade, R. D., Gagnon, J., McDonald, R. M., Granberg, R. R., and Walsh, K. A. (1977) in Solid Phase Methods in Protein Sequence Analysis: Proceedings of the Second International Conference on Solid Phase Methods in Protein Sequence Analysis, Montpelier, France, September 1977 (Previero, A., and Coletti-Previero, M. A., eds.), pp. 137–142, Elsevier/North-Holland, Amsterdam.
17. Hermann, J., Titani, K., Ericsson, L. H., Wade, R. D., Neurath, H., and Walsh, K. A. (1978) Biochemistry 17, 5672–5679.
18. Koide, A., Titani, K., Ericsson, L. H., Kumar, S., Neurath, H., and Walsh, K. A. (1978) Biochemistry 17, 5657–5672.
19. Rubinstein, M. (1979) Anal. Biochem. 98, 1–7.
20. Wilcox, P. E. (1970) in Methods in Enzymology (Perlmann, G. E., and Lorand, L., eds.), Vol. 19, 64–108, Academic Press, New York.
21. Schroeder, D. D., and Shaw, E. (1968) J. Biol. Chem. 243, 2943–2949.
22. Linke, V. R., Zwilling, R., Herbold, D., and Pfleiderer, G. (1969) Hoppe-Seyler's Z. Physiol. Chem. 350, 877–885.

Adaptation of Reverse-Phase High-Performance Liquid Chromatography for the Isolation and Sequence Analysis of Peptides from Plasma Amyloid P-Component[*,1]

JACQUELINE K. ANDERSON[2] AND JOHN E. MOLE[3]

Department of Biochemistry, University of Massachusetts Medical School, Worcester, Massachusetts 01605

Gel permeation and reverse-phase high-performance liquid chromatography were used to isolate chemical and enzymatic cleavage fragments of plasma amyloid P-component for amino acid sequence analysis. Optimal conditions for resolution of peptide mixtures were predetermined using analytical amounts (0.04–0.1 nmol) and volatile trifluoroacetic acid–acetonitrile or ammonium acetate–acetonitrile buffer systems. Thereafter, 100–200 nmol of each hydrolyzate was chromatographed on preparative columns. Size-exclusion chromatography using an acetic acid solvent containing n-propanol was found to be most useful for large-molecular-weight cyanogen bromide peptides while reverse-phase chromatography was best suited for the smaller enzymatically derived peptides. The high resolution and sensitivity of high-performance liquid chromatography using this dual approach has enabled the completion of greater than 95% of the sequence of P-component (M_r 23,500) using less than 10 mg.

P-Component is a normal plasma glycoprotein which also is associated with normal glomerular basement membrane (1,2). In addition, P-component has been identified as an integral part of the amyloid fibril in the amyloidoses (3), although no specific biological activity has been detected which would establish its role in these diseases. Preliminary NH_2-terminal sequence analysis of P-component has suggested a structural relationship between P-component and C-reactive protein $(CRP)^4$ (4), an acute-phase reactant which has many functions, among them to activate the classical pathway of complement (5–8). P-Component and CRP also share a pentagonal ultrastructure composed of 10 (9) and 5 (2) identical subunits, respectively. Furthermore, both proteins demonstrate calcium-dependent ligand binding although P-component recognizes carbohydrate polymers such as Sepharose (10) while the phosphoryl moiety promotes specific CRP-ligand interactions (11).

Until recently methods for the mapping of peptide hydrolysates and isolation of peptides in high yields for sequenator analysis have posed serious challenges for the protein chemist. In large part, this results from the fact that classical chromatographic separations are frequently hard to reproduce due

[*] This paper was presented at the International Symposium on HPLC of Proteins and Peptides, November 16–17, 1981, Washington, D. C.

[1] This work was supported in part by National Institutes of Health Grants AI 15417, AM 20614, and AM 03555, and Grants 79-156 and 80-987 from the American Heart Association.

[2] A fellow of the Arthritis Foundation.

[3] An Established Investigator of the American Heart Association.

[4] Abbreviations used: CRP, C-reactive protein; HPLC,

high-performance liquid chromatography; PTH, phenylthiohydantoin; TPCK, L-(1-tosylamido-2-phenyl)ethyl choloromethyl ketone.

to variability which exists in packing materials as well as the general variation in technique of the chromatographer. These severe limits of resolution and ability to handle small samples has made the sequenator analysis of biologically important molecules present in less-than-micromolar quantities virtually impossible. The advent and rapid development of high-performance liquid chromatography (HPLC) offers the distinct advantages of speed, high resolution, reproducibility, and high recoveries of isolated material (12–14).

The limited sequence data available for P-component and its apparent homology with CRP have prompted us to undertake studies for the complete elucidation of the primary structure of this little-understood protein. In this paper, we describe methods for the structural mapping of chemical and enzymatic hydrolyzates of P-component by HPLC as well as the isolation and subsequent analysis of specific peptides from these digests. These studies have utilized gel permeation and reverse-phase HPLC entirely in order to determine greater than 95% of the primary structure of human plasma P-component.

MATERIALS AND METHODS

Protein isolation and modification. P-Component was purified on a large scale from normal human plasma by a method previously described (15). The final product was homogeneous on polyacrylamide slab gel electrophoresis and demonstrated an amino acid analysis and NH_2-terminal sequence which were consistent with published reports (16). P-Component was reduced, [^{3}H]carboxymethylated in the presence of 6 M guanidinium chloride (17), dialyzed against 1% ammonium bicarbonate, pH 7.8, and then lyophilized. For some experiments, trypsin digestion was restricted to arginyl peptide bonds by selective modification of the ϵ-NH_2 groups of lysine with succinic an-

hydride (50 mol/mol NH_2 group) in 5 M guanidinium chloride, pH 8.0 (18). The pH was maintained during the reaction with small additions of 1 M NaOH. The completion of acetylation was verified with fluorescamine which indicated greater than 99% of the amino groups had been blocked. The succinylated, carboxymethylated protein then was dialyzed against 1% ammonium bicarbonate and lyophilized.

Chemical and enzymatic cleavages. ^{3}H-P-Component (2 mg/ml) was cleaved using a 50-fold excess (w/w) of cyanogen bromide (CNBr) at 25°C in 70% formic acid. After 4 h, the reaction was terminated by the addition of 10 vol of cold distilled water the sample was flash-frozen and immediately lyophilized.

^{3}H-P-Component and succinylated ^{3}H-P-component both were digested with TPCK-trypsin (Worthington Inc., Freehold, N. J.) in 1% ammonium bicarbonate, pH 8.0, at 37°C (1:50 ratio of enzyme/substrate) for 0.5 and 1 h, respectively. Incubation of ^{3}H-P-component with 2% staphylococcal V-8 protease (w/w) (Pierce Chemical Co., Rockford, Ill.) was performed at 37°C in 1% ammonium bicarbonate, pH 8.0, for 18 h. All enzymatic digests were terminated by the addition of 90% formic acid to a final concentration of 2% and immediately lyophilized.

High-performance liquid chromatography. Separation of the two CNBr cleavage products from uncleaved P-component was carried out using gel permeation HPLC on four 0.78 × 30.0-cm I-125 columns (Waters Assoc., Milford, Mass.) which were connected and eluted in series with 15% *n*-propanol–20% acetic acid. The flow rate was 1.0 ml/min. Absorbancy was monitored at both 280 and 254 nm. Reverse-phase chromatography was employed in order to resolve tryptic and V-8 protease peptides. Enzymatic hydrolysates were chromatographed using both 0.1% trifluoroacetic acid, pH 2.5 (System T), and 0.1% ammonium acetate, pH

6.0 (System A), as the pairing ions and the separations were monitored at both 280 and 206 nm. Standard conditions for isolation of peptides always were predetermined using 0.04–0.1 nmol of hydrolysate and a 0.38 × 30.0-cm analytical μBondapak C_{18} column. Thereafter, 100–200 nmol of each hydrolysate was chromatographed preparatively on a 0.78 × 30.0-cm μBondapak C_{18} column. For most applications, a 1-h linear gradient between 0 and 60% acetonitrile (System A or T) generally provided adequate resolution of components. Individual fractions were collected manually in order to minimize remixing of components and an aliquot of each fraction was counted to identify peptides containing carboxymethylcysteine residues. The purity of each peptide was confirmed or it was further purified by alternating with a second solvent system (T or A) on a 0.38 × 30.0-cm μBondapak C_{18} column. Each peptide was analyzed for amino acid composition following acid hydrolysis with 5.7 N HCl for 24 h at 110°C. Analyses were carried out using an updated Durrum D-500 amino acid analyzer.

Amino acid sequence determination. ^{3}H-P-Component and its enzymatically and chemically derived peptides were subjected to automated Edman degradation in the presence of polybrene using a 0.5-quadrol program (19). Phenylthiohydantoin (PTH) amino acids were identified by HPLC (20) and thin-layer chromatography (21). Confirmation of Arg and His were performed by amino acid analysis after hydrolysis of the PTH derivative to the parent amino acid with 6 N HCl containing 1% $SnCl_2$ (22,23). PTH-CmCys was confirmed also by measurement of radioactivity.

Peptide nomenclature. Each peptide was labeled according to the protease digestion from which it was derived and the solvent system used in its isolation. The prefixes Tr and SP denote cleavages with trypsin and staphylococcal V-8 protease, respectively. Tryptic peptides derived from succinylated

P-component are designated by STr. Thus, Tr9T, 2A was the ninth tryptic fraction to elute using solvent system T and which when rechromatographed on system A eluted as the second peak. This nomenclature greatly simplified the handling of peptides and their organization into a final sequence.

RESULTS AND DISCUSSION

As plasma P-component contains a single methionine residue, CNBr was investigated as a method to initially split the reduced and alkylated protein into two fragments. The separation of the two major fragments by gel permeation HPLC is shown in Fig. 1. A small amount of uncleaved P-component always chromatographed at the leading edge of the larger of the two CNBr peptides but was easily separated by manual collection of the fractions based on the uv absorbancy profile. The molecular weights of each component using sodium dodecyl sulfate gel electrophoresis were calculated as 24,000 (uncleaved P-component), 19,000 (CNBr I), and 5000 (CNBr II), respectively. Utilization of four I-125 columns in series and at a low flow rate (1.0 ml/min) permitted the separation of 2- to 5-mg samples before experiencing a loss of resolution. Recovery of the CNBr peptides was increased dramati-

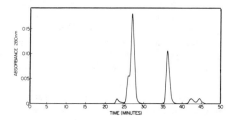

FIG. 1. Gel permeation HPLC of the CNBr peptides of P-component. A 4-h CNBr cleavage of P-component was chromatographed on four I-125 columns (0.78 × 30.0 cm) at a flow rate of 1.0 ml/min. The solvent system consisted of 15% n-propanol–20% acetic acid and absorbancy was monitored continuously at 280 nm.

cally by the inclusion of *n*-propanol into the acidic solvent system which apparently minimizes the interaction of charged groups of the sample and the free silanol residues on the column support. Automated sequence analysis (65/140 residues) established CNBr I as the amino terminus of the native molecule. The two cysteine residues known to be present in P-component both occur in CNBr I and must form a single disulfide linkage since reduction and alkylation using [³H]iodoacetic acid resulted in the incorporation of tritium only into CNBr I (1.95 mol [³H]/mol CNBr I). The 5000-dalton peptide, CNBr II, lacked homoserine or homoserine lactone in its amino acid composition and, therefore, must represent the carboxy terminus of P-component. Sequence analysis of CNBr II was successful in determining 40/45 residues for this peptide.

In order to obtain smaller peptides which would allow construction of the entire sequence of P-component, two enzymatic procedures were evaluated. The first approach involved controlled, limited digestion of ³H-P-component with TPCK-trypsin in order to generate large fragments on which extended sequences could be obtained. Thus, reverse-phase HPLC provided an initial, rapid, and exceedingly sensitive method to screen a variety of conditions for limited hydrolysis of P-component with trypsin. A 30-min incubation time at 37°C with a 1:50 ratio of trypsin/P-component was determined to be the most suitable for generating large tryptic fragments which resolved adequately using solvent system T (15). Using these identical conditions on a large C₁₈ column, a preparative separation of a tryptic digest of P-component was carried out and is presented in Fig. 2a. Since peptides separated using reverse-phase chromatograph as a function of their hydrophobicity and their ion-pairing properties, proper manipulation of the pH, choice of ion-pairing agent, and/or selection of the solvent system are all important parameters which affect resolution. The use of trifluoroacetic acid as an ion-pairing agent under acidic conditions provided excellent separation of the tryptic peptides of P-component. When a similar tryptic digest was chromatographed using ammonium acetate, pH 6.0, a completely different elution profile was observed (Fig. 2b). As expected, peptides generally eluted at a lower acetonitrile concentration as the pH of the system was increased. Approximately 70% of the lysines and arginines contained in P-component are found within the first 100 residues, resulting in an average peptide size of 7–10 residues. These short peptides proved to be very useful for the verification of sequences in the longer chemically derived peptide, CNBr I.

P-Component contains 18 potential tryptic cleavage sites (10 lysines and 8 arginines). Therefore, an attempt was made to confine hydrolysis of P-component to the arginine residues by succinylation of the ε-amino groups of lysine. This approach was expected to produce a limited number of large fragments for which overlapping lysyl sequences could be secured. Due to the sparing solubility of succinylated peptides at acidic pH values, it was necessary to fractionate succinylated peptides using system A, pH 6.0 (Fig. 2c). Interestingly, following the primary fractionation, those peptides requiring further purification were chromatographed readily using the acidic solvent system T and no apparent insolubility was noted. Comparisons of the profiles in Figs. 2b and c revealed several coincidental tryptic peptides, however, most peptides eluted at different positions within each chromatogram.

Three of the eight arginine residues were identified at positions 13, 38, and 45 of CNBr I and a fourth was located within CNBr II (Fig. 3). The placement of a fifth arginine was made from STr21A which extended the sequence for CNBr I by 18 residues. However, in order to obtain overlapping sequences for all the arginine peptides, P-component was further digested with

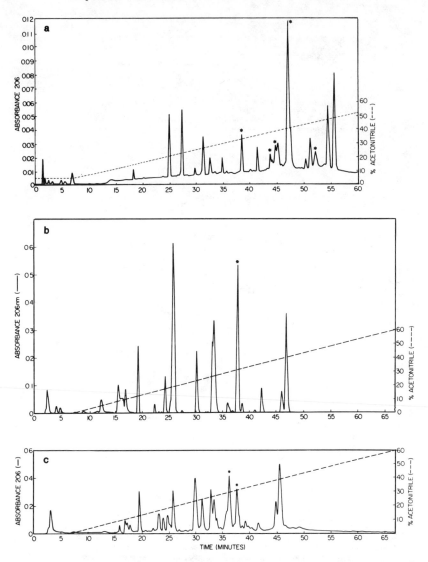

FIG. 2. Preparative reverse-phase HPLC of tryptic hydrolysates of P-component. Tryptic digests of ³H-P-component or succinylated ³H-P-component were chromatographed on a μBondapak C_{18} column (0.78 × 30.0 cm) using a 1-h linear gradient between 0 and 60% acetonitrile containing 0.1% trifluoroacetic acid, pH 2.5, or 0.1% ammonium acetate, pH 6.0, and a flow rate of 5.0 ml/min. Absorbancy measurements were recorded at 206 nm. For each chromatogram, asterisks denote tritium-labeled S-carboxymethylcysteine peptides. (a) Digest of ³H-P-component using solvent system T; (b) digest of ³H-P-component using solvent system A; (c) digest of succinylated ³H-P-component using solvent system A.

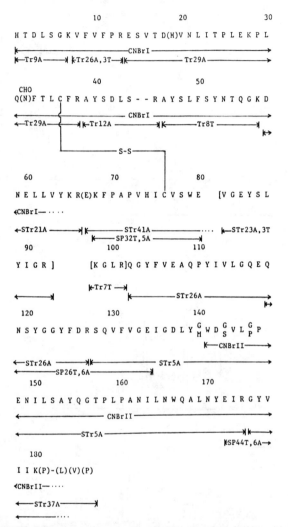

FIG. 3. Construction of the primary sequence of P-component from CNBr, tryptic, and staphylococcal V-8 protease peptides. The one letter code for the amino acids has been used. Peptides are numbered by using nomenclature based on their derivation as well as the solvent system used in their isolation. Thus, the abbreviations CNBr, Tr, STr, and SP refer to cyanogen bromide, tryptic, succinylated tryptic, and staphylococcal V-8 protease cleavages, respectively. The two solvent systems used to isolate peptides are denoted by T (0.1% trifluoroacetic acid–acetonitrile) and A (0.1% ammonium acetate–acetonitrile). For example, Tr3A,1T eluted as the third tryptic fraction using system A on the first μBondapak C$_{18}$ column and eluted as the first peak using system T and a second C$_{18}$ column. PTH amino acids were identified by thin-layer chromatography and HPLC or by HPLC only where indicated in parentheses. Peptides which are bracketed (Tr7T, STr41A, and SP32T,5A) have not been conclusively located within the sequence and were aligned based on their sequence homology to CRP.

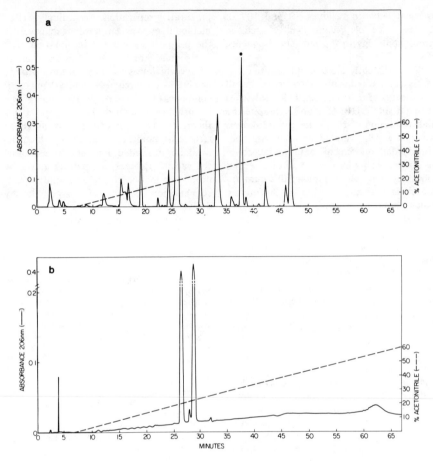

FIG. 4. Application of reverse-phase dual solvent systems for the isolation of P-component tryptic peptides. Separation of tryptic peptides was carried out using either a 0.78 × 30.0-cm or 0.38 × 30.0-cm μBondapak C$_{18}$ column and a 1-h linear gradient between 0 and 60% acetonitrile containing 0.1% trifluoroacetic acid, pH 2.5, or 0.1% ammonium acetate, pH 6.0. (a) Peptides of ^{3}H-P-component using a preparative column and system T at a flow rate of 5.0 ml/min; (b) fraction eluting at 26 min in a was rechromatographed on an analytical column using system A, at a flow rate of 2.0 ml/min.

staphylococcal V-8 protease. When the V-8 protease digest of P-component was chromatographed using solvent system T, approximately 25 well-resolved fractions were obtained. Several of these were purified further using ammonium acetate–acetonitrile, pH 6.0. One of these (SP44T, 6A) confirmed

the sequence near the single arginine in CNBr II. A second (SP26T, 6A) permitted the alignment of STr26A and STr5A. Thus, the assignment of 6/8 arginines was firmly established (Fig. 3). The remaining two arginines which occur as the COOH terminus of STr23A, 3T and Tr7T were tentatively assigned between residues 80 and 102 of the sequence of P-component, based on strong homology to the published sequence of CRP (24).

We consistently have found that one solvent system was insufficient to separate all components of most enzymatic hydrolyzates during a single HPLC analysis, despite the high resolution of the technique. However, peptides which coeluted or otherwise contained small amounts of contaminants could be isolated easily by utilization of a second solvent system. An example of this dual approach is illustrated in Fig. 4. A fraction which eluted as a symmetrical absorbance peak at 26 min using solvent system A (Fig. 4a) also appeared pure based on the results of amino acid analysis. However, sequence analysis of this sample clearly revealed two peptides in very near equimolar quantities. As seen in Fig. 4b, this fraction separated into two well-resolved components when subjected to reverse-phase HPLC using system T, pH 2.5, and each peptide had a unique polypeptide sequence.

In each chromatogram presented in Figs. 2 and 4, asterisks denote fractions that contained radioactively labeled carboxymethylcysteine. Thus, radiolabeling of the cysteine residues of P-component prior to cleavage, readily enabled the location and selective isolation of cysteine-containing peptides. It should be equally possible to prepare similar radioactive derivatives with other reactive amino acids in order to facilitate the identification and analysis of peptides containing specific amino acids.

In our studies the utilization of a second HPLC system to ascertain peptide purity prior to their amino acid or sequence analysis was a distinct advantage and offered significant savings in time and sample. The small quantity of sample required for analytical mapping (1–5 µg) is a distinct advantage over previous methods and allows the investigator to establish optimal conditions for generating and isolating peptides prior to a preparative separation. Accordingly, HPLC methodology was employed exclusively in the primary structure determination of P-component. Furthermore, the excellent recovery which was obtained for most peptides (65–95%) required less than 10 mg of P-component to elucidate greater than 95% of its primary structure. The use of HPLC, therefore, is a tremendous advantage for structural studies of proteins which are available in limited amounts. By exploiting the variables of pH, ion-pairing agent, hydrophobic mobile-phase, and solid support, subnanomole amounts of any complex hydrolyzate can be completely separated for structural analysis based entirely on HPLC methodology.

ACKNOWLEDGMENTS

Appreciation is extended to Drs. J. Claude Bennett and Ajit Bhown at the University of Alabama in Birmingham for performing many of the sequence analyses reported in this communication. The excellent technical help of Mr. Jeff Powell and Bill Lipham and the secretarial assistance of Ms. Joy Coderre are gratefully acknowledged.

REFERENCES

1. Skinner, M., Vaitukaitis, J. L., Cohen, A. S., and Benson, M. D. (1979) J. Lab. Clin. Med. 94, 633–638.
2. Dyck, R. F., Lockwood, C. M., Kershaw, M., McHugh, N., Duance, V. C., Baltz, M., and Pepys, M. B. (1980) J. Exp. Med. 152, 1162–1174.
3. Holck, M., Husby, G., Stetten, K., and Natvig, J. B. (1979) Scand. J. Immunol. 10, 55–60.
4. Osmand, A. P., Friedenson, B., Gewurz, H., Painter, R. H., Hofman, T., and Shelton, E. (1977) Proc. Nat. Acad. Sci. USA 74, 739–743.
5. Volankis, J. E., and Kaplan, M. H. (1974) J. Immunol. 113, 9–17.
6. Mortensen, R. F., Osmand, A. P., Lint, T. F., and Gewurz, H. (1976) J. Immunol. 117, 774–781.
7. Fiedel, B. A., and Gewurz, H. (1976) J. Immunol. 117, 1073–1078.
8. Mortensen, R. F., Osmand, A. P., and Gewurz, H. (1975) J. Exp. Med. 141, 821–838.
9. Pinteric, L., and Painter, R. H. (1979) Canad. J. Biochem. 57, 727–736.
10. Pepys, M. B., Dash, A. C., Munn, E. A., Feinstein,

A., Skinner, M., Cohen, A. S., Gewurz, H., Osmand, A. P., and Painter, R. H. (1977) *Lancet* **1**, 1029-1031.

11. Anderson, J. K., Stroud, R. M., and Volanakis, J. E. (1978) *Fed. Proc.* **37**, 1495.

12. Engelhardt, H., and Elglass, H. (1975) *J. Chromatogr.* **112**, 415-423.

13. Rubinstein, M., Chen-Kiang, S., Stein, S., and Udenfriend, S. (1979) *Anal. Biochem.* **95**, 117-121.

14. Jenik, R., and Porter, W. (1981) *Anal. Biochem.* **111**, 184-188.

15. Anderson, J. K., Hollaway, W., and Mole, J. E. (1981) *J. High Resolut. Chromatogr. Chromatogr. Commun.* **4**, 417-418.

16. Thompson, A. R., and Enfield, D. L. (1978) *Biochemistry* **17**, 4304-4311.

17. Ruegg, U. T., and Rudinger, J. (1977) *in* Methods in Enzymology (Hirs, C. H. W., and Timasheff, S. N., eds.), Vol. 47, pp. 111-116, Academic Press, New York.

18. Work, E., and Work, T. S. (1975) Laboratory Techniques in Biochemistry and Molecular Biology, p. 78, American Elsevier, New York.

19. Bhown, A. S., Mole, J. E., and Bennett, J. C. (1981) *Anal. Biochem.* **110**, 355-359.

20. Bhown, A. S., Mole, J. E., Hollaway, W., and Bennett, J. C. (1979) *J. Chromatogr.* **156**, 35-41.

21. Summers, M. R., Symthers, G. W., and Oroszlan, S. (1973) *Anal. Biochem.* **53**, 624-628.

22. Mendez, E., and Lai, C. Y. (1975) *Anal. Biochem.* **68**, 47-53.

23. Lai, C. Y. (1977) *in* Methods in Enzymology (Hirs, C. H. W., and Timasheff, S. N., eds.), Vol. 48, pp. 369-373, Academic Press, New York.

24. Oliveira, E. B., Gotschlich, E. C., and Liu, T.-Y. (1977) *Proc. Nat. Acad. Sci. USA* **74**, 3148-3151.

Characterization of Human Alcohol Dehydrogenase Isoenzymes by High-Performance Liquid Chromatographic Peptide Mapping*

Daniel J. Strydom and Bert L. Vallee

Center for Biochemical and Biophysical Sciences and Medicine, Harvard Medical School, 250 Longwood Avenue, Boston, Massachusetts 02115

Human liver alcohol dehydrogenase (ADH, EC 1.1.1.1) is a large and heterogeneous family of isoenzymes and the high-performance liquid chromatographic peptide mapping technique which was developed here recognizes differences and similarities between them. Isoenzymes were S-carboxymethylated, digested with trypsin, and the mixtures of tryptic peptides fractionated by reverse-phase gradient chromatography on octadecylsilane columns, using perchlorate–phosphate buffer and acetonitrile as eluants. The resultant peptide maps were reproducible, showing great similarities between the $\alpha\beta\gamma$-ADH isoenzymes (now called Class I) on the one hand and remarkable differences between these and both the π- and χ-ADH isoenzymes (now called Class II and III, respectively) on the other. This implies that these three isoenzyme groups have characteristic primary structures which correspond to their typical substrate specificities and kinetics.

Alcohol dehydrogenase (ADH)[1] (alcohol: NAD+ oxidoreductase, EC 1.1.1.1) from human liver, once thought to be a single molecular species, is actually composed of numerous molecular forms. More than 20 different isoenzymes (1–5) that exhibit a wide range of mobilities on starch-gel electrophoresis, even though they share many structural features, are now known. Thus, all ADH isoenzymes are dimeric, each monomer having a molecular weight of about 40,000, with a blocked N-terminal amino acid, containing two zinc atoms and interacting with 1 mol of NAD(H) (6). However, while they also share many functional properties, there are distinctive differences. Among these their characteristic inhibition by 4-methyl pyrazole stands out, and this has become an operational basis for their differentiation by means of chromatography. Use of the pyrazole-affinity column (7) allows the isolation of a large group of basic isoenzymes, $\alpha\beta\gamma$-ADH (1,2,7–9), to which we now refer as Class I. All are inhibited by pyrazole (10). The apparent polymorphism in this category of isoenzymes, the only ones known at the time, led Smith et al. (1) to propose a genetic model to account for 10 observed combinations of subunits attributed to three genes, α, β, and γ, the latter being polymorphic—γ_1 and γ_2. Subsequently, additional isoenzymes of Class I were found whose origin was attributed to polymorphisms in the β-locus (4,11) thought to encompass racially determined genetic variants (2,12–15).

The origin of the anodic isoenzyme of Li and Magnes (16) has eluded assignment so far. It has been isolated by virtue of its relative insensitivity to pyrazole (5,16,17); named π-ADH, it constitutes the first example of the Class II isoenzymes. Although basic, it is more anodic than Class I (17,18). The isolation of the truly anodic χ-ADH (3), the Class III isoenzymes, is based on their complete insensitivity toward pyrazole. As in the case of Class II isoenzymes, the

---

* This paper was presented at the International Symposium on HPLC of Proteins and Peptides, November 16–17, 1981, Washington, D. C.

[1] Abbreviations used: ADH, alcohol dehydrogenase; HPLC, high-performance liquid chromatography.

39

genetic origin and structural characteristics of χ-ADH remain uncertain.

The characterization of the structure, function, and origin of these three heterogeneous groups of isoenzymes now poses a major problem as does the identification of isoenzymes which might be representatives of either the whole group, of subgroups, or of others yet to be recognized. Further studies of representative isoenzymes would be designed to explore the basis and validity of genetic models, the possibility of substrate induction of isoenzymes, posttranslational modifications, and the structural basis of isoenzyme function. The separation and purification of many seemingly closely related isoenzymes taxes current chromatographic technology, particularly when the paucity of material can impose further restrictions. The task would be greatly simplified if these isoenzymes could be resolved into distinctly dissimilar subgroups of related structures. The enzymatic and electrophoretic properties which provide the basis for differentiating, e.g., Class II and III, i.e. π- and χ-ADH, respectively, from one another and from Class I, i.e. the $\alpha\beta\gamma$-forms, could be readily attributable to relatively few—if not even only one—change(s) in their amino acid sequences, as demonstrated previously for the EE, ES, and SS-isoenzymes of horse liver ADH (19).

Under prevailing circumstances conventional approaches do not detect such differences ideally. The presence of blocked amino termini of these proteins precludes aminoterminal sequencing. Amino acid analysis alone is not likely to differentiate cross-contamination among isoenzymes from genetic replacements. Peptide mapping, however, offers potential answers. Conceptually, this method is capable of simplifying resolution of the complex information contained in the primary structures of these isoenzymes. Detailed structural knowledge is not an absolute prerequisite for the interpretation of differences in peptide maps. Conventional two-dimensional paper or thin-layer mapping techniques do not yield the high level of reproducibility demanded for adequate investigation of the structures of proteins of this size.

High-performance liquid chromatography (HPLC) of peptides (20–29) has recently been applied successfully to both the isolation of peptides for sequence analysis (e.g., 30–37) and the identification of differences between related proteins (mapping) (e.g., 38–50). Chromatographic separation of peptides from a digest of any protein can now be made highly reproducible and, hence, such chromatograms become "maps" of protein structure.

Here we report the reproducible conditions to map isoenzymes of human liver ADH which serve as an aid both in their chemical characterization and in the selection of the appropriate isoenzyme groups for further studies.

MATERIALS AND METHODS

Human liver ADH isoenzymes. Isoenzymes were isolated as described previously (3,5,51).

S-Carboxymethylation of isoenzymes. S-Carboxymethylation was performed according to Crestfield *et al.* (52), except that the reducing agent was dithiothreitol and the denaturant was 6 M guanidine hydrochloride. The reaction mixtures were desalted exhaustively by dialysis against water and then lyophilized.

Digestion with trypsin. Trypsin digestion proceeded according to Jornvall (19). The protein was suspended at a concentration of 2 to 10 mg/ml in fresh 1% ammonium bicarbonate solution, 1 or 2% (w/w) of trypsin was added, and the digestion was then continued at 37°C for the specified times. At the end of the digest, 10% acetic acid was added to stop the reaction and the digests were dried under vacuum. Water was added and evaporated again to remove the last remnants of the volatile salts.

High-performance liquid chromatography. The HPLC system consisted of a

Waters μBondapak C_{18} column, two Waters Model 6000A solvent delivery systems, a Waters Model 440 absorbance monitor (at 254 nm), an LKB 2138 Uvicord S monitor with HPLC flow cell (at 206 nm), a Waters 720 data module, and a Model 730 system controller. Samples were injected automatically using a WISP 710 A injector. The solvent systems are similar to those proposed by Meek (27). Solvent A is 0.1 M perchloric acid and 0.1% o-phosphoric acid titrated to pH 2.2 ± 0.05 with solid sodium hydroxide. Solvent B is 75% acetonitrile and 25% solvent A. The column was eluted with a combination of two linear gradients, from 5 to 37% B in 20 min, and then to 85% B in 55 min. The column was reequilibrated with starting composition (5% B) by running a 5-min gradient back down to 5% B and then equilibrating for 20 min before the next injection. Flow rates were 1 ml/min throughout. Scale expansion to 0.01 absorbance units, fullscale, at 206 nm was the maximum amount possible due to baseline changes during the gradients.

The water used was HPLC grade from J. T. Baker, and the acetonitrile was HPLC grade from Waters Associates and J. T. Baker. The batches of acetonitrile were selected for low absorbance at 206 nm.

RESULTS AND DISCUSSION

Choice of Column, Buffer, and Gradient System

The literature indicates a wide range of conditions useful in the peptide mapping of proteins. The specific conditions which are adopted depend on the objectives for obtaining the maps, the type of proteins investigated, the choice of monitoring systems, and the availability of specific types of columns. In general, the use of octadecylsilane reverse-phase columns coupled with very acidic aqueous buffers and acetonitrile as organic solvent for gradient elution seems to be preferred, although higher pH values, other columns such as octyl and cyanopro-

pylsilanes and other solvents such as n-propanol and acetone, are also used successfully. Further, the presence of phosphate (22) greatly assists in successful peptide separations.

The choice among organic solvents was limited to acetonitrile, the only one with very low absorbance at 206 nm, the wavelength chosen to monitor the analytical peptide maps.

Preliminary separation studies using C_{18} (Radialpak), C_{18} (μBondapak), C_8 (Radialpak), CN (Radialpak), and phenyl (μBondapak) reverse-phase columns and employing sodium or triethylamine phosphate buffers at pH 3 did not identify any one column as markedly superior to others, except that the peaks obtained with the μBondapak columns were generally sharper. Separations at pH values above 3 were inferior.

The Radialpak C_{18} column was selected somewhat arbitrarily for studies on the effects of the gradient and pH on separations. The pH range 2.1–2.9 was investigated using triethylamine–phosphate and sodium perchlorate–phosphate buffers. The separations depended on pH and buffer, and the perchlorate–phosphate buffer at pH 2.1 to 2.3 gave the best resolution. Different gradient slopes were optimal for different parts of the separation. Optimal conditions for separations on the Radialpak column were then applied to a μBondapak C_{18} column. The peaks were not only sharper, but more numerous. The results of variation of gradient slopes paralleled those found on C_{18} Radialpak columns. The optimal separation on μBondapak C_{18} was then established as given under Materials and Methods. The separation found for Class I ADH is shown in Fig. 1.

Digestion Conditions

Samples of S-carboxymethylated Class I ADH were digested with either 1 or 2% (w/w) trypsin from 1 to 20 h and the maps of each sample were obtained. Most peaks were fully developed after 4 h of digestion. There-

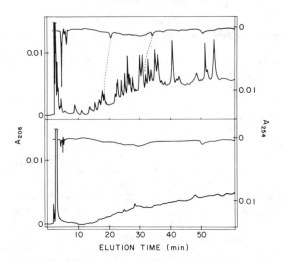

FIG. 1. Separation of a tryptic digest of Class I ADH (top) on a column of octadecylsilane (μBondapak
C_{18}) using sodium perchlorate–phosphate buffer at pH 2.2 and acetonitrile as solvent systems, as described
under Materials and Methods. The lower diagram shows the pattern obtained for the trypsin blank.

after, a few slowly decreased while others
increased in size, but the difference between
the maps obtained by digesting with 1%
trypsin for 4 h and those with 2% trypsin for
20 h was very small. A determination of the
reaction blank, i.e., the autodigestion of
trypsin in the absence of other proteins,
showed two minor peaks which were of such
low intensity that they did not interfere with
the mapping of the ADH isoenzymes (Fig.
1). To test the influence of the buffer on the
digestion process a separate digest was performed using 0.1 M N-ethylmorpholine at
pH 8.4, instead of 1% NH_4HCO_3. The resultant map did not differ appreciably. It
therefore seems that the tolerance for variation in conditions is sufficient to allow reproducible digestion. However, all of these
digests do contain some "insoluble core"
material.

Amount of Protein Required for Mapping

In preference to the experimentally more
complex postcolumn derivatization, uv absorbance was monitored to generate the
chromatograms. At 206 nm unreactive pep-

tides such as the acetylated amino-terminal
peptide of ADH are identified while those
containing tryptophan and to a lesser extent
those containing tyrosine are detected at 254
nm. In the development of such a separation
system, this dual analysis enables the recognition of a few components in the complex
mixture of unknown peptides, facilitating
the detection of changes in the chromatograms of successive experiments.

The amount of ADH needed for mapping
is quite low when readings of the eluate are
made at 206 nm. Fifty micrograms of an
individual isoenzyme generally yield a good
map with prominent peaks when the lower
limit for analysis ranges from 1 to 5 μg of
isoenzyme (approx. 10–50 pmol/subunit).
The use of highly purified acetonitrile is
mandatory when the baseline is to remain
level during gradient elution and monitoring
at 0.01 OD fullscale (206 nm). The quality
of HPLC-grade acetonitrile varies markedly
from batch to batch, even though it remains
within the manufacturers' specifications;
batches should be selected for very low absorbance at 206 nm, preferably with an A_{206}
of 0.05 or less.

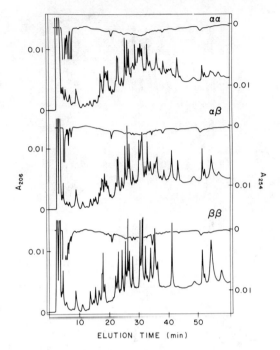

FIG. 2. Tryptic peptide maps for the $\alpha\alpha$-, $\alpha\beta$-, and $\beta\beta$-isoenzymes of Class I ADH, obtained as described under Materials and Methods.

Maps of $\alpha\beta\gamma$-Isoenzyme Fractions

The maps for the tryptic peptides of the individual $\alpha\beta\gamma$-isoenzyme fractions are closely similar (Figs. 2 and 3). The majority of peaks in these chromatograms are common to all isoenzymes of this group. Further, 60% of the areas under the chromatograms are contained under peaks with common retention times. Excluding the breakthrough fractions, they comprise the group of six peaks from 8.9 to 16.5 min; those at 17.7 and 18.5 min (the latter with a marked absorbance at 254 nm); the four peaks at 22.1, 22.7, 22.9, and 24.1 min; the peak at 25.3; the group at 26.2, 26.6, and 27.8 min; the large peaks at 30.1 and 30.9 min; as well as those at 32, 33.8, and 35.9 min. The single peak at 40.7 min, the doublet at 51.2 and 51.9 min, and the peak at 54 min are characteristic also. Jointly, these peaks constitute the general pattern which is unique to the $\alpha\beta\gamma$-isoenzymes and upon which minor differences are superimposed, reflecting close similarity of the primary structures of the $\alpha\beta\gamma$-isoenzymes. This structural resemblance underlines the likeness of their enzymatic properties, i.e., high sensitivity to pyrazole inhibition and similar kinetic constants when ethanol is the substrate.

Maps of χ-, π-, and $\beta\beta$-Isoenzymes

Figure 4 compares the maps of π- and χ-ADH isoenzymes (Classes II and III), whose enzymatic properties differ from those of the $\alpha\beta\gamma$-ADH isoenzymes, with that of $\beta\beta$-ADH, representing Class I.

The overall patterns of the former differ markedly from those of the latter. The patterns of all three groups differ markedly, strongly suggesting numerous differences in

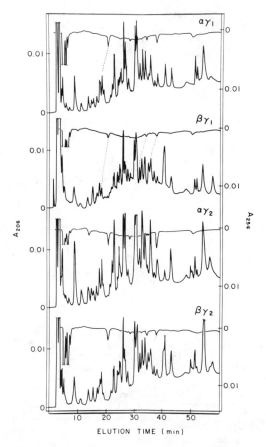

44   STRYDOM AND VALLEE

FIG. 3. Tryptic peptide maps for the $\alpha\gamma_1$-, $\alpha\gamma_2$-, $\beta\gamma_1$-, and $\beta\gamma_2$-isoenzymes of Class I ADH, obtained as described under Materials and Methods.

primary structure. This is consistent with the known enzymatic and, in the case of Class I and II ADH, immunochemical differences (53). This realization greatly facilitates the choice of isoenzymes suitable for further studies.

CONCLUSION

The present investigation was initiated to search for a method which could resolve the large and diverse group of human ADH isoenzymes, based on their sequences together

with their characteristic enzymatic properties. The Class I, II and III ADH isoenzymes do differ markedly from one another, contrasting with the high degree of similarity of the isoenzymes within Class I. Such large differences in structures of isoenzymes or isoproteins from single organs are unusual, but are reminiscent of, e.g., the differences seen between isoenzymes from different organs, such as the M- and H-forms of lactate dehydrogenase (54), or of proteins expressed at different stages of development, such as the β- and γ-chains of hemoglobin (55).

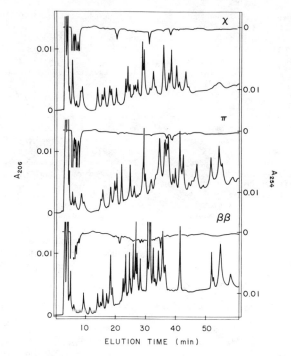

FIG. 4. Tryptic peptide maps for the χ-, π-, and ββ-isoenzymes of Classes III, II and I, respectively, of ADH, obtained as described under Materials and Methods.

Whether the simultaneous presence of these ADH isoenzymes is determined genetically and/or induced pre- or postranslationally cannot yet be stated, but will be given due consideration, as will the structural basis of the differences detected. Apart from their utility in demonstrating the *differences* between isoenzymes, the maps also reveal information regarding their *similarities*. The total pattern of peaks reflects the chemical character and therefore the identity of an isoenzyme or group of isoenzymes. Thus, the specific overall *pattern* of peaks of a map would seem to characterize the parent protein. This feature of this mapping technique may prove useful for multisubunit proteins in general, protein complexes, or groups of proteins with related but not identical functions, requiring chemical differentiation and identification.

ACKNOWLEDGMENTS

Appreciation is extended to Dr. A. Burger, Dr. M. Morelock, and Dr. D. McKenzie for the supply of isoenzymes.

REFERENCES

1. Smith, M., Hopkinson, D. A., and Harris, H. (1971) *Ann. Hum. Genet. (London)* **34**, 251–271.
2. Bosron, W. F., Li, T.-K., and Vallee, B. L. (1980) *Proc. Nat. Acad. Sci. USA* **77**, 5784–5788.
3. Pares, X., and Vallee, B. L. (1981) *Biochem. Biophys. Res. Commun.* **98**, 122–130.
4. Von Wartburg, J.-P., Papenberg, J., and Aebi, H. (1965) *Canad. J. Biochem.* **43**, 889–898.
5. Bosron, W. F., Li, T.-K., Dafeldecker, W. P., and Vallee, B. L. (1979) *Biochemistry* **18**, 1101–1105.
6. Li, T.-K. (1977) *Advan. Enzymol.* **45**, 427–483.
7. Lange, L. G., and Vallee, B. L. (1976) *Biochemistry* **15**, 4681–4686.
8. Blair, A. H., and Vallee, B. L. (1966) *Biochemistry* **5**, 2026–2034.

9. Lange, L. G., Sytkowski, A. J., and Vallee, B. L. (1976) *Biochemistry* **15**, 4687–4693.
10. Li, T.-K., and Theorell, H. (1969) *Acta Chem. Scand. Ser. B* **23**, 892–902.
11. Schenker, T. M., Teeple, L. J., and Von Wartburg, J.-P. (1971) *Eur. J. Biochem.* **24**, 271–279.
12. Stamatoyannopoulos, G., Chen, S.-H., and Fukui, M. (1975) *Amer. J. Hum. Genet.* **27**, 789–796.
13. Teng, Y.-S., Jehan, S., and Lie-Injo, L. E., (1979) *Hum. Genet.* **53**, 87–90.
14. Harada, S., Misawa, S., Agarwal, D. P., and Goedde, H. W. (1980) *Amer. J. Hum. Genet.* **32**, 8–15.
15. Azevedo, E. S., daSilva, M. C. B. O., and Tavares-Neto, J. (1975) *Ann. Hum. Genet.* **39**, 321–327.
16. Li, T.-K., and Magnes, L. J. (1975) *Biochem. Biophys. Res. Commun.* **63**, 202–208.
17. Li, T.-K., Bosron, W. F., Dafeldecker, W. P., Lange, L. G., and Vallee, B. L. (1977) *Proc. Nat. Acad. Sci. USA* **74**, 4378–4381.
18. Bosron, W. F., Li, T.-K., Lange, L. G., Dafeldecker, W. P., and Vallee, B. L. (1977) *Biochem. Biophys. Res. Commun.* **74**, 85–91.
19. Jornvall, H. (1970) *Eur. J. Biochem.* **16**, 41–49.
20. Molnar, I., and Horvath, C. (1977) *J. Chromatogr.* **142**, 623–640.
21. Rivier, J. E. (1978) *J. Liq. Chromatogr.* **1**, 343–366.
22. Hancock, W. S., Bishop, C. A., Prestidge, R. L., and Hearn, M. T. W. (1978) *Anal. Biochem.* **89**, 203–212.
23. Hearn, M. T. W., Grego, B., and Hancock, W. S. (1979) *J. Chromatogr.* **185**, 429–444.
24. O'Hare, M. J., and Nice, E. C. (1979) *J. Chromatogr.* **171**, 209–226.
25. Hancock, W. S., Bishop, C. A., Prestidge, R. L., Harding, D. R. K., and Hearn, M. T. W. (1978) *Science* **200**, 1168–1170.
26. Terabe, S., Konaka, R., and Inouye, K. (1979) *J. Chromatogr.* **172**, 163–177.
27. Meek, J. L. (1980) *Proc. Nat. Acad. Sci. USA* **77**, 1632–1636.
28. Hearn, M. T. W. (1980) *J. Liq. Chromatogr.* **3**, 1255–1276.
29. Regnier, F. R., and Gooding, K. M. (1980) *Anal. Biochem.* **103**, 1–25.
30. Larsen, B., Viswanatha, V., Chang, S. Y., and Hruby, V. J. (1978) *J. Chromatogr. Sci.* **16**, 207–210.
31. Martinelli, R. A., and Scheraga, H. A. (1979) *Anal. Biochem.* **96**, 246–249.
32. Black, C., Douglas, D. M., and Tanzer, M. L. (1980) *J. Chromatogr.* **190**, 393–400.
33. Christie, D. L., Gagnon, J., and Porter, R. R. (1980) *Proc. Nat. Acad. Sci. USA* **77**, 4923–4927.
34. Setlow, P., and Ozols, J. (1980) *J. Biol. Chem.* **255**, 8413–8416.
35. Moonen, P., Akeroyd, R., Westerman, J., Puijk, W. C., Smits, P., and Wirtz, K. W. A. (1980) *Eur. J. Biochem.* **106**, 279–290.
36. Lorimer, G. H. (1981) *Biochemistry* **20**, 1236–1240.
37. Kemp, M. C., Hollaway, W. L., Prestidge, R. L., Bennett, J. C., and Compans, R. W. (1981) *J. Liq. Chromatogr.* **4**, 587–598.
38. Rubinstein, M., Stein, S., and Udenfriend, S. (1977) *Proc. Nat. Acad. Sci. USA* **74**, 4969–4972.
39. Rubinstein, M., Chen-Kiang, S., Stein, S., and Udenfriend, S. (1979) *Anal. Biochem.* **95**, 117–121.
40. McMillan, M., Cecka, J. M., Hood, L., Murphy, D. B., and McDevitt, H. O. (1979) *Nature (London)* **277**, 663–665.
41. Fullmer, C. S., and Wasserman, R. H. (1979) *J. Biol. Chem.* **254**, 7208–7212.
42. Hughes, G. J., Winterhalter, K. H., and Wilson, K. J. (1979) *FEBS Lett.* **108**, 81–86.
43. Schroeder, W. A., Shelton, J. B., Shelton, J. R., and Powars, D. (1979) *J. Chromatogr.* **174**, 385–392.
44. Wilson, J. B., Lam, H., Pravatmuang, P., and Huisman, T. H. J. (1979) *J. Chromatogr.* **179**, 271–290.
45. Kratzin, H., Yang, C., Krusche, J. U., and Hilschmann, N. (1980) *Hoppe-Seyler's Z. Physiol. Chem.* **361**, 1591–1598.
46. Esch, F., Bohlen, P., Ling, N., Benoit, R., Brazeau, P., and Guillemin, R. (1980) *Proc. Nat. Acad. Sci. USA* **77**, 6827–6831.
47. Chaiken, I. M., and Hough, C. J. (1980) *Anal. Biochem.* **107**, 11–16.
48. Mahoney, W. C., and Hermodson, M. A. (1980) *J. Biol. Chem.* **255**, 11,199–11,203.
49. Van Der Rest, M., Bennett, H. P. J., Solomon, S., and Glorieux, F. H. (1980) *Biochem. J.* **191**, 253–256.
50. Hobbs, A. A., Grego, B., Smith, M. G., and Hearn, M. T. W. (1981) *J. Liq. Chromatogr.* **4**, 651–659.
51. Burger, A. R., and Vallee, B. L. (1981) *Fed. Proc.* **40**, 1886.
52. Crestfield, A. M., Moore, S., and Stein, N. H. (1963) *J. Biol. Chem.* **238**, 622–627.
53. Adinolfi, M., Adinolfi, M., Hopkinson, D. A., and Harris, H. (1978) *J. Immunogenet.* **5**, 283–296.
54. Kiltz, H. H., Keil, W., Griesbach, M., Petry, K., and Meyer, H. (1977) *Hoppe-Seyler's Z. Physiol. Chem.* **358**, 123–127.
55. Schroeder, W. A., Shelton, J. R., Shelton, J. B., Cormick, J., and Jones, R. T. (1963) *Biochemistry* **2**, 992–1008.

Purification of Glycopeptides of Human Ceruloplasmin and Immunoglobulin D by High-Pressure Liquid Chromatography*

Daniel Tetaert,[1] Nobuhiro Takahashi, and Frank W. Putnam[2]

Department of Biology, Indiana University, Bloomington, Indiana 47405

To facilitate structural studies of glycoproteins, reverse-phase high-pressure liquid chromatography (HPLC) methods have been developed for preparative isolation of glycopeptides and have been applied to human ceruloplasmin as an example of glycopeptides containing glucosamine (GlcN) and to human immunoglobulin D (IgD) for glycopeptides containing galactosamine (GalN). The use of RP-P columns and of trifluoroacetic acid and heptafluorobutyric acid as counterions was investigated. Various elution systems (both isocratic and programmed gradient) were used with n-propanol to assess the relative hydrophilicity of the peptides. The procedure developed for the GlcN glycopeptides of ceruloplasmin enabled purification of nine major chymotryptic peptides (ranging in size from 15 to 29 residues) and also of many minor peaks. These were characterized by amino acid and endgroup analysis, and the complete sequence of five was determined. These represent three different sites of GlcN attachment in the amino-terminal half of the ceruloplasmin chain. The procedures developed have enabled isolation of glycopeptides from ceruloplasmin having a single GlcN oligosaccharide attached; the latter are valuable for study of the structure and function of the carbohydrate groups. Separation of GalN glycopeptides from IgD was more difficult because of the high content of GalN in the hinge. Purification and sequence analysis was aided by partial removal of sugar by treatment with HF and by other methods. Four (or five) GalN oligosaccharides are attached to serine or threonine residues in the IgD hinge region, and all but one are in close proximity in the repeating sequence Ala-Thr-Thr-Ala-Pro-Ala-Thr-Thr.

Prior to development of high-pressure liquid chromatography (HPLC)[3] methods, determination of the structure of glycoproteins was difficult for four reasons: (i) the heterogeneity of the carbohydrate chains, (ii) the multiple forms of glycopeptides resulting from incomplete enzymatic digestion due to steric hindrance of the sugar chains, (iii) difficulties in purification of a series of similar large glycopeptides, and (iv) technical problems of amino acid sequence determination. These problems affected the determination of the amino acid sequence of the polypeptide chain and the sites of glycosylation and also hindered elucidation of the oligosaccharide structure. Many glycoproteins have multiple sites of attachment, and the oligosaccharides at different sites may differ in structure and function (1,2). Thus, for both the protein chemist and the carbohydrate chemist, it is desirable to isolate a glycopeptide of known sequence with a single oligosaccharide attached.

To aid the structural study of multiglycosylated proteins, we have developed reverse-phase HPLC methods for preparative isolation of glycopeptides and have applied them to human ceruloplasmin as an example

---

* This paper was presented at the International Symposium on HPLC of Proteins and Peptides, November 16–17, 1981, Washington, D. C.

[1] Present address: Unite INSERM 16, Place de Verdun, F59045 Lille Cedex, France.

[2] To whom reprint requests and correspondence should be addressed.

[3] Abbreviations used: HPLC, High-pressure liquid chromatography; GlcN, glucosamine; IgD, immunoglobulin D; GalN, galactosamine; TFA, trifluoroacetic acid; HFBA, heptafluorobutyric acid; GP, glycopeptide.

---

of GlcN glycopeptides and to human IgD as an example of GalN glycopeptides. The GlcN and GalN oligosaccharides differ in size, structure, branching, acceptor residue, and proximity of sites of attachment. Thus, the purification of the glycopeptides presented different technical problems for the two proteins. In the case of ceruloplasmin, GlcN glycopeptides representing three different sites of attachment in the 67,000-dalton amino-terminal fragment of ceruloplasmin were isolated, and their amino acid sequences were determined. One of these had not been found in earlier studies. All were linked to asparagine in the obligatory tripeptide acceptor sequence Asn-X-Ser/Thr (1,2). In the case of IgD, the hinge peptide had resisted conventional methods of enzymatic cleavage and sequence analysis. Separation of a series of GalN glycopeptides by HPLC greatly facilitated sequence analysis of the hinge and determination of the attachment sites of GalN.

MATERIALS AND METHODS

Proteins. Ceruloplasmin was purified from normal pooled human plasma (3), and the 67,000-dalton fragment was prepared from reduced and carboxymethylated ceruloplasmin by gel filtration on a column of Sephadex G-200 (4–6). IgD protein WAH was purified from the plasma of a patient with multiple myeloma, and the δ heavy chain was prepared from the reduced and aminoethylated IgD by Sephadex G-150 column chromatography (7–9). All of these proteins and fragments were pure as judged by various criteria such as amino-terminal analysis and polyacrylamide gel electrophoresis with sodium dodecyl sulfate.

Reagents. Trypsin treated with L-(1-tosylamido-2-phenol)ethylchloromethyl ketone was purchased from Worthington, and chymotrypsin was obtained from Millipore. Trifluoroacetic acid (TFA) and heptafluorobutyric acid (HFBA) of sequence grade were purchased from Pierce, and n-propanol

of chromatography grade was obtained from Burdick and Jackson Laboratories (Muskegon, Mich.).

Enzymatic digests. The 67,000-dalton fragment of ceruloplasmin (100 mg) was digested with 1 mg of chymotrypsin in 0.1 M ammonium bicarbonate for 4 h at 37°C. The chymotryptic digest was applied to a column of Sephadex G-50 (2.5 × 110 cm) equilibrated with 1 N acetic acid. The peptides were eluted at a flow rate of 25 ml/h, and the elution was monitored by absorbance at 280 nm and by reaction with fluorescamine. The same procedure was used for the separation of the peptides of the δ heavy chain (163 mg) of IgD, which had been digested with 4 mg of chymotrypsin in 0.1 M ammonium bicarbonate for 6 h at 37°C. The largest chymotryptic peptide containing galactosamine from the δ heavy chain was further digested with trypsin (10).

High-performance liquid chromatography (HPLC). Glycopeptide purification was achieved by use of an HPLC system consisting of a Beckman controller (Model 421) and pump (Model 110A) and a reverse-phase Synchropak RP-P column (0.41 × 25 cm) obtained from SynChrom (Linden, Ind.). The column was equilibrated with 0.1% TFA or 0.1% HFBA (11) as counterions and was eluted at a flow rate of 0.7 ml/min with a programmed or linear gradient of n-propanol (0–60%). The peptides were detected by measuring the absorbance at 230 nm.

Amino acid and sequence analysis. Methods for amino acid analysis with the Beckman amino acid analyzer (Model 121M) and for sequence determination with the Beckman sequencer (Model 890C) have been described (4–9). To determine the amino sugar content, the sample was hydrolyzed in 4 M HCl at 110°C for 6 h and was analyzed with the Beckman Model 121M amino acid analyzer (10).

Despite their purity, the GalN-rich glycopeptides were difficult to sequence by automatic Edman degradation, apparently because of the carbohydrate present. Thus, in

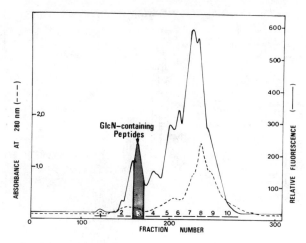

FIG. 1. Separation of the chymotryptic digest of the 67,000-dalton fragment of human ceruloplasmin by Sephadex G-50 chromatography. The fraction containing glucosamine is cross-hatched.

order to remove sugar, they were submitted to chemical reaction with HF under various conditions or to enzymatic removal of the sugar by treatment with neuraminidase or endo-α-galactosidase, as described elsewhere (10), and the sugar and the peptides were separated by HPLC.

RESULTS AND DISCUSSION

Purification of GlcN glycopeptides of ceruloplasmin. The isolation of the GlcN glycopeptides from the 67,000-dalton fragment of human ceruloplasmin was done after a prefractionation of the chymotryptic digest on a Sephadex G-50 column. The profile obtained after reaction with fluorescamine gave better definition than the absorbance at 280 nm. Ten fractions were identified (Fig. 1). Amino acid analysis revealed the presence of GlcN peptides only in fraction 3. This procedure removed uncleaved material, large peptides, and residual chymotrypsin (fractions 1 and 2), which may adversely affect the use of HPLC. Fraction 3, which eluted rapidly from the G-50 column because of its high sugar content, was ap-

plied to an RP-P column (Fig. 2). By use of a linear gradient of n-propanol (0–60%), nine main peaks were eluted; the first five peaks (GP1 to GP3) contained GlcN peptides according to the amino acid composition (Table 1).

This reverse-phase HPLC technique is very efficient for purification of glycopeptides compared to conventional techniques such as Dowex or Aminex ion-exchange chromatography or paper electrophoresis in which adsorption of GlcN peptides may occur on the matrix. In fact, very similar glycopeptides can be separated by use of reverse-phase HPLC. For example, the two glycopeptide peaks designated GP1 (under the bracket in Fig. 2) could not be differentiated by their amino acid composition and were therefore pooled and sequenced together because of the small amount of each. We assume that the two GP1 glycopeptides differed either in their carbohydrate structure or in an amide group because of deamidation during laboratory procedures. This HPLC procedure can also separate large glycopeptides that differ only by a single amino acid residue at the amino or carboxy

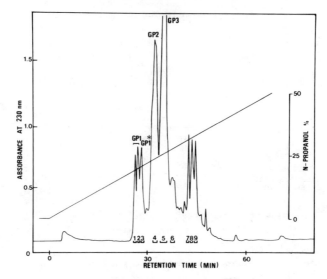

FIG. 2. Separation of the glycopeptides (GP) from Fig. 1 by HPLC on a reverse-phase Synchropak RP-P column. The peptides were eluted with a linear gradient of n-propanol containing 0.1% TFA. The peaks underlined were rechromatographed under the same conditions. The first two peaks are both designated GP1 because they appeared identical in amino acid composition and sequence (see text).

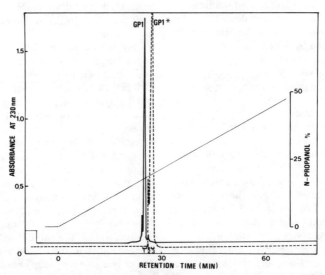

FIG. 3. Purification of glycopeptides from Fig. 2 by rechromatography. Fractions 1 and 3 from Fig. 2 were rechromatographed separately under the same conditions as in Fig. 2. The solid line gives the chromatogram of fraction 1 and the dash line shows the chromatogram of fraction 3. The peaks underlined were collected.

terminus, as is shown by rechromatography of GP1 and GP1* in Fig. 3. Thus, GP1, which has 18 amino residues, and GP1*, which has 19, were sequenced separately and the only difference found was the presence of phenylalanine as the amino terminus of GP1* (Fig. 4). Further studies have shown that two chymotryptic peptides result from the fact that the sequence in this region is -Ala-Phe-Phe-Glu-Val-. It is noteworthy that the sum of the yield of GP1 and GP1* is about the same as the yield of GP2 or GP3. Figure 3, showing the rechromatography of GP1 and GP1*, demonstrates both the high reproducibility of the chromatographic pattern and also the fact that better separation is obtained when small amounts of glycopeptides (50 nmol) are applied. However, the separation of these glycopeptides was effective even when a large amount (150 nmol) was applied, and thus the procedure is practical on a preparative scale for sequence analysis by conventional methods, instead of resorting to the microsequencing approach.

Amino acid sequence analysis of glycopeptides. Glycopeptides GP2 and GP3 from the separation shown in Fig. 2 were rechromatographed under the same conditions as for GP1 and GP1* shown in Fig. 3, and all of the peptides were submitted to sequence analysis with the results summarized in Fig. 4.

One advantage of the HPLC technique is that it enabled us to isolate and sequence a glycopeptide not previously reported in human ceruloplasmin, i.e., GP1. The amino acid sequences of three nonidentical tryptic glycopeptides from human ceruloplasmin have been reported by Rydén and Eaker (12), who proposed that the three accounted for all the carbohydrate in the protein. Their peptides were designated GP16, GP17, and GP20 because they contained 16, 17, and 20

TABLE 1

AMINO ACID COMPOSITION OF GLYCOPEPTIDES FROM THE 67,000-DALTON FRAGMENT OF HUMAN CERULOPLASMIN

Amino acid	Glycopeptide			
	GP1	GP1*	GP2	GP3
Aspartic acid	2.9	3.2	2.3	4.8
Threonine	—	—	1.9	1.9
Serine[a]	2.2	3.0	1.7	0.7
Glutamic acid	3.1	3.3	1.4	5.1
Proline	—	—	1.0	2.0
Glycine	1.0	1.2	1.1	2.0
Alanine	—	—	2.0	2.0
Half-cystine[b]	0.9	0.8	—	—
Valine	0.9	0.9	0.9	1.0
Isoleucine	0.8	1.0	—	0.8
Leucine	—	—	0.8	—
Tyrosine	—	—	—	2.8
Phenylalanine	—	0.8	0.9	1.0
Lysine	3.0	3.1	1.2	2.0
Histidine	1.0	1.0	—	1.0
Arginine	1.0	0.9	—	1.0
Total residues	18	19	15	29
Glucosamine[c]	1.7	1.6	1.6	1.1
Amino terminus	Glu	Phe	Thr	Tyr
Yield (nmol)	58	50	100	141

Note. Values are expressed as moles of residue per mole peptide.

[a] Uncorrected for destruction in acid hydrolysis. Sequence analysis gave a value of three residues for GP1.

[b] Determined as carboxymethylcysteine.

[c] The value is low because it was determined after hydrolysis for 24 h in 6 M HCl at 110°C.

FIG. 4. Amino acid sequence of the chymotryptic glycopeptides from the 67,000-dalton fragment of human ceruloplasmin. Glucosamine oligosaccharides are indicated by CHO over the residues to which they are attached.

amino acid residues, respectively. Dwulet and Putnam (6) showed that peptide GP17 is present in the sequence of the 50,000-dalton fragment and that GlcN is linked to Asn-262. GP20 corresponds to the tryptic peptide beginning at the glutamic acid in our chymotryptic peptide GP2 (Fig. 4). In work still in progress we have isolated a tryptic glycopeptide corresponding to GP20 and have confirmed the sequence of GP20. We have

also greatly extended the sequence in this area of the 67,000-dalton fragment. GP16 corresponds to the tryptic peptide beginning at the first glutamic acid in GP3 and terminating at arginine. However, where we give the sequence Ala-Ile-Tyr-Pro-Asp-Asn(CHO) Rydén and Eaker (12) give the sequence Pro-Ala-Ile-Tyr-Asn(CHO)-Asp. We also isolated the tryptic peptide corresponding to GP16 and confirmed the se-

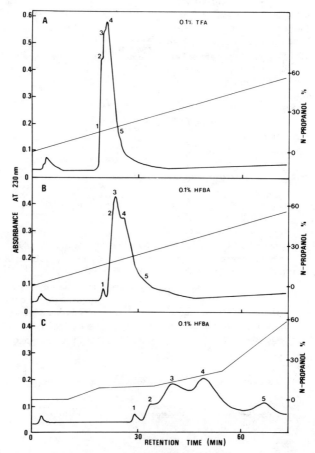

FIG. 5. Separation of the GalN glycopeptides of the hinge region of the δ chain of human IgD by use of an RP-P column under several conditions. The fraction containing galactosamine from the chymotryptic digest of the δ heavy chain was applied to the column and was eluted with a linear gradient of n-propanol containing 0.1% TFA (A) or 0.1% HFBA (B) or with a programmed gradient of n-propanol containing 0.1% HFBA (C).

quence given for GP3 in Fig. 4. Close inspection of the experimental data given by Rydén and Eaker for GP16 indicates that their results support our sequence better than their sequence. GP1, the third glycopeptide that we isolated by HPLC from the 67,000-dalton fragment, has not been reported previously. A tryptic glycopeptide that confirmed and extended the sequence at the amino terminus of GP3 was also isolated.

Purification of GalN glycopeptides of the IgD delta chain. The purification of the GalN glycopeptides of the hinge region of IgD was much more difficult than for the GlcN glycopeptides of ceruloplasmin for two reasons: (i) the nature of the oligosaccharides and (ii) the resistance to proteolytic cleavage of the GalN-rich section of the polypeptide chain of the IgD hinge region. The GalN oligosaccharides are small (probably trisaccharides with the general structure GalN-Gal-sialic acid), whereas the GlcN oligosaccharides are large and have complex structures; however, four or five GalN oligosaccharides are located close together in a stretch of some 25 residues compared to only one GlcN oligosaccharide in each of the ceruloplasmin glycopeptides. Thus, the IgD hinge glycopeptides have multiple sites of hydrophilicity. This may account for the fact that when reverse-phase HPLC was used with the normal linear gradient (n-propanol containing 0.1% TFA), several GalN glycopeptides were detected but without real separation (Fig. 5A). Substitution of HFBA as the counterion increased the separation of the GalN peptides on the RP-P column (Fig. 5B). HFBA is known to change the charge on peptides more than TFA (11), and thus the more hydrophilic GalN peptides tend to elute later. A programmed gradient of n-propanol containing 0.1% HFBA gave the best result (Fig. 5C), and five fractions were identified. The purification of the five fractions was achieved by rechromatography using various gradient conditions selected for the optimum separation of each peptide. For example, the 74-residue glycopeptide that contained all the GalN attachment sites (fraction 4) was eluted from the column at about 15% propanol in 0.1% HFBA, whereas fraction 3 (residues numbered 25–74 in Fig. 6) had one less GalN site and was eluted at 8% propanol in 0.1% HFBA. It was noted that the hinge glycopeptides eluted faster when applied in preparative amounts. The amino acid sequence of these two GalN peptides is shown in Fig. 6, which summarizes the results of sequence determination of these two glycopeptides and of other peptides obtained from other structural studies of the hinge region (10).

Structural studies and the biological role of the GlcN oligosaccharides. The HPLC method described here is useful not only to purify glycopeptides as an aid to amino acid sequence analysis but also to facilitate study of the structure and function of carbohydrates covalently linked to proteins. Samples of all four of the ceruloplasmin glycopeptides have been given to Dr. Jacques Baenziger of the School of Medicine, Washington Uni-

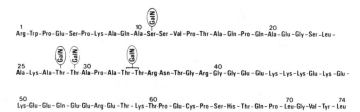

FIG. 6. Amino acid sequence of chymotryptic peptides of the IgD hinge region. Fraction 3 of Fig. 5C corresponds to the sequence of positions 25–74, and fraction 4 to positions 1–74. It is not certain whether GalN is linked to one or both of the threonine residues at positions 33 and 34.

versity of St. Louis, who is determining the structure of the oligosaccharides and also their role in endocytosis by hepatocytes. The carbohydrate structure of the GalN glycopeptides from the hinge region of IgD is likewise being determined.

ACKNOWLEDGMENTS

We thank Y. Takahashi, J. Madison, S. Dorwin, P. H. Davidson, and J. Dwulet for technical assistance. We also thank F. E. Dwulet and R. Bauman for help and advice in the work on ceruloplasmin and J. Baenziger for undertaking structural study of the carbohydrates in the GlcN and GalN glycopeptides. This work was supported by Grants AM19221 from the National Institutes of Health, IM-2G from the American Cancer Society, and CA08497 from the National Cancer Institute.

REFERENCES

1. Clamp, J. R. (1975) *in* The Plasma Proteins (Putnam, F. W., ed.), 2nd ed., Vol. 2, pp. 163–211, Academic Press, New York.
2. Kornfeld, R., and Kornfeld, S. (1976) *Annu. Rev. Biochem.* **45**, 217–237.
3. Noyer, M., Dwulet, F. E., Hao, Y., and Putnam, F. W. (1980) *Anal. Biochem.* **102**, 450–458.
4. Kingston, I. B., Kingston, B. L., and Putnam, F. W. (1980) *J. Biol. Chem.* **255**, 2878–2885.
5. Kingston, I. B., Kingston, B. L., and Putnam, F. W. (1980) *J. Biol. Chem.* **255**, 2886–2896.
6. Dwulet, F. E., and Putnam, F. W. (1981) *Proc. Nat. Acad. Sci. USA* **78**, 790–794.
7. Lin, L.-C., and Putnam, F. W. (1981) *Proc. Nat. Acad. Sci. USA* **78**, 504–508.
8. Putnam, F. W., Takahashi, N., Tetaert, D., Debuire, B., and Lin, L.-C. (1981) *Proc. Nat. Acad. Sci. USA* **78**, 6168–6172.
9. Lin, L.-C., and Putnam, F. W. (1979) *Proc. Nat. Acad. Sci. USA* **76**, 6572–6576.
10. Takahashi, N., Tetaert, D., and Putnam, F. W. (1982) *in* IVth International Conference on Methods in Protein Sequence Analysis, Brookhaven National Laboratory, Sept. 21–25, 1981 (Elzinga, M., ed.), Humana Press, Clifton, N. J., in press.
11. Bennett, H. P. J., Browne, C. A., and Solomon, S. (1980) *J. Liq. Chromatogr.* **3**, 1353–1365.
12. Rydén, L., and Eaker, D. (1974) *Eur. J. Biochem.* **44**, 171–180.

High-Performance Size-Exclusion Chromatography of Hydrolyzed Plant Proteins*

Howard G. Barth

Hercules Incorporated, Research Center, Wilmington, Delaware 19899

A high-performance size-exclusion chromatographic (SEC) procedure has been developed to determine the molecular size distribution of several sources of hydrolyzed plant proteins including soy, potato, cotton seed, corn gluten, wheat bran, and wheat germ. The SEC packing consisted of a glycerylpropylsilyl layer covalently bonded to 100-Å pore-size silica particles (10 μm). A number of mobile phases were evaluated in an attempt to reduce adsorption between hydrolyzed components and packing material. Systems containing either sodium dodecyl sulfate or methanol showed the best performances; however, adsorption could not totally be eliminated. The combination of size-exclusion and retention mechanisms of separation was valuable in establishing differences among samples. Based on dialysis of hydrolyzed soy protein, it appeared that most of the adsorbed material was of low molecular weight. Column recovery studies indicated complete elution of samples.

Hydrolyzed or water-soluble proteins from either animal or plant sources have been used extensively in the food industry as protein supplements and flavor enhancers and thickeners in soups and sauces (1,2). Because water-soluble proteins are capable of forming stable foams (2–4), they are also employed as whipping agents (5,6). Non-food-related applications of hydrolyzed proteins include their use as firefighting foams (3,7).

Commercially, proteins are hydrolyzed chemically (8,9) and/or with enzymes (10–14). The resulting mixtures are highly complex and consist of mainly proteinaceous material (proteins, peptides, and amino acids) (8) and, to a lesser extent, carbohydrates, humins, levulinic acid, formic acid (15), and a host of degradation products, the composition of which depends on hydrolysis conditions.

A knowledge of the molecular weight dis-

tribution (MWD)[1] of hydrolyzed proteins is of great importance in being able to predict the end-use characteristics of the product and for quality control. The goal in developing a size-exclusion chromatographic (SEC) system is to select a mobile phase that will eliminate partitioning and/or adsorption on the packing material. Because proteins possess a wide variety of functional groups along the chain—hydrophobic, hydrophilic, and ionic—the composition of the mobile phase must be adjusted to eliminate or reduce various nonsize-exclusion effects.

The types of solute-packing interactions and methods to eliminate or reduce these effects are given in Table 1. Briefly, the ionic strength of the mobile phase must be sufficiently large to eliminate ion-inclusion and ion-exclusion effects. Depending on the charge density of the polymer and packing, an ionic strength greater than 0.01 M may

[1] Abbreviations used: MWD, molecular weight distribution; SEC, size-exclusion chromatography; HPLC, high-performance liquid chromatography; SDS, sodium dodecyl sulfate; RI, refractive index; $\bar{M}_r$, weight-average molecular weight.

* This paper was presented at the International Symposium on HPLC of Proteins and Peptides, November 16–17, 1981, Washington, D. C.

TABLE 1

METHODS TO REDUCE OR ELIMINATE NONSIZE-
EXCLUSION EFFECTS IN AQUEOUS SEC OF PROTEINS

Effect	Method
Electrostatic interactions	
Ion exchange	Increase ionic strength; adjust pH.
Ion exclusion	Increase ionic strength.
Ion inclusion	Increase ionic strength.
Adsorption	
Coulombic	Increase ionic strength; adjust pH.
Hydrogen bonding	Add urea or guanidine · HCl.
Hydrophobic	Add organic moderator (SDS, glycol, alcohol); lower ionic strength.

be adequate. Ion exchange may also occur between basic proteins ($pI > 7$) and acidic groups on the packing. To eliminate this interaction, the ionic strength can be increased or the pH of the mobile phase decreased to suppress ionization of the acidic groups. Finally, hydrophobic interactions between protein and packing may be reduced or eliminated by either decreasing the ionic strength or polarity of the mobile phase or by adding an ionic surfactant.

Previously, conventional SEC techniques using crosslinked dextran or polyacrylamide gels have been employed for the determination of the MWD of hydrolyzed proteins (11,13,14). During the past several years, high-performance SEC packings for proteins have become commercially available (16–22). Because of their high efficiencies and short analysis times, these supports are superior to conventional packings.

In this report, we describe the development of a high-performance SEC method for the determination of the MWD of hydrolyzed plant proteins employing a silica-based packing chemically modified with a glycerylpropylsilyl coating.

MATERIALS AND METHODS

Both commercial and laboratory samples of chemically hydrolyzed proteins were analyzed by SEC. Typical protein content of

these samples were soy, 50%; corn gluten, 62%; potato, 74%; cotton seed, 58%; wheat germ, 26%; and bran, 17%. Samples of enzymatically (pepsin) hydrolyzed soy protein isolates (peptone, albumen, and D100WA) were obtained from A. E. Staley Manufacturing Company.

Protein standards used were bovine serum albumin, chymotrypsinogen A, ribonuclease A (Pharmacia Fine Chemicals), insulin chain B, bacitracin, bradykinin (Serva Fine Biochemicals), and glycylglycylglycyl-glycine (United States Biochemical Co.). Dextran Blue T2000 was obtained from Pharmacia and D_2O was purchased from Aldrich Chemical Company. All other chemicals were reagent grade. Dialysis tubing employed was Spectrapor No. 6 (2000 molecular weight cut-off) from Spectrum Medical Industries.

HPLC apparatus. The high-performance liquid chromatography (HPLC) system consisted of a Varian 8500 pump and a Rheodyne injection valve equipped with a 20-μl injection loop. A uv spectrophotometric detector (Shoeffel Model SF770P-A1) and a Waters 401 refractometer were employed. Analysis was done at a flow rate of 0.25 ml/min at a chart speed of 1 in./min.

The packing material consisted of a glycerylpropylsilyl layer covalently bonded to LiChrospher silica particles (10 μm) and was purchased prepacked in 25 cm × 4.6 mm i.d. (4.1 mm i.d. for earlier columns) stainless-steel columns from SynChrom. The nominal pore size of this packing was 100 Å, which has a molecular weight exclusion limit of about 3×10^5 with respect to proteins. To prevent contamination build-up on the analytical column, a guard column, 4 cm × 4.6 mm i.d. packed with 100-Å Synchropak packing, obtained from Brownlee Laboratory, was placed before the column. The efficiency of a newly installed column was about 7000 theoretical plates at a flow rate of 0.25 ml/min determined with D_2O. Typical column lifetimes were about 6–9 months, after which time increased adsorption was noted.

TABLE 2

RECOVERY STUDIES

	Mobile-phase composition[a]			
Sample	0.05 M Na$_2$SO$_4$	0.01 M Na$_2$SO$_4$ with 30% MeOH[b]	0.05 M Na$_2$SO$_4$ with 1% SDS	0.05 M Na$_2$SO$_4$ buffered to pH 6 with AcONa/AcOH[c] containing 1% SDS
Hydrolyzed soy protein (lab preparation)	41%	96%	113%	101%
Hydrolyzed soy protein (commercial preparation)	54%	83%	98%	100%

[a] Column by-passed with capillary tube. Areas obtained at 254 nm and measured with a planimeter. Chromatographic conditions: 100-Å Synchropak column (4.1 mm i.d. × 25 cm), 0.25-ml/min flow rate, 20-μl injection volume, 2.8-mg/ml sample concentration.
[b] MeOH, methanol.
[c] AcONa/AcOH, sodium acetate/acetic acid.

Mobile phase. The recommended mobile phase consisted of a 0.05 M sodium acetate/acetic acid, pH 6.0, buffer containing 1% sodium dodecyl sulfate (SDS). To facilitate sample preparation, double-strength mobile phase was prepared from a 0.095 M sodium acetate solution containing 0.005 M acetic acid. The solution was filtered under vacuum through a 0.22-mm Millipore filter (Type GS, 47 mm diameter). After filtration, SDS was added at a concentration of 2% solution. The mobile phase was prepared by diluting this solution 1:1 with filtered, distilled water.

Sample preparation. To avoid mobile-phase mismatch, samples were diluted with equal volumes of double-strength mobile phase and further diluted with mobile phase to give protein concentrations of 2.5 to 3.0 mg/ml. Solid samples were dissolved directly in the mobile phase with stirring for 1 h. Before injection, all samples were filtered through a 0.65-mm Millipore filter (Type DA, 13 mm diameter) using a Swinny adapter.

RESULTS AND DISCUSSION

Mobile Phase Selection

Recovery studies were carried out using commercial and laboratory preparations of chemically hydrolyzed soy protein. The results were calculated by comparing the peak area of injected samples obtained through the chromatographic column with the area obtained in which the column was replaced with a capillary tube. The percentage recoveries, shown in Table 2, indicate that there was a significant amount of protein adsorption with 0.05 M Na$_2$SO$_4$ mobile phase. With the addition of 30% methanol, recoveries increased significantly, indicating that hydrophobic interactions were the dominant source of adsorption. Quantitative recoveries were obtained by the use of 1% SDS in either unbuffered or buffered (pH 6.0) Na$_2$SO$_4$ solutions.

TABLE 3

COMPARISON OF RECOVERIES BASED ON UV AND RI MEASUREMENTS

	Recovery (%)	
Sample	RI	UV
Hydrolyzed soy protein (lab preparation)	43	41
Hydrolyzed soy protein (commercial preparation)	50	54

Note. Mobile phase 0.05 M Na$_2$SO$_4$. See Table 2 for other chromatographic conditions.

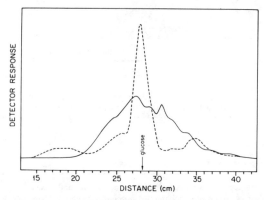

FIG. 1. Size-exclusion chromatography of hydrolyzed soy protein (lot A) using 0.01 M Na₂SO₄ containing 30% methanol for the mobile phase; column, Synchropak 100 Å (4.1 mm i.d. × 25 cm) with guard column (4.6 mm i.d. × 4 cm); flow rate, 0.25 ml/min; chart speed, 1 in./min; injection volume, 20 μl; concentration, 2.8 mg/ml; detectors: ——, uv, 254 nm, 0.1 AUFS; - - -, RI, ×8.

It should be noted that area recoveries obtained from uv measurements are based on absorptivities and not, necessarily, on weight percentage. For example, if a high-uv-absorptivity peptide were adsorbed, the recovery results would be underestimated; the converse is also true. However, with recoveries approaching 100%, this error should be small. Area measurements based on a refractive index (RI) detector would be more reliable. Recovery studies using the RI detector are shown in Table 3, in which glucose

was injected as a control. As indicated, both uv and RI results were comparable. Unfortunately, refractive index measurements could not be employed with samples eluted with mobile phases containing either methanol or SDS because of the occurrence of negative peaks partly caused by mobile-phase mismatch.

Typical chromatograms of hydrolyzed soy protein using various mobile phases are shown in Figs. 1–3. Both uv and RI detector tracings are given. The arrow indicates the

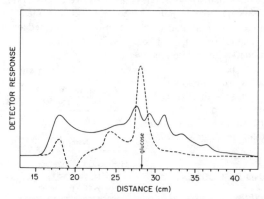

FIG. 2. Size-exclusion chromatography of hydrolyzed soy protein (lot A) using 0.05 M Na₂SO₄ containing 1% SDS; ——, uv; - - -, RI; see Fig. 1 for other conditions.

retention time of glucose, which was used to estimate the totally permeated volume of the column for these analyses. The material eluting before glucose was separated on the basis of size exclusion; components eluting after were retained because of either ion-exchange or hydrophobic interactions. Attempts to eliminate adsorption completely by the use of lower pH mobile phases, increased ionic strength, or increased methanol content were unsuccessful due to protein precipitation.

The RI peak profile in Fig. 1 showed the presence of a high-molecular-weight peak which was not observed with the uv detector. This suggests the presence of nonproteinaceous material. With the addition of SDS, there was a significant change in the MWD profile as well as in peak area, as shown in Fig. 2. The alkyl portion of SDS binds to the hydrophobic regions of the protein, thereby reducing its hydrophobic properties. In addition, its conformation is also altered from a globular to a more elongated shape resulting in an increased hydrodynamic volume.

In order to obtain reproducible chromatographic profiles, a buffered mobile phase was employed to eliminate pH variability among samples. This system was initially composed of 1% SDS in 0.05 M Na_2SO_4 and 0.19 M

buffer (pH 6.0). A chromatogram obtained with this mobile phase is shown in Fig. 3. As compared to Fig. 2, the elution volume of the high-molecular-weight material increased. Ostensibly, the increased ionic strength of this mobile phase was responsible for the observed protein contraction. The occurrence of negative peaks (caused by the presence of SDS) in the RI traces cannot be readily explained.

It should be emphasized that there were significant retention time variations among several columns that were evaluated. Presumably this was caused by differences in stationary phase loadings which affected hydrophobic interactions. Thus, in later experiments, the ionic strength of the mobile phase was decreased by eliminating the Na_2SO_4 and lowering the sodium acetate/acetic acid content to 0.05 M in order to eliminate the influence of stationary phase differences among columns.

Because these samples included a wide range of components of different chemical compositions, it was not possible to eliminate nonsize-exclusion effects totally. Because the object of this work was to develop a method to examine the total composition of these mixtures, protein-, and nonprotein-related components, the combination of both size-

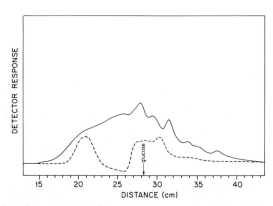

FIG. 3. Size-exclusion chromatography of hydrolyzed soy protein (lot A) using 0.05 M Na_2SO_4/0.19 M sodium acetate buffer (pH 6.0) containing 1% SDS; ——, uv; - - -, RI; see Fig. 1 for other conditions.

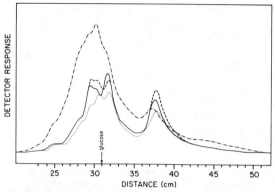

FIG. 4. Size-exclusion chromatography of hydrolyzed soy protein (lot B) at different wavelengths: — · —, 230 nm; ——, 254 nm; – – –, 265 nm; · · ·, 280 nm; see Fig. 3 for other conditions.

exclusion and retention mechanisms of separation was nevertheless valuable in discerning compositional differences among samples.

Detector Wavelength Selection

Figure 4 represents a composite chromatogram of a laboratory preparation of hydrolyzed soy protein monitored at 230, 254, 265, and 280 nm. The 254-nm wave-length was selected for routine analysis because of its universal use in HPLC photometers. Also, there was no need to use the lower, more sensitive wavelengths since detection limit was not a problem.

Sample Dialysis

To determine the molecular weight range of adsorbed components, hydrolyzed soy protein samples were dialyzed against

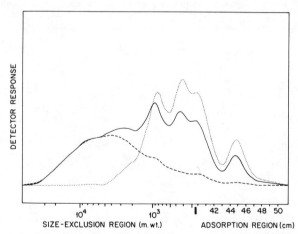

FIG. 5. Molecular weight distribution of hydrolyzed soy protein (lot C) dialyzed for 24 h; ——, original sample; · · ·, dialysate; – – –, dialyzed sample; mobile phase, 0.05 M sodium acetate buffer (pH 6.0) containing 1% SDS; see Fig. 1 for other conditions.

water using a 2000-molecular-weight-cut-off membrane. Figure 5 represents a composite chromatogram of the original hydrolyzed sample and the dialysate and dialyzed portions after 24 h of dialysis. Based on these results and the composite chromatograms, adsorbed species consisted mainly of lower molecular weight (<2000) components.

Column Calibration

The recommended protein standards for column calibration are listed in Table 4, and a typical calibration curve is given in Fig. 6. Other peptides that were examined but were found to adsorb using the 0.05 M acetate buffer/1% SDS mobile phase were glycyltyrosine (K_d = 1.06) and trypsin inhibitor (lung) which did not elute. Bradykinin, bacitracin, and insulin B gave rather broad peaks indicating impure preparations. Gly-gly-gly-glycine was detected at 220 nm; all other compounds were determined at 254 nm. For molecular weight calculations, D_2O served as a calibration to determine the totally permeated volume of the column (23); for this compound an RI detector was employed.

Sample $\bar{M}_r$ calculations were based on the assumption that the uv absorptivities of all species were similar. Because of possible absorptivity differences, the $\bar{M}_r$'s were used to reflect relative molecular weight changes during hydrolysis of a given protein rather than absolute values.

Because both size-exclusion and adsorption separation processes are present in this system, $\bar{M}_r$ calculations ignore the presence of adsorbed species. However, as demonstrated from dialysis experiments (Fig. 5), adsorbed components appear to be of low molecular weight.

Applications

With this high-performance SEC method, the MWD of hydrolyzed plant proteins can be determined within 25 min. Because hydrolyzed samples contain a complex array

TABLE 4

RECOMMENDED SEC CALIBRATION STANDARDS

Standard	Molecular weight	$K_d{}^a$
Dextran blue	2×10^6	0
Bovine serum albumin	67×10^3	0.02
Ovalbumin	43×10^3	0.06
Chymotrypsinogen A	25×10^3	0.12
Ribonuclease	13.7×10^3	0.26
Insulin chain B	3.5×10^3	0.54
Bacitracin	1450	0.78
Bradykinin	1060	0.82
Glycylglycylglycyl-glycine	240	0.95
Glucose	180	0.96
Deuterium oxide	20	1.0

^{a} Distribution coefficient, $K_d = (V_r - V_0)/(V_t - V_0)$, where V_0 is the column exclusion volume and V_t is the totally permeated volume. Chromatographic conditions: 100-Å Synchropak column (4.6 mm i.d. × 25 cm), 0.05 M sodium acetate buffer (pH 6) mobile phase, 20-μl injections, 0.25-ml/ml flow rate.

of proteins, peptides, amino acids, oxidized products, and nonproteinaceous material (mainly carbohydrates), the resulting chromatograms contain both size-excluded and retained species. However, this method has been used to characterize a number of different hydrolyzed proteins for quality control purposes and to establish relationships between composition and end-use performance.

The SEC profile of a laboratory preparation of chemically hydrolyzed soy protein is shown in Fig. 7. The rather complex chromatogram contains numerous components which are both size-excluded and adsorbed. A composite chromatogram of three commercial samples of soy protein isolates, enzymatically hydrolyzed with pepsin, is shown in Fig. 8. Both the albumen and peptone fractions have similar chromatographic features; however, D100WA shows higher molecular weight components and the absence of retained material. Finally, the SEC method was applied to a number of chemically hydrolyzed plant proteins including

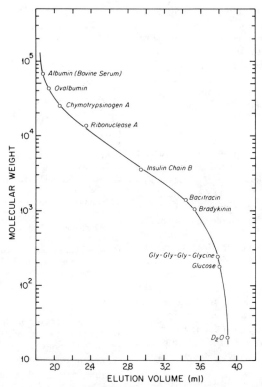

FIG. 6. Protein calibration curve. Mobile phase, 0.05 M sodium acetate buffer (pH 6.0) containing 1% SDS; column, Synchropak 100 Å (4.6 mm i.d. × 25 cm) with guard column; see Fig. 1 for other conditions.

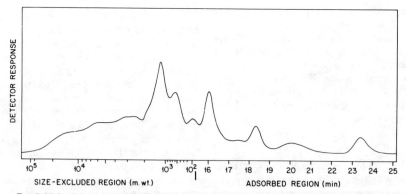

FIG. 7. Molecular weight distribution of chemically hydrolyzed soy protein (lot D); 2 mg/ml; see Fig. 6 for other conditions.

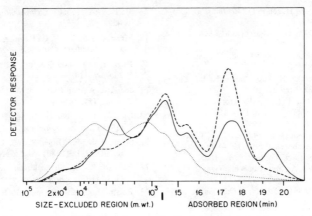

FIG. 8. Molecular weight distribution of enzymatically (pepsin) hydrolyzed soy protein isolates; · · ·, D100 WA; - - -, soy peptone; ——, soy albumen; 5 mg/ml concentrations, see Fig. 6 for conditions.

potato, cotton seed, wheat bran, wheat germ, and corn gluten.

CONCLUSIONS

Chemically hydrolyzed protein consists of a complex array of proteins, peptides, oxidized components, amino acids, and nonproteinaceous material. A mobile-phase system was developed to minimize interactions between sample components and the chromatographic packing. Although a number of different mobile phases were examined, the recommended mobile phase consisted of a 0.05 M sodium acetate/acetic acid buffer (pH 6.0) containing 1% SDS. The sodium acetate was required to prevent ion-exclusion and ion-inclusion effects and to reduce ion exchange between sample and residual silanol groups on the packing. The SDS was needed to reduce hydrophobic interactions. With this mobile phase, adsorption could not be totally eliminated. Because of the large chemical heterogeneity of the sample, further mobile-phase adjustment was not possible without the onset of sample precipitation. Nevertheless, the entire elution profile could be used for "fingerprinting" samples. The method is also applicable for hydrolyzed animal protein and purified protein isolates.

Furthermore, the $\bar{M}_r$ and MWD of the size-excluded portion of the chromatogram can be determined wth an appropriate protein calibration curve.

ACKNOWLEDGMENTS

The author gratefully acknowledges the helpful comments from Walter J. Freeman, William W. Maslanka, William T. McNabola, and Gavin G. Spence of Hercules Incorporated and Fred E. Regnier of Purdue University. The excellent technical assistance of and discussions with David Allen Smith are appreciated. A portion of this paper was written during the author's faculty appointment at Northeastern University, Boston, in the fall of 1981.

REFERENCES

1. Olsman, H. (1979) *J. Amer. Oil Chem. Soc.* **56,** 375–376.
2. Kinsella, J. E. (1979) *J. Amer. Oil Chem. Soc.* **56,** 242–258.
3. Akers, R. J. (1976) Foams, Academic Press, London.
4. Bikerman, J. J. (1973) Foams, Springer-Verlag, New York.
5. Mansvelt, J. W. (1979) *J. Amer. Oil Chem. Soc.* **56,** 350–352.
6. Gunther, R. C. (1979) *J. Amer. Oil Chem. Soc.* **56,** 345–349.
7. Grove, C. S., Wise, G. E., Marsh, W. C., and Gray, J. B. (1951) *Ind. Eng. Chem.* **43,** 1120–1122.

8. Perri, J. M., and Hazel, F. (1946) *Ind. Eng. Chem.* **38**, 549–554.

9. Gunther, R. C., U. S. Patent 3,814,816, June 4, 1974.

10. Mori, K., Kotaka, H., and Maeda, T., Jap. Patent 1978-58981, May 27, 1978.

11. Mori, K., Kotaka, H., and Maeda, T., Jap. Patent 1978-58982, May 27, 1978.

12. Fikushima, M., Fukube, T., Horiuchi, T., and Okayasu, M., Jap. Patent 1974-109551, October 18, 1974.

13. Myers, D. V., Ricks, E., Myers, M. J., Wilkinson, M., and Iacobucci, G. A. (1974) *Proc. IV Int. Congr. Food Sci. Technol.* **5**, 96–102.

14. Constantinides, A., and Adu-Amankwa, B. (1980) *Biotechnol. Bioeng.* **22**, 1543–1565.

15. Fukushima, D. (1979) *J. Amer. Oil Chem. Soc.* **56**, 357–362.

16. Barth, H. G. (1980) *J. Chromatogr. Sci.* **18**, 409–429.

17. Pfannkoch, E., Lu, K. C., Regnier, F. E., and Barth, H. G. (1980) *J. Chromatogr. Sci.* **18**, 430–441.

18. Regnier, F. E., and Gooding, K. M. (1980) *Anal. Biochem.* **103**, 1–25.

19. Jenik, R. A., and Porter, J. W. (1981) *Anal. Biochem.* **111**, 184–188.

20. Diosady, L. L., Bergen, I., and Harwalker, V. R. (1980) *Milchwissenschaft* **35**, 671–674.

21. Schmidt, D. E., Giese, R. W., Conron, D., and Karger, B. L. (1980) *Anal. Chem.* **52**, 177–182.

22. Barford, R. A., Sliwinski, B. J., and Rothbart, H. L. (1979) *Chromatographia* **12**, 285–288.

23. Barth, H. G., and Regnier, F. E., manuscript in preparation.

The Isolation of Peptides by High-Performance Liquid Chromatography Using Predicted Elution Positions*

C. A. BROWNE,[1] H. P. J. BENNETT, AND S. SOLOMON

Endocrine Laboratory, Royal Victoria Hospital, and Departments of Medicine and Biochemistry, McGill University, Montreal, Quebec H3A 1A1, Canada

This paper describes the derivation and use of predictive retention coefficients for the reversed-phase high-performance liquid chromatography of peptides. The use of predicted elution positions in the isolation of peptides is illustrated by two examples where peptides, whose existence was postulated from cDNA sequence data, have been successfully isolated. The combination of the powerful chromatographic technology and the ability to predict the elution positions of peptides based on their composition provides a very potent method for the isolation of peptides from biological tissues.

The use of reversed-phase high-performance liquid chromatography (RP-HPLC)[2] in the separation of peptides has been rapidly increasing in the last few years (1–4). RP-HPLC has been used for a variety of different applications including the separation of peptide hormones (5,6), the separation of proteolytic digests for peptide mapping (7,8), and the separation of globin chains (9,10). The preparative applications of RP-HPLC have so far been mostly limited to the final steps of an isolation procedure (11–13). This approach has been useful, perhaps most notably in the recent isolation of corticotropin releasing factor (14).

In our laboratory, much effort has gone into the development of a method for the complete isolation of peptides by the combination of a reversed-phase extraction procedure (15,16) and RP-HPLC (17,18). This isolation method depends upon the different properties of two hydrophobic ion-pairing reagents, trifluoroacetic acid (TFA), and heptafluorobutyric acid (HFBA). During a study of the potential use of various perfluorinated carboxylic acids as hydrophobic ion-pairing reagents in RP-HPLC of peptides, it was observed that the elution order of a series of standard peptides was highly dependent on the nature of the perfluorinated acid used (19). Thus, it became apparent that the ability to predict the elution positions of peptides under different chromatographic conditions would be extremely useful.

An attempt to correlate elution positions in HPLC with amino acid composition was reported by Meek (20), who derived retention coefficients for individual amino acid residues from the observed elution positions of a series of peptides using aqueous acetonitrile gradients at both pH 2.1 and 7.4. More recently, the same author has refined his retention coefficients and has reported the results of a study of other factors, such as flow rate and gradient rate, on elution

---

* This paper was presented at the International Symposium on HPLC of Proteins and Peptides, November 16–17, 1981, Washington, D. C.

[1] Present address: Department of Physiology, Monash University, Clayton, Victoria 3168, Australia.

[2] Abbreviations used: RP-HPLC, reversed-phase high-performance liquid chromatography; TFA, trifluoroacetic acid; HFBA, heptafluorobutyric acid; NIL, neurointermediary lobe; ACTH, adrenocorticotropin; β-LPH, β-lipotropin; α-MSH, α-melanotropin; CLIP, corticotropin-like intermediary lobe peptide.

TABLE 1

RETENTION COEFFICIENTS FOR AMINO ACIDS
AND SOME POST-TRANSLATIONAL
MODIFICATION FUNCTIONS

	Retention coefficient (0.1% TFA)	Retention coefficient (0.13% HFBA)
Trp	16.3	17.8
Phe	19.2	14.7
Ile	6.6	11.0
Leu	20.0	15.0
Tyr	5.9	3.8
Met	5.6	4.1
Val	3.5	2.1
Pro	5.1	5.6
Thr	0.8	1.1
Arg	−3.6	3.2
Ala	7.3	3.9
Gly	−1.2	−2.3
His	−2.1	2.0
Cys	−9.2	−14.3
Lys	−3.7	−2.5
Ser	−4.1	−3.5
Asn	−5.7	−2.8
Gln	−0.3	1.8
Asp	−2.9	−2.8
Glu	−7.1	−7.5
Amino	4.2	4.2
Carboxyl	2.4	2.4
N-Acetyl	10.2	7.0
Amide	10.3	8.1
O-Phospho	−2.4	−4.1
N-Glyco	−8.0	−6.5

position (21). During attempts to apply the reported retention coefficients (for pH 2.1) to peptide elution positions observed in our laboratory, it became clear that the retention coefficients as described by Meek (20) or Meek and Rossetti (21) were not completely compatible with the data obtained by our chromatographic systems. Furthermore, it was clear that the retention coefficients had to change on going from our TFA system to the HFBA system, in order to account for the observed changes in the relative elution order of a series of peptides.

This paper describes the derivation of retention coefficients for the TFA and HFBA systems employed in our isolation procedure,

and gives two examples of the usefulness of such an approach.

MATERIALS AND METHODS

The HPLC system used was from Waters Associates and the uv detector was a Perkin–Elmer LC 75 set at 210 nm. All aspects of the chromatography system have been fully described elsewhere (22). All of the chromatography described here was performed using a Waters C_{18} μBondapak column which was eluted with linear gradients of aqueous acetonitrile containing either 0.1% TFA or 0.13% HFBA throughout. The rate of increase of the gradient was always 20% acetonitrile per hour and the flow rate was always 1.5 ml/min. The extraction procedure for the neurointermediary lobes (NIL) of the rat pituitaries has been described elsewhere (23). All standard peptides were obtained as described (19,22).

The elution positions of the major peptide components of the rat NIL were determined in both the TFA and HFBA solvent systems, by chromatographing an extract of five rat NILs in each system. The peptides were identified by a combination of radioimmunoassays and peptide mapping as previously described (22–25).

The elution position of each peptide was expressed as an acetonitrile concentration. These estimates take into account the total dead volume of our HPLC system. A reproducible uv baseline disturbance at 210 nm has previously been observed which corresponds to the time when the solvent containing acetonitrile first mixes with the aqueous solvent. Using this as the true starting point for the gradient shows that the dead volume accounts for a 5.1-min delay between starting the chart recorder and commencement of the acetonitrile gradient. The 10-s delay between the spectrophotometer and the fraction collector was also taken into account. All 25 peptides which were used to determine the retention coefficients have also been characterized by amino acid analysis (manuscript in preparation).

The retention coefficients for the 20 amino acid residues and the six other functions (Table 1) were calculated by iterative linear regression analysis as described by Meek (20). The set of retention coefficients for pH 2.1 reported by Meek (20) was used as the starting set, together with estimates for the two functions (O-phospho = -2.0; CHO = -4.0) not included in Ref. (20), based on their observed effect on retention time in the TFA system. Values for each amino acid were changed by 0.1 until maximum correlation was achieved. No attempt was made to "normalize" the retention coefficients by making the slopes of the correlation lines approach 1.0, nor was any attempt made to make the correlation lines intersect with the origin. The correlation analysis was performed on a Hewlett–Packard HP 9831 desktop computer, equipped with 8K of RAM memory, a 9866B termal printer, and

a 9827A matrix plotter. A listing of the program is available from the author on request.

RESULTS

During the last few years, we have been isolating and characterizing peptides from the neurointermediary lobe of the rat pituitary (22–25). For peptides related to α-MSH, ACTH, and β-endorphin, we have relied heavily on radioimmunoassays as a means of identification. However, from the cDNA sequences that were obtained for the bovine (26) and rat (27) precursors for ACTH/β-LPH, it was clear that other peptides should exist whose sequences and amino acid compositions were predicted from the nucleotide sequence data. Two such peptides were γ_3-MSH and an acidic linking or "hinge" peptide between γ_3-MSH and ACTH (26,27). As we lacked radioimmu-

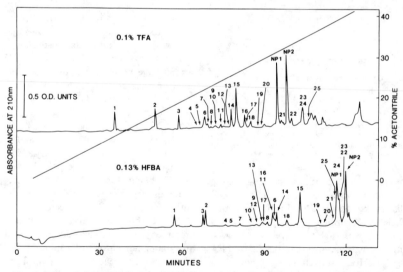

FIG. 1. Reversed-phase HPLC of two ODS-silica extracts of the neurointermediary lobes from five rat pituitaries. In the upper panel, the column was eluted over 3 hs from 1.6 to 61.6% acetonitrile containing 0.1% TFA throughout and in the lower panel the column was eluted with the same acetonitrile gradient, this time containing 0.13% HFBA throughout. The uv absorbance at 210 nm was monitored for both separations. Twenty-five peptides were identified by a combination of radioimmunoassay and amino acid analysis and are numbered as in Table 2. NP₁ and NP₂ are rat neurophysins I and II, which are not included in the data for the correlation analyses.

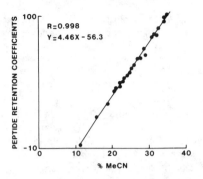

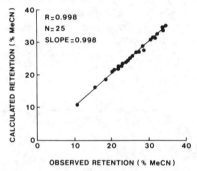

FIG. 2. Correlation of predicted and observed retention values for the 0.1% TFA system. The upper panel shows the correlation between the calculated peptide retention coefficients calculated using the data in Table 1 for the TFA system. The lower panel shows the correlation between the calculated and observed peptide elution positions using the data in Table 2.

noassays for these peptides, we attempted to predict their elution position using the data published by Meek (20,21) for retention coefficients for amino acids at pH 2.1. However, when we tried to correlate observed retention positions in either the 0.1% TFA or 0.13% HFBA solvent systems with predicted retentions based either on the Meek (20) or the Meek and Rossetti (21) data for low pH conditions, we found rather poor correlations. Furthermore, it was clear that we needed two different sets of values, one for the TFA system and one for the HFBA system.

As can be seen in Fig. 1, a series of 25 peptides from the rat NIL, which had been identified and characterized fully, could be chromatographed by identical acetonitrile gradients on the same column, in one case with 0.1% TFA as the hydrophobic counter ion throughout, and in the other case, with 0.13% HFBA throughout. After identifying the elution positions of all 25 peptides in both systems we attempted to generate a set of retention coefficients for all 20 amino acids, and 4 modifying groups, by maximizing the correlation coefficient in an iterative linear regression analysis. This was done for both the TFA and the HFBA systems, using the Meek (20) values as the starting point. For the O-phospho and N-glyco functions, initial values of -2.0 and -4.0 were used. For both the TFA (Fig. 2) and the HFBA (Fig. 3) sets of data, it was possible to obtain a very good correlation between the observed retention times and the predicted values. In this study we have chosen to express the retention value as an acetonitrile concentration, rather than an elution time, so that the data would be more widely useful. The intrinsic dead volume volumes of our HPLC system have been taken into account in calculating acetonitrile concentrations (see Materials and Methods).

If a comparison of the observed and predicted elution positions in the TFA and HFBA systems is made (Fig. 1 and Table 2), it is clear that there is a very close correlation for all 25 peptides, the largest deviation being for peptide 18, CLIP$_{18-39}$, which differs from the predicted position in both systems by 1.4% acetonitrile. This would represent a difference between the predicted and observed elution time of 4.7 min under our chromatographic conditions. The mean difference between the observed and predicted elution positions was 0.28% acetonitrile or 0.9 min for the TFA system and was 0.48% acetonitrile or 1.6 min for the HFBA system.

The calculated retention times were derived by using the appropriate $Y = MX + C$ relationship, where, for the TFA system,

M was 4.46 and C was −56.3, and for the HFBA system, where M was 4.59 and C was −88.9 (Fig. 2). It is difficult to compare the two sets of derived retention coefficients directly, or to compare them to the Meek (20) or the Meek and Rossetti (21) values. This is partly because the values from the literature are calculated as retention times, whereas our values are in terms of percentage acetonitrile. Secondly, the general effect of the HFBA in retaining peptides longer than the TFA system is mostly reflected in the differences in the C values (−88.9 and −56.3, respectively).

DISCUSSION

We have made use of retention coefficients for predicting the elution position of peptides to isolate two peptides. In our initial work on pituitary peptides (22–25) all of the gradients started at 20% acetonitrile. Using the retention coefficients originally generated by Meek (20), we attempted to predict the retention position of the acidic linking peptide Pro ACTH/β-LPH 77-95 which might be expected to be present in the neurointermediary lobe of the rat. We obtained a negative (−20.4) value for the retention position of this peptide. When it was calculated using the Meek and Rossetti retention coefficients, a small positive value of 19.8 was obtained. Although these values were of little help in pinpointing the exact position of the desired peptide, they did indicate that it would elute from the column very rapidly, and that we should be using a lower initial concentration of acetonitrile for our chromatograms. When we did use a gradient starting from 1.6% acetonitrile as in Fig. 1, a peak appeared at about 11% acetonitrile (Peak 1). This peptide was repurified in the HFBA system and, on amino acid analysis, was found to correspond in composition to the sequence Pro ACTH/β-LPH 77-94. This peptide was therefore located by application of predictive retention coefficients, even though the absolute elution position was not determined. We had similarly

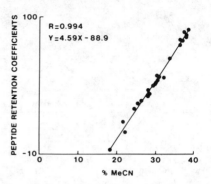

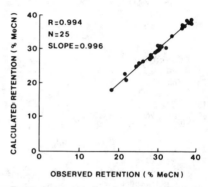

FIG. 3. Correlation of predicted and observed retention values for the 0.13% HFBA system. The upper panel shows the correlation between the calculated peptide retention coefficients using the data in Table 1 for the HFBA system. The lower panel shows the correlation between the calculated and observed peptide elution positions using the data in Table 2.

isolated Lys-γ_3-MSH from rat neurointermediary lobes (28) by using a modification of the retention values of Meek. In this case, we used the cDNA information for the predicted sequence, and used the knowledge that the peptide should contain tryptophan as an aid in its location in the chromatogram. In this case, we were perhaps fortunate to succeed, as the isolated peptide differed in composition from the γ_3-MSH by having an extra lysine residue and a carbohydrate group. These observations led us to the conclusion that while this approach was of great

TABLE 2

COMPARISON OF CALCULATED AND OBSERVED ELUTION POSITIONS FOR 25 RAT
NEUROINTERMEDIARY LOBE PEPTIDES[a]

No.	Peptide	Observed TFA elution position	Calculated TFA elution position	Δ	Observed HFBA elution position	Calculated HFBA elution position	Δ
1	Pro-ACTH-β-LPH 77–94	11.0	10.8	−0.2	18.2	18.0	−0.2
2	Arg-Vasopressin 1–9	15.8	16.1	+0.3	22.0	21.1	−0.9
3	Oxytocin 1–9	18.8	18.6	−0.2	21.6	22.8	+1.2
4	Phospho-glyco-CLIP 18–38	20.6	21.0	+0.4	24.5	25.1	+0.6
5	Glyco-CLIP 18–38	21.0	21.5	+0.5	25.3	26.0	+0.7
6	Lys-γ_3-MSH 1–25	22.0	21.9	−0.1	30.5	30.9	+0.4
7	Phospho-CLIP 18–38	22.2	22.8	+0.6	26.3	26.5	+0.2
8	Desacetyl α-MSH 1–13	22.8	22.5	−0.3	29.8	29.5	−0.3
9	α-Endorphin 1–16	23.2	24.4	+1.2	28.3	27.7	−0.6
10	CLIP 18–38	24.0	23.3	−0.7	28.0	27.4	−0.6
11	Monoacetyl α-MSH 1–13	24.6	23.9	−0.7	30.3	30.1	−0.2
12	Phospho-glyco-CLIP 18–39	25.2	25.3	+0.1	28.2	28.3	−0.1
13	Glyco-CLIP 18–39	25.2	25.8	+0.6	29.3	29.2	+0.1
14	Diacetyl α-MSH 1–13	25.8	25.2	−0.6	30.8	31.7	+0.9
15	γ-LPH 1–38	26.8	27.0	+0.2	33.6	33.9	+0.3
16	Phospho-CLIP 18–39	27.6	27.1	+0.5	30.0	29.7	−0.3
17	γ-Endorphin 1–17	28.6	28.9	+0.3	30.2	31.0	−0.8
18	CLIP 18–39	29.0	27.6	−1.4	32.0	30.6	−1.4
19	Phospho-ACTH 1–39	30.8	31.0	+0.2	36.3	36.3	0
20	ACTH 1–39	31.4	31.5	+0.1	36.5	37.1	+0.6
21	β-Endorphin 1–31	32.0	31.4	−0.6	37.0	37.0	0
22	Acetyl β-endorphin 1–31	32.6	32.7	+0.1	38.5	37.6	−0.9
23	Acetyl δ-endorphin 1–27	34.0	34.7	+0.7	38.5	38.8	+0.3
24	Posterior pituitary glycopeptide 1–39	34.0	33.9	−0.1	38.0	37.8	−0.2
25	Acetyl-δ-endorphin 1–26	34.8	35.2	+0.4	37.5	38.4	+0.9

[a] Values are expressed as percentage acetonitrile (see Materials and Methods).

potential, it was necessary for us to evolve a set of retention coefficients that would be more appropriate for our chromatographic conditions. The Meek values were generated using a sodium perchlorate–phosphoric acid–acetonitrile mobile phase and a Bio-Rad ODS column. We have observed that this column behaved significantly differently from the C_{18} μBondapak columns which we use, and we knew that the nature of the hydrophobic counterion has a profound effect on the elution order of peptides from reversed-phase columns (19).

Strictly speaking, the retention coefficients that we have calculated are applicable only to the chromatographic systems we have used and, in particular, to our HPLC column. For readers to obtain an accurate set of coefficients, we suggest that they generate their own coefficients using their own chromatographic conditions and reversed-phase column. However, over the past few years we have found that one Waters Associates μBondapak C_{18} column has much the same characteristics as another. According to the manufacturer they all have similar carbon-loading. The chromatographic properties of these columns indicate that they have been efficiently "end-capped" and this gives rise to reasonably pure reversed-phase chromatography. There appears to be little polar interactions of peptides with under-ivitized silica. We have successfully applied our own retention coefficients to the behavior of peptides on other μBondapak C_{18} columns. In order to do this, all controllable variables must be kept constant, i.e., concentration of counterion (0.1% TFA or 0.13% HFBA), use of acetonitrile as the low polarity solvent, gradient rate (20% acetonitrile per hour), and solvent flow rate (1.5 ml/min.). Under these standard conditions, the column is first "calibrated" with an extract of neurointermediary lobes from five rat pituitaries (22). The theoretical elution position of putative peptides are calculated using our original data (Table 1) and their positions are then determined relative to the natural peptides in Fig. 1 and Table 2. The theoretical elution positions using the new column can be estimated relative to the same natural peptides in the chromatographic profile of the calibration mixture. This maneuver is possible because the relative positions of the peptides in the calibration mixture do not change very much from column to column even though the acetonitrile concentration at which each peptide elutes may vary considerably.

By using natural peptides instead of synthetic peptides as the data base for our derived retention coefficients, we have the extra advantage of obtaining values for the post-translational modifications, N-acetyl, O-phospho, and N-glyco, which are important constituents of a variety of natural peptides, and which clearly have a large effect on elution positions in HPLC (Fig. 1).

The approach described in this paper, whereby the elution positions of peptides can be predicted from their amino acid composition and post-translational modifications, should be generally applicable to the isolation of any small peptide (i.e., less than 50 amino acids) from tissue extracts, and would also be of use in predicting the fragmentation pattern obtained from proteolytic digestion of peptides and proteins. It would also be useful in the selection of the appropriate conditions for achieving separations of peptides which only differ in composition by one or two amino acids, or by the presence of a post-translational modification.

ACKNOWLEDGMENTS

The authors would like to thank Isabel Lehmann and Susan James for their expert technical assistance. We would especially like to thank Dr. J. Meek of St. Elizabeth's Hospital, National Institute of Mental Health, Washington, D. C., for generously making a preprint for Ref. (21) available to us prior to publication. This work was supported by the Medical Research Council of Canada, Grants MT-1658 and MA-6733, Fond de la recherche en santé du Québec, Grant 800208, and U. S. Public Health Service, Grant HD 04365. H.P.J.B. is a recipient of a scholarship from the Fond de la recherche en santé du Québec.

REFERENCES

1. Rivier, J. E. (1978) *J. Liquid Chromatogr.* **1**, 343–366.
2. Hancock, W. S., Bishop, C. A., Prestidge, R. L., Harding, D. R. K., and Hearn, M. T. W. (1978) *Science* **200**, 1168–1170.
3. Rubinstein, M., Stein, S., and Udenfriend, S. (1977) *Proc. Nat. Acad. Sci. USA* **74**, 4969–4972.
4. Bennett, H. P. J., Hudson, A. M., McMartin, C., and Purdon, G. E. (1977) *Biochem. J.* **168**, 9–13.
5. O'Hare, M. J., and Nice, E. C. (1979) *J. Chromatogr.* **171**, 209–226.
6. Lewis, R. V., Stein, S., and Udenfriend, S. (1979) *Int. J. Peptide Protein Res.* **13**, 493–497.
7. Hancock, W. S., Bishop, C. A., Prestridge, R. L., and Hearn, M. T. W. (1978) *Anal. Biochem.* **89**, 203–212.
8. Fullmer, C. S., and Wasserman, R. H. (1979) *J. Biol. Chem.* **254**, 7208–7212.
9. Congote, L. F., Bennett, H. P. J., and Solomon, S. (1979) *Biochem. Biophys. Res. Commun.* **89**, 851–858.
10. Petrides, P. E., Jones, R. T., and Böhlen, P. (1980) *Anal. Biochem.* **105**, 383–388.
11. Ling, N., Burgus, R., and Guillemin, R. (1976) *Proc. Nat. Acad. Sci. USA* **73**, 3942–3946.
12. Rubenstein, M., Stein, S., Gerber, D., and Udenfriend, S. (1977) *Proc. Nat. Acad. Sci. USA* **74**, 3052–3055.
13. Speiss, J., Rivier, J. E., Rodfrey, J. A., Bennett, C. D., and Vale, W. (1979) *Proc. Nat. Acad. Sci. USA* **76**, 2974–2978.

14. Vale, W., Speiss, J., Rivier, C., and Rivier, J. (1981) *Science* **213**, 1394–1397.

15. McMartin, C., Bennett, H. P. J., Hudson, A. M., and Purdon, G. E. (1977) *J. Endocrinol.* **73**, 14P–15P.

16. Bennett, H. P. J., Hudson, A. M., Kelly, L., McMartin, C., and Purdon, G. E. (1978) *Biochem. J.* **175**, 1139–1141.

17. Bennett, H. P. J., Browne, C. A., Brubaker, P. L., and Solomon, S. (1980) *in* Biomedical Applications of Liquid Chromatography (Hawk, G. L., ed.), pp. 197–210, Dekker, New York.

18. Bennett, H. P. J., Browne, C. A., Goltzman, D., and Solomon, S. (1979) *Proc. Amer. Pept. Symp. 6th 1979*, 121–124.

19. Bennett, H. P. J., Browne, C. A., and Solomon, S. (1980) *J. Liquid Chromatogr.* **3**, 1353–1365.

20. Meek, J. L. (1980) *Proc. Nat. Acad. Sci. USA* **77**, 1632–1636.

21. Meek, J. L., and Rossetti, Z. L. (1981) *J. Chromatogr.* **211**, 15–28.

22. Bennett, H. P. J., Browne, C. A., and Solomon, S. (1981) *Biochemistry* **20**, 4530–4538.

23. Browne, C. A., Bennett, J. P. J., and Solomon, S. (1981) *Biochemistry* **20**, 4530–4546.

24. Browne, C. A., Bennett, H. P. J., and Solomon, S. (1981) *Proc. Amer. Pept. Symp. 7th 1981*, 509–512.

25. Bennett, H. P. J., Browne, C. A., and Solomon, S. (1981) *Proc. Amer. Pept. Symp. 7th 1981*, 785–788.

26. Nakanishi, S., Inoue, A., Kita, T., Nakamura, M., Chang, A. C., Cohen, S. N., and Numa, S. (1979) *Nature (London)* **278**, 423–427.

27. Drouin, J., and Goodman, H. M. (1980) *Nature (London)* **288**, 610–612.

28. Browne, C. A., Bennett, H. P. J., and Solomon, S. (1980) *Biochem. Biophys. Res. Commun.* **100**, 336–343.

The Application of High-Performance Liquid Chromatography for the Resolution of Proteins Encoded by the Human Adenovirus Type 2 Cell Transformation Region[1]

MAURICE GREEN AND KARL H. BRACKMANN

*Institute for Molecular Virology, St. Louis University Medical Center,
3681 Park Avenue, St. Louis, Missouri 63110*

The human adenovirus 2 (Ad2) transformation genes are located in early region E1a (map position (mp) 1.3–4.5) and E1b (mp 4.6–11.2) on the linear duplex Ad2 DNA genome of M_r 23 × 10⁶ (viral DNA is divided into 100 map units). E1b codes for three major proteins of apparent molecular weights 53,000 (53K), 19K, and 20K; smaller quantities of 21K, 22K, and 23K proteins that are related to 53K are also synthesized in Ad2-infected cells. Because the resolution and purification of these Ad2 candidate transformation proteins proved very difficult by conventional protein purification methods, the applicability of high-performance liquid chromatography (HPLC) methodology was examined. Starting with a crude cytoplasmic S100 fraction of Ad2-infected human cells, the resolution of the Ad2 E1b-coded 19K, 20K, 21K, 22K, and 23K proteins by reverse-phase HPLC using a C_8 column and a linear 0–60% 1-propanol gradient in 0.5 M pyridine formate was achieved. E1b proteins purified under these conditions retained their immunological reactivity. By anion-exchange HPLC using a linear 10 mM to 1 M NaCl gradient in 10 mM 4-(2-hydroxyethyl)-1-piperazineethanesulfonic acid buffer, pH 7.6, the same five Ad2 E1b-coded 19K–23K proteins were separated, with improved resolution of the 19K protein. Based on these findings, protocols for the extensive purification of the E1b-19K and E1b-20K proteins have been developed. These results illustrate the potential of HPLC methodology for the rapid purification of biologically interesting proteins from complex cellular mixtures of proteins.

Human adenovirus type 2 (Ad2)[2] is a DNA tumor virus which has been developed into an excellent model for understanding the molecular biology of mammalian cells and the molecular basis of cell transformation. Ad2 has a linear duplex DNA genome of 35,000 base pairs (1). Upon infection of permissive human cells, the viral genome is expressed in two major stages, an early stage

which precedes viral DNA replication, followed by a late stage. The early genes lie in four noncontiguous regions designated E1, E2, E3, and E4. The transformation genes, which are integrated and expressed in all Ad2-transformed cell lines thus far examined (2), are located in the left 11% of the Ad2 DNA molecule in early region E1 (3). E1 is subdivided into two regions, E1a and E1b. E1a encodes five proteins of apparent molecular weights of 28,000 (28K) to 53K, whereas E1b encodes major proteins of apparent molecular weights of 19K, 20K, and 53K, as well as additional proteins of molecular weights 21K, 22K, and 23K (4–15).

The great promise of tumor virology is that understanding the functions of the virus-coded transformation proteins will help

[1] This paper was presented at the International Symposium on HPLC of Proteins and Peptides, November 16–17, 1981, Washington, D. C.

[2] Abbreviations used: Ad2, adenovirus serotype 2; HPLC, high-performance liquid chromatography; MEM, Eagle's minimal essential medium; AraC, arabinosyl cytosine; PMSF, phenylmethylsulfonyl fluoride; NP-40, Nonidet P-40; SDS, sodium dodecyl sulfate; Hepes, 4-(2-hydroxyethyl)-1-piperazineethanesulfonic acid.

elucidate the nature of cell growth control and neoplasia. The isolation of the viral transformation proteins in a purified state is necessary for many studies on biological function. We are interested in purifying the Ad2 E1b-coded proteins, since several studies suggest that they play an important role in Ad2-induced cell transformation. The purification of the Ad2 candidate transformation proteins and of viral transformation proteins in general has proven to be difficult. Viral transformation proteins are often "sticky," tend to aggregate, and spread into broad peaks during column purification. In addition, viral transformation proteins are usually present in low concentrations in infected and transformed cells.

We have found that recently developed HPLC procedures, in particular reverse-phase HPLC and anion-exchange HPLC, are very useful for the purification of the Ad2 E1b-coded candidate transformation proteins. We describe in this report the resolution of proteins of molecular weights 19K, 20K, 21K, 22K, and 23K that are encoded by the Ad2 E1b transformation region.

EXPERIMENTAL PROCEDURES

Preparation of [^{35}S]methionine-labeled cytoplasmic S100 fraction from Ad2 early infected cells. The [^{35}S]methionine-labeled Ad2 early S100 fraction was prepared for use in purification studies with HPLC as follows. Three to 20 liters of human KB cells, growing exponentially in suspension culture (4×10^5 cells/ml) in Eagle's minimal essential medium (MEM), were centrifuged at 300g at room temperature and infected in 1/20 of the initial volume of MEM with 500 plaque-forming units/cell of Ad2. After 1 h of adsorption at 37°C, cells were diluted to one-half the initial volume with MEM containing 5% horse serum and 25 μg/ml of cycloheximide. At 5 h postinfection, AraC was added to 20 μg/ml and the cells were centrifuged and washed twice with warm methionine-free MEM containing 5% horse

serum and 20 μg/ml of AraC. Cells were resuspended at 1.0×10^6 cells/ml in methionine-free MEM containing 5% horse serum and 20 μg/ml of AraC, and labeled with 5 mCi of [^{35}S]methionine (900–1100 Ci/mmol) from 6 to 21 h postinfection. The following steps were performed at 0 to 4°C. Cells were collected by centrifugation and washed twice with phosphate-buffered saline containing 1 mM PMSF. The cell pellet was incubated for 20 min in 10 vol of buffer A (10 mM Tris–HCl, pH 8.5/250 mM sucrose/3.7 mM CaCl$_2$/12 mM MgCl$_2$/1 mM PMSF/1% NP-40). Nuclei were removed by centrifugation at 800g for 10 min at 4°C. The cytoplasmic fraction (combined supernatants) was centrifuged at 100,000g in the Beckman Ti50.2 rotor after the addition of KCl to 0.4 M and sodium deoxycholate to 1%. The supernatant is the S100 fraction and contained about 2×10^9 cpm of trichloroacetic acid-insoluble radioactivity.

Immunoprecipitation analysis of infected cell fractions. Portions of column fractions containing [^{35}S]methionine-labeled proteins were immunoprecipitated with antiserum to F17 cells (an Ad2-transformed cell line that contains only Ad2 E1 genetic information). The immunoprecipitate was analyzed by sodium dodecyl sulfate–polyacrylamide gel electrophoresis and autoradiography (10).

Two-dimensional peptide maps. Bands of [^{35}S]methionine-labeled polypeptides, identified on dried gels by autoradiography, were excised, digested with trypsin, and analyzed by two-dimensional electrophoresis (pH 1.7) thin-layer chromatography (10).

Fractionation of Ad2 E1b-coded proteins by DE-53 cellulose chromatography. All steps were performed at 0 to 4°C. The S100 fraction (containing 1.5–2 × 10^9 cpm) prepared as described above from about 7×10^9 [^{35}S]methionine-labeled Ad2 early infected cells was loaded on a 2.5 × 22-cm DE-53 column equilibrated with buffer B (10 mM Tris, pH 8.5/1 mM dithiothreitol/1 mM PMSF/0.05% NP-40/10% glycerol) containing 15 mM NaCl. After washing with

200 ml of buffer B containing 15 mM NaCl, the column was eluted with 600 ml of a 15- to 800-mM NaCl gradient in buffer B. Fractions (6 ml) were collected, and 25- and 500-μl aliquots were used for measuring radioactivity and for immunoprecipitation with F17 antiserum, respectively.

Purification of Ad2 E1b proteins by reverse-phase HPLC. Reverse-phase HPLC was performed with a C_8 column (Altex ultrasphere-octyl or SynChrom SynChropak RP-P) (4.6 × 250 mm) equilibrated in 0.5 M pyridine formate, pH 4.1. The Ad2 S100 fraction (2–6 ml, 1–3 × 10^8 cpm) was loaded on the C_8 column and eluted with a linear 100-min gradient of 0 to 60% 1-propanol in 0.5 M pyridine formate, pH 4.1, at a flow rate of 0.33 ml/min. These conditions are similar to those described by Kimura *et al.* (16). Fractions (1.0 ml) were collected and aliquots were analyzed for radioactivity and for immunoprecipitation with F17 antiserum. Protein was monitored continuously by an automated fluorescence detection system (17). Recovery of radioactive protein ranged from 10 to 30%; the proteins not eluted from the column represented mainly high molecular weight species, i.e., above 30K.

Purification of Ad2 E1b proteins by anion-exchange HPLC. Anion-exchange HPLC was performed with a SynChropak AX300 column (SynChrom) (4.6 × 250 mm) equilibrated in buffer C (10 mM Hepes, pH 7.6/0.1% NP-40/1 mM PMSF/1 mM dithiothreitol/5% glycerol) containing 10 mM NaCl. The Ad2 S100 fraction (2–6 ml, 1–3 × 10^8 cpm) was dialyzed against buffer C containing 10 mM NaCl and loaded on the column. After it was washed with 20 ml of buffer C containing 10 mM NaCl, the column was eluted with a linear 70-min gradient of 10 mM to 1 M NaCl in buffer C at a flow rate of 1.0 ml/min. Fractions (1.0 ml) were collected and analyzed for total radioactivity, for protein content (fluorescence), and for immunoprecipitable Ad2 E1b proteins, as described above. Recovery of radioactive protein ranged from 90 to 100%.

RESULTS

Attempts to Fractionate Ad2 E1b Proteins in the Cytoplasmic S100 Fraction by Chromatography on DE-53 Cellulose

Ad2 early infected cells were treated with cycloheximide from 1 to 5 h postinfection to enhance the synthesis of early proteins (18) and were labeled with [^{35}S]methionine from 6 to 18 h postinfection in the presence of Ara C (to block the late stages of infection). The crude cytoplasmic S100 fraction, which contains approximately 80% of the total cellular protein, was isolated as described under Experimental Procedures. As we have recently shown by immunoprecipitation and peptide mapping (19,20), the S100 fraction contains the Ad2 E1b-coded major 53K, 19K, and 20K proteins, as well as smaller amounts of 21K, 22K, and 23K proteins that are related to E1b-53K.

We tested the suitability of DE-53 cellulose chromatography for the resolution of the Ad2 E1b proteins. The distribution of [^{35}S]methionine-labeled proteins after chromatography of the S100 fraction from Ad2 early infected cells on a DE-53 column is shown in Fig. 1 (top). An aliquot of every third fraction was analyzed by immunoprecipitation with rat antiserum to the Ad2-transformed cell line F17, which contains antibodies against Ad2 E1b proteins. SDS–polyacrylamide gel electrophoresis analysis of the immunoprecipitates is presented in Fig. 1 (bottom). As visualized in Fig. 1 (bottom) and summarized in Fig. 1 (top), the E1b-20K eluted first from the column. However, the Ad2 E1b-coded 53K, 19K, 21K, 22K, and 23K proteins were found to coelute, and were not resolved by DE-53 cellulose chromatography, under the experimental conditions used.

Resolution of Ad2 E1b Proteins in the Cytoplasm by Reverse-Phase HPLC

The cytoplasmic S100 fraction, isolated from [^{35}S]methionine-labeled Ad2 early in-

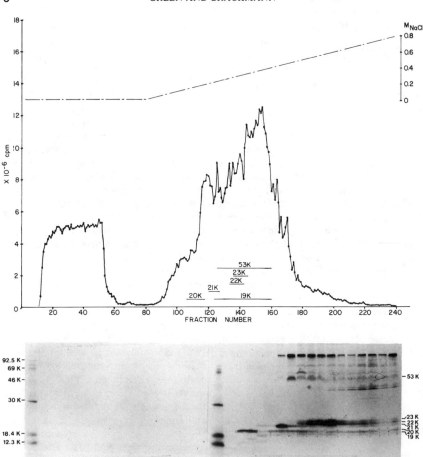

FIG. 1. DE-53 cellulose chromatography of [^{35}S]methionine-labeled S100 fraction isolated from Ad2 early infected cells. (Top) Fractionation of Ad2 E1b proteins in cytoplasm on DE-53 cellulose. The S100 fraction from [^{35}S]methionine-labeled Ad2 early infected cells was chromatographed on a DE-53 column as described under Experimental Procedures. Each fraction was analyzed for total radioactivity, and every third fraction was analyzed by immunoprecipitation with F17 antiserum. (Bottom) Immunoprecipitation. The immunoprecipitates from above were analyzed by SDS–polyacrylamide gel electrophoresis and autoradiographed. The M_r markers are [Me-^{14}C]-labeled phosphorylase b (92.5K), bovine serum albumin (69K), ovalbumin (46K), carbonic anhydrase (30K), lactoglobulin A (18.4K), and cytochrome c (12.3K).

fected cells, were dialyzed against buffer C containing 10 mM NaCl and was fractionated by reverse-phase HPLC. A 0 to 60% 1-propanol gradient in 0.5 M pyridine formate, pH 4.1, was used. The distribution of proteins, as determined by fluorescence analysis (17), is shown in Fig. 2 (top). Sixteen pools of the first 55 fractions were prepared,

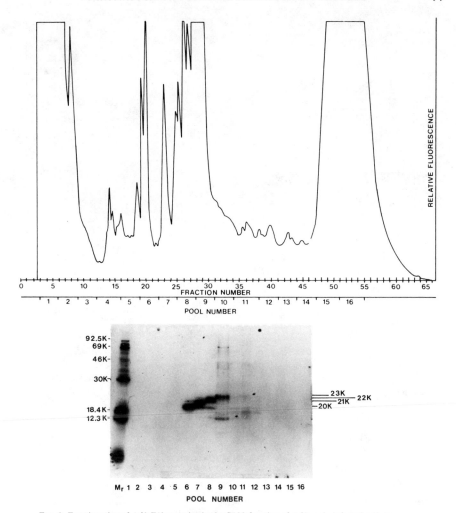

FIG. 2. Fractionation of Ad2 E1b proteins in the S100 fraction of Ad2 early infected cells by reverse-phase HPLC. (Top) Distribution of Ad2 E1b 20K, 21K, 22K, and 23K from cytoplasm by reverse-phase HPLC. An S100 fraction from [^{35}S]methionine-labeled Ad2 early infected cells was fractionated by reverse-phase HPLC on a C$_8$ column as described under Experimental Procedures. The protein content (relative fluorescence) is plotted against fraction number. (Bottom) Immunoprecipitation of pools from reverse-phase HPLC of Ad2 cytoplasm. Fractions were pooled, as shown in the abscissa in Fig. 2 (top), lyophilized, and analyzed by immunoprecipitation with F17 antiserum. Each immunoprecipitate was resolved by SDS–polyacrylamide gel electrophoresis and autoradiographed. M_r markers are as given in legend of Fig. 1.

lyophilized, and analyzed by immunoprecipitation. As shown by the SDS–gel electrophoresis in Fig. 2 (bottom) and confirmed by peptide-map analysis (not shown), E1b-20K, 21K, 22K, and 23K were resolved in pools 6, 7, 8, and 9, respectively. Small

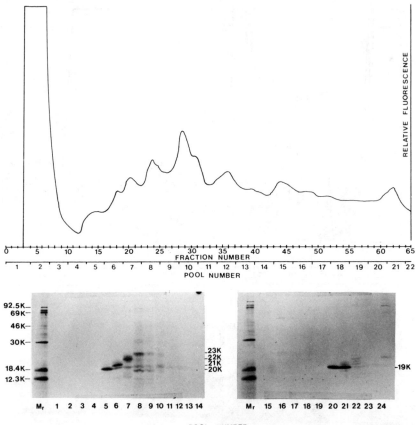

FIG. 3. Fractionation of Ad2 E1b proteins in the S100 fraction of Ad2 early infected cells by anion-exchange HPLC. (Top) Distribution of Ad2 E1b 20K, 21K, 22K, 23K, and 19K from cytoplasm by anion-exchange HPLC. An S100 fraction from [^{35}S]methionine-labeled Ad2 early infected cells was fractionated by anion-exchange HPLC, as described under Experimental Procedures. The protein content (relative fluorescence) is plotted against fraction number. (Bottom) Immunoprecipitation of pools from anion-exchange HPLC of Ad2 cytoplasm. Fractions were pooled, as indicated in the abscissa to Fig. 3 (top), and immunoprecipitated with F17 antiserum. Each immunoprecipitate was resolved by SDS–polyacrylamide gel electrophoresis and autoradiographed. M_r markers are as given in legend of Fig. 1.

amounts of 19K were found in pools 9 through 12 (fractions 32–39) (not shown).

Resolution of Ad2 E1b Proteins in the Cytoplasm by Anion-Exchange HPLC

As shown above, the five Ad2 E1b-coded proteins of M_r 19K to 23K can be resolved by reverse-phase HPLC. However, the use of pH 4.1 buffer and 1-propanol may be detrimental to the unknown biological activity of the Ad2 E1b proteins. To develop additional purification steps for E1b proteins under milder conditions, we examined the usefulness of anion-exchange HPLC. A [^{35}S]methionine-labeled S100 fraction from

Ad2-infected cells was dialyzed against buffer C (pH 7.6) containing 10 mM NaCl and loaded on the column. A 10-mM to 1-M NaCl gradient in buffer C was used to develop the column. The distribution of proteins eluted from the column in shown in Fig. 3 (top). The distribution of the E1b proteins, as determined by immunoprecipitation with F17 serum and SDS–gel electrophoresis (and confirmed by peptide-map analysis), is shown in Fig. 3 (bottom). The E1b-20K, 21K, 22K, and 23K proteins were resolved in pools 5, 6, 7, 8, respectively. Of practical use for purification, E1b-19K eluted in pools 20 and 21, well separated from E1b-20K, 21K, 22K, and 23K. It is interesting to note that the resolution of [^{35}S]methionine-labeled Ad2 E1b proteins by anion-exchange HPLC was superior to that found by DE-53 chromatography (see Fig. 1 (top)).

DISCUSSION

Improved methodologies that are suitable for the purification of proteins present in low concentrations in complex biological mixtures are needed for the isolation of viral transformation proteins. Our experiences attempting to purify the Ad2 E1b-coded proteins have taught us that conventional column procedures are inadequate. The large-size columns that are necessary for large amounts of starting material require long operation times, i.e., 10–30 h, which often result in large losses of Ad2 E1b-coded proteins. A major advantage of HPLC technology is the short operation time necessary for an HPLC column, i.e., 1–2 h. Thus, labile proteins can be rapidly recovered with minimal loss of activity. An additional advantage is that HPLC purified proteins are eluted in small volumes, in contrast to the large volumes eluted by conventional column procedures. The higher protein concentration in the HPLC eluates helps retain the biological activity of proteins and avoids a time-consuming (possibly denaturing) concentration step.

Our reverse-phase HPLC protocol uses low pH (pH 4.1) and 1-propanol to resolve the Ad2 E1b proteins. These conditions are possibly detrimental to biological activity, although the immunological reactivity of the proteins, i.e., immunoprecipitability, was retained. The identification of more "physiological" mobile phases for reverse-phase HPLC would be valuable. We show here that anion-exchange HPLC can resolve the five E1b-coded 19K to 23K proteins, using neutral pH (pH 7.6) and aqueous solvents. Of interest, conventional anion-exchange column chromatography (DE-53) was less effective in resolving these same proteins.

The standard commercially available HPLC columns (4.6 × 250 mm) can be loaded with up to 50 mg of proteins. The Ad2 E1b transformation proteins are present at a concentration of only about 0.04% of the total protein of the infected cell. Thus, to purify 1 mg of one of these proteins, assuming 100% yield, would require 2.5 g of cell protein as starting material. Thus, before the HPLC technology can be applied, a preliminary purification step is necessary. We have found that chromatography on DE-53 cellulose or phosphocellulose can serve this function. Using DE-53 cellulose chromatography followed by reverse-phase HPLC, we recently have been able to purify the Ad2 E1b-20K protein to near homogeneity (19). Using phosphocellulose chromatography followed by anion-exchange HPLC and then reverse-phase HPLC, we have been able to purify extensively the Ad2 E1b-19K (our unpublished data).

Further developments in HPLC technology should greatly increase the scope of its application to the purification of biologically important proteins. The availability of cation-exchange supports for HPLC should provide an additional dimension to protein purification. The availability of shorter chains, e.g., C_3 to C_6, for reverse-phase HPLC may be especially helpful since highly hydrophobic proteins appear to stick irreversibly to C_8 columns. For example, we have not been able to elute the E1b-53K from a C_8 column. The methodical exploration of mobile phases in reverse-phase HPLC of proteins would be of great benefit. The ability to purify interesting cellular and viral proteins may spark important advances in our understanding of the role of such proteins in cell growth and development.

ACKNOWLEDGMENTS

This work was supported by U. S. Public Health Service Grants 5 RO1 CA-21824, 5 RO1 CA-29561, and 5 T32 CA-09222 from the National Institutes of Health and by Specialized Cancer Center (CORE) Grant 5 P30 CA-25843 from the National Cancer Institute. Dr. Green is the recipient of Research Career Award 5 K06 AI-04739 from the National Institutes of Health. We thank Maria Cartas, Leann Young, Rose Epler, and Eve Dake for excellent technical assistance, and Carolyn Mulhall for editorial assistance. We thank Dr. W. Wold for critical reading of the manuscript.

REFERENCES

1. Green, M., Piña, M., Kimes, R., Wensink, P. C., MacHattie, L. A., and Thomas, C. A. (1967) *Proc. Nat. Acad. Sci. USA* **57**, 1302.
2. Green, M., Wold, W. S. M., and Büttner, W. (1981). *J. Mol. Biol.* **151**, 337.
3. Van der Eb, A. J., Mulder, C., Graham, F. L., and Houweling, A. (1977) *Virology* **99**, 372.
4. Gilead, Z., Jeng, Y., Wold, W. S. M., Sugawara, K., Rho, H. M., Harter, M. L., and Green, M. (1976) *Nature (London)* **264**, 263.
5. Levinson, A. D., and Levine, A. J. (1977) *Cell* **11**, 871.
6. Lewis, J. B., Atkins, J. F., Baum, P. R., Solem, R., Gesteland, R. F., and Anderson, C. W. (1976) *Cell* **7**, 141.
7. Harter, M. L., and Lewis, J. M. (1978) *J. Virol.* **26**, 736.
8. Halbert, D. N., Spector, D. J., and Raskas, H. J. (1979) *J. Virol.* **31**, 621.
9. Wold, W. S. M., and Green, M. (1979) *J. Virol.* **30**, 297.
10. Green, M., Wold, W. S. M., Brackmann, K. H., and Cartas, M. A. (1979) *Virology* **97**, 275.
11. Schrier, P. I., Van den Elsen, P. H., Hertoghs, J. J. L., and Van der Eb, A. J. (1979) *Virology* **99**, 372.
12. Brackmann, K. H., Green, M., Wold, W. S. M., Cartas, M., Matsuo, T., and Hashimoto, S. (1980) *J. Biol. Chem.* **255**, 6772.
13. Ross, S. R., Flint, S. J., and Levine, A. J. (1980) *Virology* **100**, 419.
14. Esche, H., Mathews, M. B., and Lewis, J. B. (1980) *J. Mol. Biol.* **142**, 399.
15. Lupker, J. H., Davis, A., Jochemsen, H., and Van der Eb, A. J. (1981) *J. Virol.* **37**, 524.
16. Kimura, S., Lewis, R. B., Gerber, L. B., Brink, L., Rubinstein, M., Stein, S., Udenfriend, S. (1979) *Proc. Nat. Acad. Sci. USA* **76**, i756.
17. Böhlen, P., Stein, S., Dairman, W., and Udenfriend, S. (1973) *Arch. Biochem. Biophys.* **155**, 213.
18. Harter, M. L., Shanmugan, G., Wold, W. S. M., and Green, M. (1976) *J. Virol.* **19**, 232.
19. Green, M., Brackmann, K. H., Cartas, M. A., and Matsuo, T. (1982) *J. Virol.* **42**, 30.
20. Matsuo, T., Wold, W. S. M., Hashimoto, S., Rankin, A. R., Symington, J., and Green, M. (1982) *Virology* **118**, 456.

The Importance of Silica Type for Reverse-Phase Protein Separations[*,1]

JAMES D. PEARSON, NAN T. LIN, AND FRED E. REGNIER

Department of Biochemistry, Purdue University, West Lafayette, Indiana 47907

Various large-pore-diameter silicas have been coated with n-alkylchlorosilanes and tested for efficacy in protein separation. The optimal silica has been characterized for loading and column-length effects by means of resolution, load capacity, and desorption tests. The mechanism of interaction between protein and stationary phase is discussed. Theoretical plate values determined for small, unretained molecules are found to be noncorrelative to protein resolution. A test mixture is proposed for comparing the ability of commercial columns to resolve proteins.

Optimization of high-performance reverse-phase chromatography for proteins and peptides has been approached by two routes: mobile phase manipulation on a given reverse-phase packing or stationary phase selection on a specific silica. Most reports have been on the former approach (1–5). The use of trifluoroacetic acid (TFA)[2] as a mobile phase ion-pairing agent has been an especially important advance in reverse-phase high-performance liquid chromatography (RP-HPLC) because of its volatility and effectiveness as a solubilizing agent (5–8). Other perfluorocarboxylic acids, such as heptafluorobutyric acid, have also been found to have desirable selectivity (5).

The importance of utilizing large-pore-diameter silicas (>100 Å) for large, denatured cyanogen bromide fragments (7,9) and protein (10) separations has recently

emerged. It has also been suggested that in addition to the mobile phase, alkyl ligand, and pore size, the silica matrix itself may play a significant role in the performance of a support in protein–peptide separations (9). This report examines the contribution of various large-pore-diameter silica matrices to resolution and recovery of proteins.

A true evaluation of various silicas cannot be accomplished by evaluating packed columns from a series of vendors. Manufacturers vary coating and packing conditions to such an extent that definitive statements about silica contribution to support selectivity and resolution are impossible. To circumvent this problem we have coated, packed, and tested silicas under uniform conditions.

After selection of an optimal silica, the influence of different bonding procedures on resolution and recovery were studied. Columns were also evaluated for resolution as a function of loading capacity, length, and column efficiency. Additionally, loading capacity vs length was examined.

MATERIALS AND METHODS

LiChrosorb and LiChrospher silicas were purchased from E. Merck (Darmstadt, West Germany). Nucleosil 100-5 was obtained from Macherey-Nagel and Company (Duren, West Germany). Hypersil was pur-

* This paper was presented at the International Symposium on HPLC of Proteins and Peptides, November 16–17, 1981, Washington, D. C.

[1] This is Journal Paper No. 000 from the Purdue University Agricultural Experimental Station.

[2] Abbreviations used: TFA, trifluoroacetic acid; RP-HPLC, reverse-phase high-performance liquid chromatography; BSA, bovine serum albumin; OVA, ovalbumin; αCB-2 and -3, 44- and 65-amino-acid-residue fragments of fetal globin α-chain formed by cleavage with cyanogen bromide; SiMe₂ClC₈, n-octyldimethylchlorosilane; SiCl₃C₈, octyltrichlorosilane.

chased from Shandon Southern Instruments (Sewickley, Pa.). Spherisorb SG30F was obtained from Harwell Ceramics Centre (Berkshire, England). Spherosil XOBO75 was purchased from Rhone-Poulenc Fine Chemicals (France). Vydac TP was obtained from The Separations Group (Hesperia, Calif.) or purchased from SynChrom (Linden, Ind.) under the name of Silica 2362. Partisil ODS-3 and C_8 columns were obtained from Whatman (Clifton, N. J.). The Bio-Rad ODS-5S column was obtained from Bio-Rad Laboratories (Richmond, Calif.). The LiChrosorb-RPC column was purchased from Alltech Associates (Deerfield, Ill.). n-Alkylchlorosilanes were purchased from Petrarch Systems (Levittown, Pa.). Trifluoroacetic acid (TFA) was obtained from Pierce (Rockford, Ill.). 2-Propanol was purchased from Fisher Scientific Company (Fair Lawn, N. J.). Bovine serum albumin (BSA) No. A-7511, ovalbumin (OVA) No. A-5503, and glycyl-L-tyrosine were purchased from Sigma Chemical Company (St. Louis, Mo.). CNBr fragments of fetal globin were a gift from Dr. Mark A. Hermodson, Dr. Peter E. Nute, and Dr. Walter C. Mahoney.

CNBr globin fragments. Globin fragments were prepared as previously described (9). The cyanogen bromide fragments were numbered consecutively from those containing the amino-terminal sequence to those containing the carboxyl-terminal sequence of the α-chain. Thus, the three α-chain fragments were designated αCB-1 (residues 1–32), αCB-2 (residues 33–76), and αCB-3 (residues 77–141).

Preparation of stationary phases in Table 1. Octyltrichlorosilane was bonded to silicas via siloxane formation and then endcapped with trimethylchlorosilane using the methanolysis technique of Evans *et al.* (11).

Preparation of stationary phases in Table 2. Four grams of n-octyldimethylchlorosilane were added to 40 g of CCl_4 (10% w/w solution), mixed, and degassed. Silica samples (100 mg) were added to 5-g aliquots of the 10% (w/w) solution, mixed thoroughly, sonicated briefly, and degassed. Reaction occurred at room temperature for 24 h. Samples were filtered on medium Büchner funnels, washed three times with 10 ml CCl_4, and three times with 10 ml acetone. Samples were dried overnight in an oven at 40°C before microanalysis.

Preparation of stationary phases for mass action study (Table 4). A 20% (w/v) stock solution of n-octyldimethylchlorosilane in CCl_4 was prepared and then serial dilutions of 10, 5, 1, and 0.1% were made. One-gram samples of Vydac TP (5.2 μm) were added to each silylating solution, mixed, sonicated, and degassed. Solutions were heated to 65°C for 3 h and then remained at room temperature for 3 days. Supports were washed and dried as described.

Preparation of stationary phases for thermal study (Table 5 and Figs. 2 and 3). Nine 1.0-g samples of Vydac TP (5.01 μm) silica were subjected to temperatures of 24, 120, 200, 300, 400, 500, 600, 700, and 1000°C for a period of 2 h in a Thermolyne Model 1400 furnace. Samples were cooled in a dessicator containing drierite and phosphorous pentoxide (12) until the silylation step. These "thermally pretreated silicas" were then added to 10-ml aliquots of 10% (w/v) n-octyldimethylchlorosilane in CCl_4, mixed, sonicated, and degassed. Solutions were heated to 65°C for 2.5 h and then left at room temperature for 4 days. Samples were filtered on Büchner funnels and washed three times with 10 ml CCl_4 and three times with 10 ml acetone. Samples were dried overnight in a 40°C oven before packing. This thermal pretreatment process was repeated for a second series of samples heated to 24, 100, and 200°C (Fig. 3).

Column packing. Silicas were slurry packed (2% w/v) into columns with 2-propanol at 8000 to 8500 psi by means of a pneumatic pump, (Haskel, Burbank, Calif.). All column diameters were 0.41 cm and, unless otherwise stated, 5.0 cm long. The packing time alotted was 1 min/cm column

TABLE 1

EFFECTS OF SILICA MATRIX ON PROTEIN SEPARATIONS

Silica type	Particle size (μm)	Pore diameter (Å)	$R_{S_{BSA/OVA}}$	$Rec_{BSA/OVA}$	$R_{S_{\alpha CB_2/CB_3}}$	Rec_{CBpeps}	N
Nucleosil 100-5	5	100	1.39	1.97	0.75	5.02	—
Hypersil	5	120	0.43	0.70	0.99	2.62	1208
Spherosil XOBO75	5	300	Poor	Poor	Poor	Poor	1151
Spherisorb SG30F	10	300	1.80	4.33	1.81	6.65	860
LiChrospher Si 300	10	300	1.72	3.80	1.00	6.25	1472
Vydac TP	5	330	2.38	5.70	2.37	6.59	1535
LiChrospher Si 1000	10	1000	1.54	4.17	0.79	—	1643

Note. Resolution values for proteins were determined by conditions as in Fig. 2. Peptides were chromatographed under conditions in Fig. 1, except that the linear gradient was 0–60% in 60 min. Recovery values were determined on a relative scale using a known amount of glycyl-L-tyrosine (GT) or acetophenone as internal standard for protein and peptide systems, respectively. Theoretical plate values were determined with GT in 0.1% (v/v) TFA at 0.7 ml/min.

length. Some silicas crushed at 8000 psi; subsequent investigation established back-pressure limits for packing these materials (Table 3).

Microanalysis. Carbon and hydrogen analyses were performed by C. S. Yeh, Chemistry Department, Purdue University. Accuracy was ±0.2%.

High-performance liquid chromatography. Analyses by HPLC were done using a Varian Vista system (Varian Associates, Walnut Creek, Calif.). The program employed allowed the microprocessor-controlled data system to tabulate resolution and recovery measurements directly.

RESULTS

Comparing Large-Pore-Diameter Silicas

Various silicas were coated with octyl trichlorosilane (SiCl$_3$C$_8$) by a standard procedure (see Materials and Methods) to ascertain the importance of pore size and silica type. Smaller pore-size silicas (<120 Å) were found to be inferior to large-pore-diameter supports for resolution and recovery of peptides and proteins (Table 1). Two such silicas were tested in this study and three others were similarly found to be inferior in a previous report (9). Pore sizes of 300 Å apparently offer the best resolution for pep-

TABLE 2

SILICA TYPE VS LIGAND DENSITY

Silica coated with SiMe$_2$ClC$_8$	%C	%H	Surface area[a] (m^2/g)	Ligand density (μmol C$_8$/m^2)
LiChrospher Si 100	7.81	1.93	256	2.84
LiChrospher Si 300	6.13	1.83	250	2.24
Vydac TP	3.32	1.15	100	2.90
Spherosil XOBO75	4.01	1.73	100	3.54
LiChrospher Si 500	1.91	0.71	45	3.63
LiChrospher Si 1000	0.79	0.26	19	3.50

[a] Surface area values for LiChrospher silicas are from Unger (13); others are from the manufacturers if available.

TABLE 3

PRESSURE STABILITY OF POROUS SILICAS

Silica[a]	Particle size (μm)	Pore diameter (Å)	Surface area (m²/g)	Pore volume (ml/g)	Packing pressure limit (psi)
LiChrosorb Si 60	5	60	475	0.76	>8000
LiChrosorb Si 100	10	100	278	1.02	>8000
Nucleosil 100-5	5	100	300	1.0	>8000
Hypersil	5	120	200	—	>8000
LiChrospher Si 100	10	120	256	1.2	>8000
LiChrospher Si 300	10	300	250	2.0	2000
Spherosil XOBO75	5	300	50–100	0.7	>8000
Spherosorb SG30F	10	300	—	—	1000
Vydac TP	5	330	100	—	>8000
LiChrospher Si 500	10	500	45	0.88	5000
LiChrospher Si 1000	10	1000	19	0.72	>8000
LiChrospher Si 4000	10	3850	4.7	0.78	>8000

[a] Pore diameter, surface area, and pore volume values for LiChrosorb and LiChrosphers are from Unger (13); others are from manufacturers if available.

tides and proteins (Table 1). It is also apparent that 300-Å pores do not necessarily portend a silica will be successful in every case. Within the series in Table 1, four silicas of approximately 300 Å were studied to determine whether selectivity was based on the

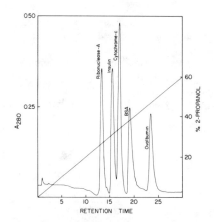

FIG. 1. Column (0.41 × 5 cm) packed with SiMe₂ClC₈-coated Vydac (5.2 μm). Solvent system A, 0.1% (v/v) TFA in H₂O; Solvent B, 0.1% (v/v) TFA in 2-propanol. Gradient was 0–80% Solvent B in 40 min. Flow, 0.7 ml/min; sample load, 630 μg (mass ratio, in order of elution, 2:1.2:1:1.8:1.8); injection volume, 100 μl; ambient temperature.

siliceous matrix. Even though Spherosil XOBO75 was n-alkylated as completely as the other silicas (Table 2), it was not found to be a satisfactory support for reverse-phase separations. This indicated that pore size was not a singular factor for optimizing resolution; although Spherosil XOBO75 itself was found to be durable (withstood packing pressures above 8000 psi) and packed well (more than 30% better efficiency than Spherisorb SG30F), it would not resolve BSA from OVA. Spherisorb SG30F and LiChrospher Si 300 were both found to be acceptable silicas in our study. The main disadvantage of these materials was inherent fragility; both silicas crushed easily when back pressures approached 2000 psi (Table 3). This is a severe handicap, especially in light of the recent popularity of propanol as the mobile phase. The high viscosity of propanol often induces operating pressures of 1500 psi or greater. Vydac TP was found to be the best silica for both protein (Table 1, Fig. 1) and peptide (Table 1 and Ref. (9)) separations. The structural integrity of Vydac TP allowed packing pressures in excess of 8000 psi without signs of particle collapse. From experience gained in this study it has become apparent that pore volume rather than pore diameter is a critical physical parameter when approximating maximum pressures a silica can tolerate (Table 3). The various 300-Å silicas tested ran the gamut of packing stability from only 1000 to greater than 8000 psi.

LiChrospher Si 1000 was superior to the smaller pore Nucleosil 100-5 or Hypersil silicas for protein resolution, but not necessarily for peptides (Table 1). While this study indicated that Vydac TP offered better selectivity than LiChrospher Si 1000, it must be recognized that in the case of extremely large proteins (>100,000 daltons) large-pore LiChrospher-based supports sometimes worked better (14).

Although the carbon loading (%) was lowest for Vydac TP among the 300-Å silicas, surface area must be considered in determining ligand density (13,15). After this normalization (16), it was apparent that alkyl-chain density of the Vydac-based support lay between that of LiChrospher Si 300 and Spherosil XOBO75 (Table 2). The ligand density calculations for LiChrospher Si 500 and Si 1000 silicas were subject to significant error because of the ±0.2% accuracy in elemental microanalysis. The remainder of the study concentrated on investigating the suitability of various techniques of applying the bonded phase and identifying

characteristics of Vydac-based reverse-phase columns that were most suitable for protein separations.

Optimizing n-Alkylation and Thermal Treatment

Two variables were investigated in this study: (i) percentage (w/v) n-octyldimethylchlorosilane ($SiMe_2ClC_8$) in CCl_4 during the coating process and (ii) thermal pretreatment of silica before n-alkylation. The separation of BSA from OVA according to conditions in Fig. 1 was used to monitor protein resolution.

Serial dilutions from 20 to 0.1% $SiMe_2ClC_8$ in CCl_4 were used for reaction with silica (see Materials and Methods). Table 4 shows that protein resolution was affected by this manipulation in support preparation. Interestingly, the data from elemental analyses (Table 4) did not reveal a significant change in the total amount of carbon load for 5% and higher solutions, but resolution values of packed columns were characteristically better from solutions > 5%. The %C values were corrected for inherent carbon in silica prior to n-alkylation. One blank silica, Spherosil XOBO75, contained 1.1% C prior to n-alkylation. It is advisable when reporting %C data that ambient CO_2 be dispelled

TABLE 4

THE INFLUENCE OF ORGANOSILANE MONOMER CONCENTRATION ON PROTEIN RESOLUTION AND LIGAND DENSITY

Organosilane[a] concentration (% w/v)	Microanalysis		$R_{OVA/BSA}$[b]	Mass action ratio[c]
	%C	%H		
0.1	0.58	0.66	0.80	0.12
1	2.53	0.71	3.06	1.2
5	3.22	1.25	2.57	6
10	3.00	1.28	3.47	12
20	2.89	1.20	3.43	24

[a] $SiMe_2ClC_8$.

[b] R_s values were determined under conditions as in Fig. 2.

[c] Mass action ratio is the number of $SiMe_2ClC_8$ molecules in the reaction mixture divided by the number of reactive silanols available on the silica surface (4 $\mu mol/m^2$ (16)).

TABLE 5

THE INFLUENCE OF THERMALLY PRETREATING SILICA ON PROTEIN RETENTION

Temperature[a] (°C)	Microanalysis		Relative protein recovery[b,c]	t_R		Column back pressure[d] (psi)
	%C	%H		BSA (min)	OVA (min)	
24	2.46	1.40	2.57	11.7	14.2	725
120	2.40	1.40	2.63	11.6	14.7	700
200	1.48	0.79	2.59	11.7	14.9	640
300	1.47	0.80	2.56	11.7	14.1	550
400	1.42	0.97	—	—	—	—
500	1.37	0.76	2.34	11.7	13.8	490
600	0.92	0.52	2.18	11.1	12.5	440
700	0.91	0.52	2.19	10.9	12.2	410
1000	0.74	0.30	—	10.0	12.3	380

[a] Vydac TP was heated to the temperature indicated for 2 h prior to silylation with $SiMe_2ClCl_8$.
[b] Relative recovery refers to the protein ovalbumin with GT as the internal standard.
[c] For elution conditions and additional data for protein resolution, see Fig. 2.
[d] Column back pressure was measured at 1 ml/min using 0.1% TFA at 27°C.

thermally and possible background carbon, inherent to the manufacturing process, be subtracted.

The thermal study indicated that preheating Vydac TP to 200°C prior to n-alkylation increased protein resolution with a concomitant sharp drop in %C loading (Table 5). Compared to initial resolution, values following 80 h of gradient-flow conditioning fell disproportionally among the six test columns in Fig. 2. The precipitous drop in resolution values (R_s) was characteristically observed during the first few hours of column usage. To verify this phenomenon and quan-

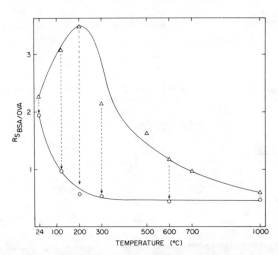

FIG. 2. Protein resolution vs thermal pretreatment of silica before and after successive blank gradients (0–100% in 20 min, 0.1% (v/v) TFA into 2-propanol at 0.7 ml/min) for 80 h. △, Resolution values of new columns; ○, values after 80 h of use.

titate resolution decline, a second experiment monitored column usage for silicas preheated to 24, 100, and 200°C prior to n-alkylation, with R_s values periodically determined as a function of time (Fig. 3). As mentioned above, it must be remembered that determining R_s values for new columns can only be taken as approximate (minimum), whereas values measured after many hours of gradient usage are reliable figures. Therefore, the rate of descending slopes and not the initial R_s values themselves are to be noted in Fig. 3. From Figs. 2 and 3 it was apparent that thermally induced surface alterations prior to n-alkylation increased protein resolution, but the enhancement was ephemeral and overall undesirable. Over the range of 100 to 500°C pretreatment, the %C was relatively unchanged (Table 5), while R_s values dropped progressively for the initial measurement and then leveled off at similar low levels after 80 h (Fig. 2). This indicated that initially the 200–500°C thermal treatment offered different selectivity. According to retention times listed in Table 5, it appeared that more change occurred with OVA than BSA over that temperature range. After 80 h of gradient elution, resolving power for each 200–500°C column was normalized to a lower level. It is known that hydrogen-bonded water is released above 200°C and that surface vicinal silanols condense to form siloxanes ($\equiv$Si$-$OH + HO$-$Si$\equiv$ $\rightarrow$ $\equiv$Si$-$O$-$Si$\equiv$ + H_2O) above 450°C (17). It is possible that augmented resolution occurred when the surface concentration of accessible free silanols was temporarily reduced. It would follow that surface siloxane groups, aided by a putative acid catalysis,[3] hydrolyzed as a result of exposure to water, thereby increasing silanol concentration (13) and diminishing protein resolution. Scott (18) has reported that siloxane bonds are stable in distilled water. Qualitative evidence of surface siloxane bond reactivity in acidic medium, to our knowledge, has not been investigated. Since the

[3] TFA, 0.1% (v/v) in H_2O has a pH of about 1.9.

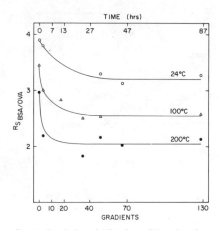

FIG. 3. Resolution of BSA from OVA plotted as a function of the number of prior gradient elutions (40-min linear gradient, 0–80% 0.1% (v/v) TFA in H_2O into 2-propanol with 0.1% (v/v) TFA at 0.7 ml/min); 130 gradients = 87 h of usage. Silica pretreated to 24, 100, and 200°C as indicated by plots O, $\triangle$, and $\bullet$, respectively.

majority of peptide and protein elution protocols reported in the literature are under acidic conditions, this matter seems to be an important variable in column performance. When a comparison is made of the three columns after 85 h of usage in Fig. 3, it should be noted that BSA and OVA retention times were constant throughout for the column at 24°C, whereas times decreased 14 and 19% for the columns at 100 and 200°C, respectively (data not shown). This could suggest an initial unstable bonded phase for the latter two columns. All three columns were subject to increased surface silanol concentration when exposed to acidified aqueous phase. In the case of the column at 24°C, a more stable C_8-bonded-phase–silica interface could explain why the R_s degeneration curve was comparatively less severe in Fig. 3.

Column Loading, Length, and Mechanism

Resolution as a function of sample loading with a BSA/OVA mixture declined in a linear relationship to the point where the column was saturated with protein (Fig. 4).

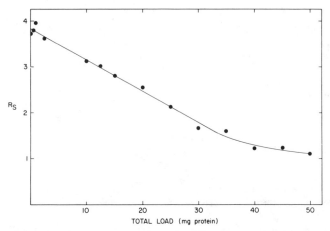

FIG. 4. Separation of BSA from OVA (9:11 ratio) on a 0.41 × 25-cm column Vydac coated with SiMe$_2$ClC$_8$. Injection volume, 1 ml; gradient conditions as in Fig. 1.

When the loading of a column surpassed its saturation limit, resolution approached a lower limit. Column saturation was indicated when solute eluted as a breakthrough peak. Thus, unlike what may occur in size-exclusion chromatography (19), overloading a reverse-phase column does not result in total loss of resolution (Fig. 5). An interesting sidelight found in Fig. 5 was the optimized loading capacity for OVA prepeaks. When the column was overloaded with the major components of the BSA/OVA mixture, total resolution of the minor components was still possible. Breakthrough peaks were omitted in Fig. 5 for clarity.

Loading also affected solute retention times. For loading levels of 1, 10, and 25 mg in Fig. 5, retention times of all solutes in the BSA/OVA mixture decreased. As an example, the first OVA prepeak eluted in 21.8, 21.6, and 21.2 min., respectively. The mechanism of this phenomenon is thought to involve availability of stationary phase sites for protein binding during the elution process (20). As loading increases, the number of free sites decreases until a state of "overloading" or total depletion of binding sites exists. It would follow that an overloaded column results in some protein being rapidly eluted as a breakthrough peak because of nonavailability of additional binding sites. Longer columns introduced additional binding sites and thus afforded greater loading capacities. A 0.41 × 25-cm column could load 21 mg of BSA/OVA mixture with a single injection, whereas a 0.41 × 5-cm column loaded only 2.5 mg (Fig. 6). The loading capacity per centimeter was found to be more than additive as column length increased. This was most likely due to radial dispersion of solute within a column. When solute is introduced at a point source on the head of the column, a conical-shape pattern of solute dispersion is generated. Depending on the size of particles in the bed and column diameter, solute will eventually reach the column walls (21). As a consequence, not all of the stationary phase surface will be exposed to solute, but as column length increases, the fraction of unavailable material within a column will decrease. The loading study in this work was based on single, 1-ml injections of varying protein concentration. This represented loading under dynamic conditions because protein was introduced in a plug and traveled down the column be-

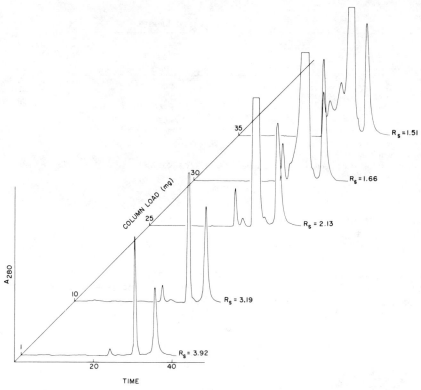

FIG. 5. A representative series of chromatograms used for plotting data in Fig. 4. Elution conditions are the same as in Fig. 1. The first major peak is BSA, the second is OVA. Small peaks prior to BSA peak are from OVA.

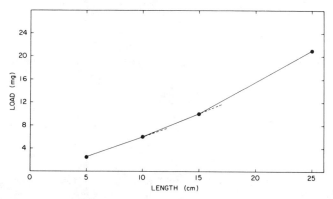

FIG. 6. Maximum protein load vs column length. Same packing material and elution conditions as in Fig. 5, except column lengths of 5, 10, 15, and 25 cm were employed.

TABLE 6

BSA/OVA SEPARATION VS COLUMN LENGTH

Column[a] length (cm)	Loading (mg)[b]	Avg. Δt^c between peaks (min)	% of 25-cm Column Δt
5	0.05 → 2.5	4.08	87
10	0.05 → 7.5	4.28	91
15	0.05 → 10	4.55	97
25	0.05 → 21	4.69	100

[a] All columns were gradient eluted as in Fig. 1.
[b] Limits of nonsaturated loading.
[c] Δt, the average difference in retention time between BSA and OVA peaks over a series of nonsaturating loading conditions as determined in Fig. 6.

fore attaining full stationary phase–solute equilibrium. Although this is the common mode of protein introduction onto a column, we have recently found that loading can be substantially increased when protein is introduced under less dynamic conditions (22). Instead of a single injection of solute onto a column prior to initiating the gradient, loading via several smaller injections several minutes apart enabled a column to accept more protein.

Column length was not found to be important for protein separations. As shown in Table 6, a fivefold increase in column length increased the time between BSA and OVA by only 13%. This represented a substantial deviation from what would have been expected if a multiple-step partitioning process

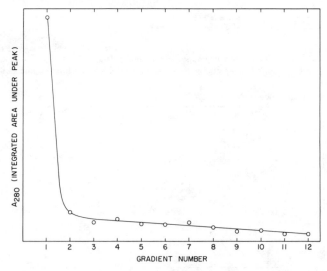

FIG. 7. Ovalbumin desorption vs number of gradients. Several injections of OVA were applied to a C_8-coated Vydac column (0.41 × 25 cm) until the column was saturated. Consecutive desorbing linear gradients of 0 to 80% 0.1% (v/v) TFA in H_2O into 0.1% (v/v) TFA in 2-propanol were run at a flow rate of 0.7 ml/min.

is assumed. The small difference in relative retention as a function of column length would be more consistent with a mechanism involving a small number of adsorption–desorption steps rather than the multiple-step partitioning process associated with the chromatography of small molecules. It was also observed that solute desorption was not complete in a single gradient elution cycle when columns were used in the preparative mode. OVA was chosen as a probe in Fig. 7 because of its relatively high degree of lipophilicity in a TFA/2-propanol system (cf. Fig. 1). OVA (21 mg) was loaded onto the column. Of the total protein eluted, 60% was recovered in the first gradient cycle while the remaining 40% was in the next 11. The desorption process is therefore not an all-or-none response to TFA/propanol gradients.

Column Efficiency vs Protein Resolution

Column efficiency is commonly expressed by Eq. [1] (23), where t_R is the retention time of a small, unretained solute, t_w is the width of the peak, and N is the theoretical plate number for the column. N is related to H, the height equivalent to a theoretical plate and the column length (L). Determining N and solving for H in Eqs. [1] and [2] and then substituting into Eq. [3] gives the value of h, the reduced height of a theoretical plate (24). Reduced plate height (h) can be used to compare packing efficiencies of columns regardless of length or particle size, d_p.

$$N = 16\left(\frac{t_R}{t_w}\right)^2 \qquad [1]$$

$$N = \frac{L}{H} \qquad [2]$$

$$h = \frac{H}{d_p} \qquad [3]$$

$$R_s = \frac{1}{4}(\alpha - 1)\sqrt{N}\left(\frac{k'}{1+k'}\right) \qquad [4]$$

According to Eq. [4] (23)—the popular resolution equation relating resolution to α, selectivity factor; k', capacity factor; and N, plate value—resolution is thought to be proportional to the square root of the N value. It would follow that high N values would be indicative of high R_s values for any given column. An important point to note is that Eqs. [1]–[4] apply only to the isocratic elution mode. A common misconception is that a column of high efficiency (high N value) will necessarily connote superior protein resolution. This may not be true. Proteins are large, tenaciously retained, and usually require a gradient mode for elution (parameters not accounted for in Eqs. [1]–[4]). To examine the relationship between column efficiency and actual protein resolution, four commercial columns and a column prepared in this laboratory with a $SiMe_2ClC_8$-bonded phase on Vydac (desig-

TABLE 7

PROTEIN RESOLUTION VS THEORETICAL PLATES

Column BSA/OVA	Column dimensions (cm), particle size (μm)	N^a	h	R_s^b
Partisil ODS-3	0.46 × 25, 10	4050	6.17	2.03
LiChrosorb-RPC	0.46 × 25, 10	3918	6.38	Poor
Partisil C8	0.46 × 25, 10	2397	10.43	1.52
Bio-Rad ODS-5S	0.4 × 25, 5	2330	21.46	1.74
Vydac-test-C_8	0.41 × 25, 5	830	60.24	2.90

[a] Theoretical plate values were determined with $NaNO_3$ in H_2O at 0.5 ml/min.

[b] Proteins were chromatographed by the linear gradient conditions as in Fig. 4. BSA/OVA loads were 10 mg (9:11 ratio) in 1-ml injection.

nated Vydac-test-C_8) were compared. Columns were isocratically eluted with H_2O employing $NaNO_3$ as the solute (25) to obtain h, H, and N. The data in Table 7 indicated that the less efficiently packed Vydac-test-C_8 column outperformed commercially packed columns in protein resolution. In addition, an interesting point about relative selectivity can be found in Table 5. Because all columns were run under identical conditions and the retention time of BSA was constant in most cases, differences in the retention time of OVA were responsible for intercolumn variation in resolution. Resolution was more a function of the relative selectivity of packing materials (23) than the number of plates (N) in the column. Therefore, when columns are compared for protein resolution, the R_s value of a protein mixture such as BSA and OVA is a more realistic indicator of column utility than plate value.

DISCUSSION AND CONCLUSION

Large-pore-diameter silicas have recently emerged as the premier type to use when large peptides (>30 residues) need to be resolved from one another. These silicas are also best suited for protein separations. Of the several silicas compared in this study, it may be concluded that Vydac TP affords the most utility as a support for reverse-phase chromatography of large peptide fragments and proteins (Table 1). When C_8-coated Vydac was compared to various commercially prepared columns, it was observed that its unique selectivity, inherent in the siliceous matrix, overshadowed our relatively inferior packing techniques to achieve the best resolution of BSA from OVA (Table 7).

It is generally thought that it is beneficial to preheat silica prior to n-alkylation to reduce surface water and thereby slightly increase the carbon loading (13,17). In the case of Vydac TP for protein separations, preheating silica may have an overall deleterious effect on resolution (Figs. 2 and 3).

This conclusion was reached as the result of monitoring protein resolution during usage, not just initial performance. The C_8 supports prepared from preheated Vydac performed well initially, but resolution deteriorated substantially during the first few hours of operation (Fig. 3). On the other hand, C_8 columns prepared without thermal treatment of silica retained resolution for a longer time.

No relationship was found between the performance of columns in the resolution of proteins and the theoretical plate data obtained from the chromatography of small, nonretained solutes. Using BSA, OVA, and cyanogen bromide fragments of human globin, it was possible to evaluate the performance of a variety of commercial and experimental columns (Tables 1 and 7).

Longer columns increased protein loading capacity considerably (Fig. 6), but not necessarily resolution (Table 6). The fact that a column five times longer increased separation between BSA and OVA peaks by only 13% (Table 6) coincides with what was previously reported for large peptides (9). This same effect has been observed for both cation- and anion-exchange supports with high charge-to-surface-area ratios (26,27). Chang (28) calculated that the density of charge on an anion exchanger and a typical protein surface area were compatible with a multisite cooperative binding process. We suspect the same process occurs for proteins in reverse-phase chromatography. Given that the ligand density of C_8 chains on Vydac was 2.90 $\mu mol/m^2$ (Table 2), a C_8 ligand will be located within every 64 Å^2 according to Berendsen's (16) method of calculation. Taking the geometry of a typical protein such as BSA (60 Å in diameter and 45 Å high (29)), its dimensions theoretically allow it to have as many as 44 sites of interaction with C_8 stationary phase ligands.

The selectivity of Vydac TP is apparently not due to a unique ligand density as indicated in Table 2. The carbon percentage determined by microanalysis showed that specific ligand density (μmol/m^2) for Vydac TP appeared not to differ significantly from other silicas. These values are in agreement with values of 2 to 4 μmol/m^2 calculated for supports in other studies (30).

The order of elution in Fig. 1 did not correspond to increasing molecular weights of proteins. This suggested that density or intensity of lipophilic sites vary between proteins. In fact, on the basis of lipophilic sites/unit volume, OVA is 33% more hydrophobic than BSA (31). The concept of multisite interaction stems from a series of papers by Jennissen (32–37), who used glycolytic enzymes. Jennissen reported that with alkylated agarose the mechanism of separation for small molecules was site independent, while for proteins the mechanism was site dependent. Small molecules followed a simple lock-and-key-type mechanism. The adsorption–desorption of proteins, on the other hand, followed a cooperative, multisite binding pattern. Jennissen's kinetic data showed that there were between three and nine binding sites for a series of five proteins (37). This type of multisite requirement for desorption could explain the phenomenon observed by Lewis (10) (cf. Fig. 2): BSA isocratically eluted on a C_8 column in approximately 13 min with a mobile phase of 0.5 M formic acid–0.4 M pyridine (pH 4.0) containing 34% 1-propanol, but when the mobile phase was 32% in 1-propanol, the peak would not elute. It was interesting to note in Lewis' data that even after 40 min of noneluting 32% propanol flow, the peak base width, when finally eluted with 34% propanol, was not more than 10–15% greater than the original 34% propanol isocratic run. The tenacity with which OVA was retained on a column in Fig. 7 even after being subjected to numerous gradient elution cycles may be due to multiple site binding.

Characterization of C_8 bonded phase coated Vydac TP indicated that resolution decreased with increasing BSA/OVA load. As sample mass increased, retention times for all peaks decreased. Column length was important when considering loading capacity but not resolution. The advantage of shorter columns is that back pressures are significantly less and column life is prolonged. These data suggest that a preparative column should be short and of large diameter.

In summary, the influence of column efficiency (theoretical plates), column length, carbon loading, pore size, thermal conditioning, organosilane bonding procedures, and type of silica for resolution of proteins in reverse-phase chromatography was examined. The key to protein resolution was apparently not the theoretical plate value or length of a column, nor was it exclusively manifest in the carbon loading or thermal conditioning of silica. Although macroporosity was an essential feature of a high-resolution column, it did not determine superiority. The single most profound determinant of resolution was the silica support itself.

ACKNOWLEDGMENTS

We are grateful to Mr. Mears Mitchell for determining column efficiencies for this study. We are indebted to Dr. George Vanecek for many helpful discussions on protein interactions with large-pore-diameter supports. This work was supported by NIH Grant 0685-51-11535.

REFERENCES

1. Regnier, F. E., and Gooding, K. M. (1980) *Anal. Biochem.* **103**, 1–25.
2. Hearn, M. T. W., and Hancock, W. S. (1978) *TIBS* **4**, 58–62.
3. Rivier, J. E. (1980) *J. Chromatogr.* **202**, 211–222.
4. Rubinstein, M. (1979) *Anal. Biochem.* **98**, 1–7.
5. Bennett, H. P. J., Browne, C. A., and Soloman, S. (1980) *J. Liq. Chromatogr.* **3**(9), 1353–1365.
6. Dunlap III, C. E., Gentleman, S., and Lowney, L. I. (1978) *J. Chromatogr.* **160**, 191–198.
7. Van Der Rest, M., Bennett, H. P. J., Soloman, S., and Glorieux, F. H. (1980) *Biochem. J.* **191**, 253–256.
8. Mahoney, W. C., and Hermodson, M. A. (1980) *J. Biol. Chem.* **255**, 11,199–11,203.
9. Pearson, J. D., Mahoney, W. C., Hermodson, M. A., and Regnier, F. E. (1981) *J. Chromatogr.* **207**, 325–332.
10. Lewis, R. V., Fallon, A., Stein, S., Gibson, K. D.,

and Udenfriend, S. (1980) *Anal. Biochem.* **104,** 153–159.

11. Evans, M. B., Dale, A. D., and Little, C. J. (1980) *Chromatographia* **13,** 5–10.

12. Scott, R. P. W., and Kucera, P. (1979) *J. Chromatogr.* **171,** 37–48.

13. Unger, K. K. (1979) Porous Silica: Its Properties and Use as Support in Column Liquid Chromatography, Elsevier, New York.

14. Vanecek, G., and Regnier, F. E. (1982) *Anal. Biochem.,* **121,** 156–169.

15. Colin, H., and Guiochon, G. (1977) *J. Chromatogr.* **141,** 289–312.

16. Berendsen, G. E., Pikaart, K. A., and de Galan, L. (1980) *J. Liq. Chromatogr.* **3**(10), 1437–1464.

17. Scott, R. P. W. (1980) *J. Chromatogr. Sci.* **18,** 297–306.

18. Scott, R. P. W., and Trainman, S. (1980) *J. Chromatogr.* **196,** 193–205.

19. Vivlecchia, R. V., Lightbody, B. G., Thimot, N. Z., and Quin, H. M. (1977) *J. Chromatogr. Sci.* **15,** 424–433.

20. Regnier, F. E., manuscript in preparation.

21. Knox, J. H., Laird, G. R., and Raven, P. A. (1976) *J. Chromatogr.* **122,** 129–145.

22. Pearson, J. D., and Regnier, F. E., manuscript in preparation.

23. Snyder, L. R., and Kirkland, J. J. (1979) Introduction to Modern Liquid Chromatography, 2nd ed., Chap. 2, Wiley, New York.

24. Giddings, J. C. (1965) Dynamics of Chromatography, Chap. 2, Dekker, New York.

25. Wells, M. J. M., and Clark, C. R. (1981) *Anal. Chem.* **53,** 1341–1345.

26. Gupta, S., and Regnier, F. E., personal communication.

27. Vanecek, G., and Regnier, F. E. (1980) *Anal. Biochem.* **109,** 345–353.

28. Chang, S. H., Gooding, K. M., and Regnier, F. E. (1976) *J. Chromatogr.* **120,** 321–333.

29. Slayter, E. M. (1965) *J. Mol. Biol.* **14,** 443–452.

30. van de Venne, J. L. M., Rindt, J. P. M., Coenen, G. J. M. M., and Cramers, C. A. M. G. (1980) *Chromatographia* **13,** 11–17.

31. Fisher, H. F. (1964) *Proc. Nat. Acad. Sci.,* Wash. **51,** 1285.

32. Jennissen, H. P., and Heilmeyer, Jr., L. M. G. (1975) *Biochemistry* **14,** 754–760.

33. Jennissen, H. P. (1976) *Biochemistry* **15,** 5683–5692.

34. Jennissen, H. P. (1978) *J. Chromatogr.* **159,** 71–83.

35. Jennissen, H. P. (1979) *Protides Biol. Fluids, Proc. Colloq.* 26th, 657–660.

36. Jennissen, H. P. (1980) *Protides Biol. Fluids, Proc. Colloq.* 27th, 765–770.

37. Jennissen, H. P. (1977) *Hoppe-Seyler's Z. Physiol. Chem.* **358,** 255.

Variables in the High-Pressure Cation-Exchange Chromatography of Proteins*

CHARLES A. FROLIK, LINDA L. DART, AND MICHAEL B. SPORN

Laboratory of Chemoprevention, National Cancer Institute, Building 37, Room 3C02, Bethesda, Maryland 20205

The use of cation-exchange high-pressure liquid chromatography for the separation of proteins has been investigated. Several factors, including solvent composition, pH, flow rate, and temperature, were examined for their effects on the resolution of protein standards (insulin, β-lactoglobulin, and carbonic anhydrase B; molecular weight range, 6000 to 30,000 and pI range, 5.3 to 6.5). An initial comparison was made of the recovery of these proteins from three commercially available columns (Whatman Partisil SCX, Separation Industry CM silica, and MCB Reagents Lichrosorb KAT). In general, under the conditions employed, the SCX column gave the highest recovery of applied protein. Based on this recovery data, the Partisil SCX column was chosen for subsequent examination of chromatographic parameters that would optimize protein resolution. An increase in temperature decreased retention and resolution but increased recovery, with some proteins being affected more than others. A decrease in pH in the final eluant or an increase in pH in the initial eluant caused an increase in retention times. For some proteins, the decrease in pH resulted in a greater total recovery of protein. This information has been applied to the purification by cation exchange high pressure liquid chromatography of transforming growth factors from a human tumor cell line.

High-pressure liquid chromatography (HPLC)[1] is rapidly becoming an increasingly useful technique in the separation of peptides and proteins (1-3). In the past, most work has concentrated on the use of reverse-phase HPLC. The application of ion-exchange HPLC to peptide purification has received less attention. There are, however, a limited number of reports concerning the use of either anion-exchange (4-11) or cation-exchange HPLC (4,8,12-17) for the isolation of proteins. Most of these separations employed nonvolatile buffers to elute the compounds of interest. When isolating compounds of biological importance, the use of such buffers often necessitates an additional

dialysis or gel filtration step to remove the high concentration of salt prior to assay in a biological system. Recently, Radhakrishnan *et al.* (13) described a method that utilized volatile buffers to successfully chromatograph proteins up to 24,000 daltons on a cation-exchange HPLC column. It is the purpose of this report to examine the effect of volatile solvent composition, flow rate, pH, and temperature on the recovery and resolution of standard proteins on a commercially available HPLC cation-exchange column and to apply this information to the isolation of polypeptide growth factors from human tumor cells.

MATERIALS AND METHODS

Reagents. Nonradioactive insulin, β-lactoglobulin, and carbonic anhydrase B were obtained from Sigma Chemical Company, St. Louis, Missouri. Monoiodinated insulin

* This paper was presented at the International Symposium on HPLC of Proteins and Peptides, November 16-17, 1981, Washington, D. C.

[1] Abbreviations used: HPLC, high-pressure liquid chromatography; TGF, transforming growth factor; R_s, resolution.

(100 μCi/μg), β-lactoglobulin (21.9 μCi/ mg), and carbonic anhydrase (13 μCi/mg), both [methyl-^{14}C]methylated, were purchased from New England Nuclear, Boston, Massachusetts. Pyridine and acetonitrile were supplied by Burdick and Jackson, Muskegon, Michigan, trifluoroacetic acid (sequanal grade) by Pierce Chemical Company, Rockford, Illinois, and HPLC grade water by J. T. Baker Chemical Company, Phillipsburg, New Jersey. Ammonium acetate, ACS grade, was purchased from Eastman Kodak Company, Rochester, New York.

Instrumentation. All chromatographic runs were performed on a Model 5060 HPLC (Varian, Walnut Creek, Calif.) equipped with a universal column heater. Cation-exchange HPLC columns tested included a 10-μm Partisil-10-SCX (4.6 mm $\times$ 25 cm) column (Whatman, Clifton, N. J.), a 10-μm CM Silica (4.6 mm $\times$ 30 cm) column (Separation Industries, Orange, N. J.) and a 10-μm Lichrosorb KAT (4.0 mm $\times$ 25 cm) column (MCB Mfg. Chemists, Cincinnati, Ohio). Solvent systems for these columns consisted of pyridine–ammonium acetate mixtures with the pH adjusted by glacial acetic acid. For determination of optimal conditions for protein separation by cation-exchange HPLC, standard protein solutions were prepared in 4 m$_M$ hydrochloric acid or in water that contained

^{125}I-insulin (8.8 $\times$ 10^5 dpm/mg), [^{14}C]-β-lactoglobulin (2.5 $\times$ 10^5 dpm/mg), or [^{14}C]-carbonic anhydrase (2.5 $\times$ 10^6 dpm/mg). Twenty-microliter aliquots containing either 17 μg of the individual proteins or a total of 15 μg of a mixture of these proteins were applied to the column. Fractions (0.4 ml) were collected in small vials (Bio-Vial, Beckman Instruments, Inc., Palo Alto, Calif.); 2 ml of Aquasol (New England Nuclear) were added and the samples was counted on a Model 5500 gamma counter (Beckman Instruments) for detection of ^{125}I and/or on a Tri-Carb 460 CD liquid scintillation counter (Packard Instrument Co., Downers Grove, Ill.) for determination of ^{125}I and ^{14}C.

Purification of TGF. TGF was extracted from human rhabdomyosarcoma cells (A673) by an acid–ethanol procedure (18). The lyophilized extract was redissolved in 1 M acetic acid and applied to a Bio-Gel P30 column, 100–200 mesh, 2.6 $\times$ 90 cm. The column was eluted with 1 M acetic acid at a flow rate of 10 ml/h. Fractions (2.5 ml) were collected and aliquots assayed for protein using an assay which is based on the dye-binding procedure of Bradford (19) (Bio-Rad Laboratories, Richmond, Calif.), and for transforming activity in a soft agar growth assay (18). Briefly, samples to be tested were suspended in 0.3% agar (Difco, Noble agar) in Dulbecco's modified Eagles

TABLE 1

PROTEIN RECOVERY FROM CATION-EXCHANGE RESINS

		Percentage recovery					
		Insulin		β-Lactoglobulin		Carbonic anhydrase	
Column	Temperature (°C)	Isocratic	Linear gradient	Isocratic	Linear gradient	Isocratic	Linear gradient
SCX	25	87	77	78	74	78	32
	55	91	75	88	90	88	72
CM Silica	55	74	—	75	—	66	—
Lichrosorb KAT	55	76	—	75	—	14	—

Note. Flow, 0.5 ml/min; isocratic, 0.05 M pyridine, 2.0 M ammonium acetate, pH 4.0; linear gradient, 0.05 M pyridine, 0.01 M ammonium acetate, pH 2.5, to 0.05 M pyridine, 2.0 M ammonium acetate, pH 4.0 (0.04 M/min).

medium (GIBCO, Grand Island, N. Y.) supplemented with 10% calf serum (GIBCO), penicillin (100 units/ml), and streptomycin (100 μg/ml) containing 3×10^3 normal rat kidney fibroblasts (clone 49F) (20) per milliliter. A portion (0.7 ml) of the mixture was pipetted onto a 0.7-ml base layer (0.5% agar in the supplemented medium) in a 35-mm petri dish. After 1 week at 37°C in a humidified 5% CO_2 atmosphere, the cells were stained by layering over the agar 0.7 ml of a solution of 2-(p-iodophenyl)-3-(p-nitrophenyl)-5-phenyl tetrazolium chloride (0.5 mg/ml in water) and continuing the incubation for 24 h as described by Schaeffer and Friend (21). The number and size of the colonies were determined using a Bausch & Lomb Omnicon image analysis system

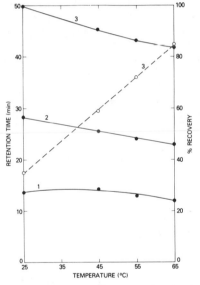

FIG. 1. Dependence of retention time (——) and recovery (- - -) on temperature. Standard proteins (1, insulin; 2, β-lactoglobulin; 3, carbonic anhydrase) were chromatographed on a Partisil-10-SCX column at a flow rate of 0.5 ml/min using a linear gradient starting at 0.05 M pyridine, 0.01 M ammonium acetate, pH 2.5, and finishing at 0.5 M pyridine, 0.01 M ammonium acetate, pH 4.0 (0.008 M pyridine/min) at the temperature indicated.

TABLE 2

EFFECT OF FLOW RATE ON PROTEIN RESOLUTION

Flow (ml min⁻¹)	Protein	Elution volume (ml)	R_s
0.2	Insulin	7.0	
	β-Lactoglobulin	13.4	1.4[a]
	Carbonic anhydrase	24.3	1.6[b]
0.5	Insulin	6.9	
	β-Lactoglobulin	13.0	1.6[a]
	Carbonic anhydrase	23.8	2.1[b]
2.1	Insulin	6.8	
	β-Lactoglobulin	12.7	1.3[a]
	Carbonic anhydrase	22.0	1.3[b]

Note. Column, Partisil-10-SCX; temperature, 55°C, linear gradient, 0.05 M pyridine, 0.01 M ammonium acetate, pH 2.5, to 0.5 M pyridine, 0.01 M ammonium acetate, pH 4.0 (0.016 M pyridine/ml).
[a] R_s between insulin and β-lactoglobulin.
[b] R_s between β-lactoglobulin and carbonic anhydrase.

(22). The fractions containing transforming activity (molecular weight, approximately 20,000) were combined and lyophilized to dryness. The sample was redissolved in 500 μl 25:75:0.1, acetonitrile:water:trifluoroacetic acid, pH 2.0, and applied to a 10-μm μBondapak C_{18} (Waters Associates, Milford, Mass.) (7.8 mm × 30 cm) reverse-phase HPLC column and chromatographed as described in the legend to Fig. 3. The major region showing transforming activity from this column (eluting at 10 min) was combined, lyophilized to dryness, redissolved in 250 μl of 0.05 M pyridine, 0.01 M ammonium acetate, pH 2.5, and applied to a Partisil-10-SCX column. This column was developed as described in the legend to Fig. 4.

RESULTS AND DISCUSSION

Recovery. In order to evaluate various commercially available HPLC cation-exchange columns for applicability to the purification of polypeptide growth factors, standard proteins were chosen that had molecular weights and isoelectric points that

TABLE 3

EFFECT OF pH ON RETENTION TIME

pH		Retention time (min)		
Initial	Final	Insulin	β-Lacto-globulin	Carbonic anhydrase
2.5	4.0	14	23	39
2.5	6.0	10	18	28
4.0	4.0	25	31	40

Note. Column, Partisil-10-SCX; temperature, 55°C; flow rate, 0.5 ml/min; linear gradient, 0.05 M pyridine, 0.01 M ammonium acetate to 0.05 M pyridine, 2.0 M ammonium acetate (0.04 M/min) at the pH indicated.

were similar to the compounds of interest. In addition, due to the nature of the chromatographic solvents used, detection of proteins by uv absorption was not practical. For this reason, radioactively labeled compounds were chosen for the standards.

Before examining the various parameters that influence the resolution of proteins on cation-exchange HPLC columns, the recovery of the standard proteins from several commercially available columns was determined. As is shown in Table 1, at a temperature of 55°C and using a solvent com-

position that eluted all three proteins in the void volume of the column, the SCX column gave the highest recovery of applied protein. When a gradient system that resolved the three proteins was employed with the SCX column, the recovery of insulin and carbonic anhydrase decreased while the recovery of β-lactoglobulin remained constant. More importantly, when the temperature of the column was decreased from 55 to 25°C, the recovery of the proteins was diminished with carbonic anhydrase showing the greatest affect (Table 1, Fig. 1). Based on this recovery data, the Partisil SCX column at a temperature of 55°C was chosen for subsequent examination of the chromatographic parameters that would optimize protein resolution.

Temperature. Because of the increase in protein recovery with an increase in temperature, it was decided to first examine the effect of temperature on resolution. As is shown in Fig. 1, there is a slight decrease in the retention time for all three proteins with an increase in column temperature. This decrease in the retention time is also reflected in a decrease in the resolution of these proteins. At 25°C, β-lactoglobulin and carbonic anhydrase had an R_s value of 1.5. At 55°C

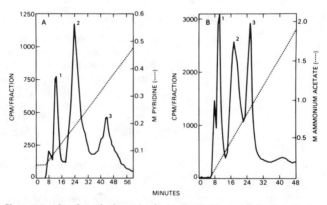

FIG. 2. Chromatography of standard proteins (1, insulin; 2, β-lactoglobulin; 3, carbonic anhydrase) on a Partisil-10-SCX column at a temperature of 55°C, flow rate of 0.5 ml/min, and linear gradients (A) from 0.05 M pyridine, 0.01 M ammonium acetate, pH 2.5, to 0.5 M pyridine, 0.01 M ammonium acetate, pH 4.0 (0.008 M pyridine/min), and (B) from 0.05 M pyridine, 0.01 M ammonium acetate, pH 2.5, to 0.05 M pyridine, 2.0 M ammonium acetate, pH 4.0 (0.04 M ammonium acetate/min).

this value had decreased to 1.0. It has previously been observed that the capacity factor for cytochrome *c* chromatographed on an acrylic cation exchanger increased with an increase in temperature (16) while the capacity factor for hemoglobin decreased (15). With an anion-exchange column, resolution of ovalbumin was found to increase with a change of temperature from 4 to 25°C (9). It therefore appears that the effect of temperature on the selectivity of ion-exchange separations varies depending on the protein mixture being examined. In the present study, while an increase in temperature reduced resolution, it also decreased peak width, reduced analysis time and, as previously mentioned, increased protein recovery. For these reasons, 55°C was chosen as the standard temperature in all further resolution studies.

Flow rate. Previous investigations have shown that a decrease in the mobile-phase velocity substantially increases resolution in both anion- (9) and cation-exchange (14) chromatography. Under the experimental conditions employed in the present study, decreasing the flow rate from 2.1 to 0.5 ml/min did result in a small increase in resolution (Table 2). However, a further reduction in mobile-phase velocity to 0.2 ml/min did not yield any further improvement. Therefore, for the Partisil-10-SCX column, it appears that 0.5 ml/min is the optimum flow rate under the gradient conditions employed.

pH. Normally in cation-exchange chromatography, a decrease in the pH of the mobile phase causes an increase in sample retention due to an overall increase in the positive charge of the sample. For proteins, the conformation of the molecule may also be important in determining which groups are available for interaction with the cation-exchange resin and this conformation may change with pH. As is shown in Table 3, an increase in the pH of the initial mobile phase from 2.5 to 4.0 caused an increase in the retention of the three proteins chromato-

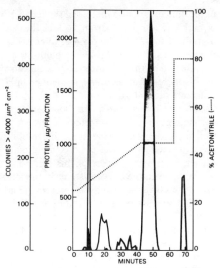

FIG. 3. Chromatography of human TGF on a reverse-phase μBondapak C_{18} column. TGF was extracted from A673 cells and chromatographed on a Bio-Gel P30 column as described under Materials and Methods. The fractions containing transforming activity were combined and chromatographed on a μBondapak C_{18} column at a temperature of 25°C, a flow rate of 1.2 ml/min and a 40-min linear gradient (· · ·) from 25:75:0.1, acetonitrile:water:trifluoroacetic acid, pH 2.0, to 45:55:0.1, acetonitrile:water:trifluoroacetic acid, pH 2.0, followed by an 80:20:0.1, acetonitrile:water:trifluoroacetic acid, pH 2.0, strip at 60 min. Fractions (1.2 ml) were collected and aliquots (200 μl) removed for determination of protein (shaded region) and transforming activity.

graphed. This would appear to indicate that at the higher pH there were more positive charges available on the protein due to conformational changes for interaction with the resin or that the number of available ionic binding sites on the resin had increased due either to an increase in the ionization of the sulfonic acid functional groups on the resin or to ionization of the silica gel backbone (17). A decrease in the pH of the final mobile phase from 6.0 to 4.0 was accompanied by the expected increase in sample retention, a result consistent with other studies employing cation-exchange HPLC (8,16). It was noted, however, that lowering the pH of the final eluant also increased the recovery of carbonic anhydrase from 24% in the pH 6 gradient to 72% in the pH 4 gradient. The recovery of insulin and β-lactoglobulin were not affected by the pH values tested. Because of this increase in recovery, subsequent pH gradients did not go above 4.0.

Solvent composition. In ion-exchange chromatography, peak retention is controlled by the ionic strength of the mobile phase as well as by the pH. As mentioned earlier, a primary purpose of this study was to evaluate volatile buffer systems in the ion-exchange chromatography of proteins. The pyridine–acetate buffer utilized by Radhakrishnan et al. (13) was, therefore, taken as the starting point. As demonstrated in Fig. 2A, at a constant concentration of ammonium acetate, a pyridine gradient from 0.05 to 0.48 M was sufficient to separate the three standard proteins with insulin being resolved from β-lactoglobulin ($R_s = 1.6$) which in turn was resolved from carbonic anhydrase

($R_s = 2.1$). The proteins could also be resolved by an ammonium acetate gradient at a constant pyridine concentration (Fig. 2B). As anticipated, decreasing the ammonium acetate gradient slope as a function of time from 0.04 to 0.02 M/min increased the resolution between β-lactoglobulin and carbonic anhydrase ($R_s = 1.0$ at 0.04 M/min and 1.5 at 0.02 M/min). Omission of 0.05 M pyridine in the ammonium acetate gradient resulted in an increase in retention times of all three proteins with carbonic anhydrase not eluting even with 2.0 M ammonium acetate, pH 4.0. It therefore appears that the pyridine–ammonium acetate solvent system has sufficient resolving power to separate insulin, β-lactoglobulin, and carbonic anhydrase using either a pyridine or an ammonium acetate gradient.

Application. The information provided in the previous sections was applied to the practical problem of purification of an unknown protein. Transforming growth factors (TGFs)

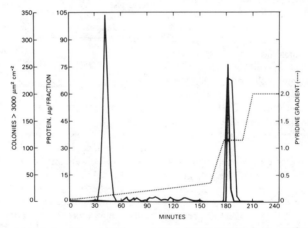

FIG. 4. Chromatography of human TGF on a cation-exchange HPLC column. The peak of transforming activity that eluted from the reverse-phase column at 9 to 10 min (Fig. 3) was applied to a Partisil-10-SCX column and the column developed at a flow rate of 0.5 ml/min at 25°C with a linear gradient ($\cdot\ \cdot\ \cdot$) of 0.05 M pyridine, 0.01 M ammonium acetate, pH 2.5, to 0.35 M pyridine, 0.01 M ammonium acetate, pH 3.5 (0.002 M/min). After 162 min the column was eluted for 32 min with 1.2 M pyridine, 0.01 M ammonium acetate, pH 4.0, and then for an additional 32 min with 2.0 M pyridine, 0.01 M ammonium acetate, pH 6.0, using a 10-min linear gradient to change each solvent. Fractions (1.5 ml) were collected and aliquots (400 μl) removed for determination of protein (shaded region) and transforming activity.

are a family of acid-stable polypeptides that are able to confer a transformed phenotype on normal rat kidney fibroblasts and to induce these cells to grow in soft agar (18,23–26). When human tumor cells, grown in culture, are extracted using an acid–ethanol extraction procedure (18) and the residue obtained is chromatographed on a Bio-Gel P30 gel filtration column in 1 M acetic acid, two regions of the column effluent display the ability to transform normal cells. When the higher molecular weight TGF region (18–20,000) is pooled and chromatographed on a reverse-phase HPLC column (Fig. 3) one prominent transforming peak (elution at 10 min) is found that separates from the majority of the protein. When this peak is chromatographed on the SCX cation-exchange column (Fig. 4) it is divided into two regions, one that elutes at 42 min and is separated from the bulk of the protein and a second peak that is eluted in the 1.0 M pyridine strip along with a major protein peak. Because there was an insufficient amount of protein in the first peak to do further characterization studies, efforts are now underway to obtain more of this intracellular human TGF by the same purification scheme. It is, however, apparent that cation-exchange HPLC in combination with reverse-phase HPLC can be a useful tool in the purification of peptides and proteins.

ACKNOWLEDGMENTS

The authors are indebted to Dr. George J. Todaro for supplying the A673 cells used for obtaining intracellular human TGF, to Mr. Das of Separation Industries for supplying the CM-Silica column for testing, and to Ellen Friedman for the typing of this manuscript.

REFERENCES

1. Rivier, J. E. (1978) J. Liquid Chromatogr. 1, 343–366.
2. Rubinstein, M. (1979) Anal. Biochem. 98, 1–7.
3. Regnier, F. E., and Gooding, K. M. (1980) Anal. Biochem. 103, 1–25.
4. Chang, S. H., Noel, R., and Regnier, F. E. (1976) Anal. Chem. 48, 1839–1845.
5. Chang, S. H., Gooding, K. M., and Regnier, F. E. (1976) J. Chromatogr. 125, 103–114.
6. Kudirka, P. J., Schroeder, R. R., Hewitt, T. E., and Toren, E. C., Jr. (1976) Clin. Chem. 22, 471–474.
7. Alpert, A. J., and Regnier, F. E. (1979) J. Chromatogr. 185, 375–392.
8. Barford, R. A., Sliwinski, B. J., and Rothbart, H. L. (1979) J. Chromatogr. 185, 393–402.
9. Vanecek, G., and Regnier, F. E. (1980) Anal. Biochem. 109, 345–353.
10. Hanash, S. M., and Shapiro, D. N. (1981) Hemoglobin 5, 165–175.
11. Lenda, K. (1981) J. Liquid Chromatogr. 4, 863–869.
12. McMurtrey, K. D., Meyerson, L. R., Cashaw, J. L., and Davis, V. E. (1976) Anal. Biochem. 72, 566–572.
13. Radhakrishnan, A. N., Stein, S., Licht, A., Gruber, K. A., and Udenfriend, S. (1977) J. Chromatogr. 132, 552–555.
14. Jones, B. N., Lewis, R. V., Paabo, S., Kojima, K., Kimura, S., and Stein, S. (1980) J. Liquid Chromatogr. 3, 1373–1383.
15. Schifreen, R. S., Hickingbotham, J. M., and Bowers, G. N., Jr. (1980) Clin. Chem. 26, 466–472.
16. van der Wal, S. J., and Huber, J. F. K. (1980) Anal. Biochem. 105, 219–229.
17. Hansen, S. H. (1981) J. Chromatogr. 212, 229–233.
18. Roberts, A. B., Lamb, L. C., Newton, D. L., Sporn, M. B., De Larco, J. E., and Todaro, G. J. (1980) Proc. Nat. Acad. Sci. USA 77, 3494–3498.
19. Bradford, M. M. (1976) Anal. Biochem. 72, 248–254.
20. De Larco, J. E., and Todaro, G. J. (1978) J. Cell. Physiol. 94, 335–342.
21. Schaeffer, W. J., and Friend, K. (1976) Cancer Lett. 1, 259–262.
22. Roberts, A. B., Anzano, M. A., Lamb, L. C., Smith, J. M., and Sporn, M. B. (1981) Proc. Nat. Acad. Sci. USA 78, 5339–5343.
23. De Larco, J. E., and Todaro, G. J. (1978) Proc. Nat. Acad. Sci. USA 75, 4001–4005.
24. Todaro, G. J., Fryling, C., and De Larco, J. E. (1980) Proc. Nat. Acad. Sci. USA 77, 5258–5262.
25. Ozanne, B., Fulton, R. J., and Kaplan, P. L. (1980) J. Cell. Physiol. 105, 163–180.
26. Moses, H. L., Branum, E. L., Proper, J. A., and Robinson, R. A. (1981) Cancer Res. 41, 2842–2848.

Application of High-Performance Liquid Chromatography to Competitive Labeling Studies: The Chemical Properties of Functional Groups of Glucagon*

STEPHEN A. COCKLE,†,1 HARVEY KAPLAN,‡ MARY A. HEFFORD,‡
AND N. MARTIN YOUNG†,2

†Division of Biological Sciences, National Research Council of Canada, Ottawa K1A 0R6, Canada; and
‡Department of Biochemistry, University of Ottawa, Ottawa K1N 6N5, Canada

A competitive-labeling study of glucagon was carried out using [³H]- and [¹⁴C]-1-fluoro-2,4-dinitrobenzene to determine simultaneously the chemical properties of the α-amino and imidazole groups of the N-terminal histidine residue, and the lysine and tyrosine residues, under conditions where glucagon is in its physiologically active monomer form. The dinitrophenyl derivatives of these groups were purified by high-performance liquid chromatography which greatly simplified the separation steps of the procedure. The results showed the α-amino and tyrosine groups to have relatively normal behavior, with pK values of 7.98 and 10.22, respectively, while the lysine had a low pK of 8.46. The imidazole function had an apparent pK of 7.84, substantially higher than previous estimates. This difference may be accounted for by the effect of the charged form of the adjacent α-amino group on the nucleophilicity of the imidazole group.

Competitive labeling (1) is a method for studying the chemical properties of particular residues in proteins or peptides, to elucidate their functional roles. The method, which has been reviewed by Bosshard (2) under its alternative name of differential chemical modification, involves the controlled use of reagents such as 1-fluoro-2,4-dinitrobenzene (Dnp-F),³ acetic anhydride, or formaldehyde. In a typical experiment with Dnp-F, aliquots of a mixture of a protein, such as an immunoglobulin (3), and an

internal standard are prepared at various pH values. The aliquots are first treated with [³H]Dnp-F in very low amounts, so that the reactive groups on the protein compete for the available reagent on the basis of their pK values and second-order rate constants. The samples are subsequently saturated with [¹⁴C]Dnp-F. The resulting mixtures must then be chemically treated, hydrolyzed, and separated in order to obtain in pure forms both the labeled internal standards and the Dnp derivatives from the protein such as S-Dnp-cysteine, O-Dnp-tyrosine, ε-Dnp-lysine, imidazolyl-Dnp-histidine, and the α-amino Dnp derivatives of the amino-terminal residues. Some of these derivatives may represent unique residues, but in general additional steps of enzymatic or chemical fragmentation of the modified protein samples and separation of the resulting peptides will be necessary to characterise residues uniquely.

---

* This paper was presented at the International Symposium on HPLC of Proteins and Peptides, November 16–17, 1981, Washington, D. C.

¹ Present address: Department of Chemistry, University of Winnipeg, 515 Portage Ave., Winnipeg, MB, R3B 2E9, Canada.

² To whom reprint requests should be addressed.

³ Abbreviations used: Dnp, 2,4-dinitrophenyl; Dnp-F, 1-fluoro-2,4-dinitrobenzene; HPLC, high-performance liquid chromatography.

These separation procedures are the most laborious of the steps involved in competitive-labeling studies. This led us to investigate high-performance liquid chromatography (HPLC) as an aid in such work. We have successfully used HPLC in a study of the peptide hormone glucagon, and we report the determination of the chemical properties of both the α-amino and imidazole moieties of its N-terminal histidine residue and of its lysine and tyrosine residues.

MATERIALS AND METHODS

The [^{3}H]Dnp-F and [^{14}C]Dnp-F were obtained from Amersham Corporation. Samples of N^α-Dnp-histidine were prepared by thiolysis of di-Dnp-histidine (4) and imidazolyl-Dnp-histidine by reaction of leucylhistidine or valyl-histidine (Bachem, Inc.) or acetyl-histidine (Sigma Chemical Co.) with Dnp-F, followed by hydrolysis and purification by HPLC. S-Dnp-cysteine was prepared similarly from N-acetyl-cysteine. Other Dnp amino acid derivatives were obtained from Sigma Chemical Company, as was the glucagon and the dipeptide alanyl-alanine used as the internal standard.

Competitive labeling. A mixture of the alanyl-alanine internal standard and glucagon, each ~1 mM, was prepared in 0.1 M KCl, 5 mM K-borate buffer. The exact proportions were determined by amino acid analysis. Portions of the mixture were adjusted to pH 6.5 and 10.5, respectively, and treated with 1 mM [^{14}C]Dnp-F, at 20°C for 24 h. The reactions were then quenched by addition of taurine and the portions were combined. In this way, it was possible to ensure a high incorporation of Dnp groups into the more reactive amino acids such as tyrosine without producing much of the di-Dnp derivative of the N-terminal histidine, or excessive byproducts. The proportions of the [^{14}C]Dnp derivatives were determined by hydrolysis and HPLC separation (see below). The remainder of the glucagon–internal standard mixture was diluted 1000-fold into a buffer solution containing 0.1 M KCl, 2 mM acetic acid, 2 mM K-phosphate, and 2 mM K-borate, and divided into 5-ml aliquots each containing approximately 5 nmol of glucagon. Each aliquot was adjusted to the desired pH and reacted at 20°C for 36 h with 6.24 nmol [^{3}H]Dnp-F added in 25 μl of acetonitrile. To one-half of each aliquot was added 1.0 ml of the above [^{14}C]Dnp-glucagon solution, containing approximately 50 nmol of glucagon. One-half of each mixture was then taken to dryness and hydrolyzed in 0.5 ml 6 N HCl for 18 h at 110°C. The hydrolysates were again evaporated to dryness, redissolved in 0.5 ml of water, and the buffer salts were removed by use of Sep-Pak cartridges (Waters Scientific Ltd.) which contain a C18 silica medium. The hydrolysates were applied to the cartridges, the salts were removed by rinsing with 4 ml of water, and the Dnp derivatives were recovered by elution with 3 ml of acetonitrile, and again evaporated to dryness. For HPLC injection, the samples were redissolved in 200 μl of the running solvent, which had additional unlabeled imidazolyl-Dnp-histidine in it to provide carrier material.

HPLC procedures. The separations of the Dnp derivatives were carried out with a 5-μm reverse-phase C18 silica column (Beckman Ultrasphere-ODS, 0.46 × 25 cm) preceded by a 10-μm C18 precolumn (3.2 × 40 mm) both being maintained at 62°C by means of water-jackets and a circulating water-bath. The HPLC equipment was from Beckman-Altex and the eluate was analyzed with a Model 155 variable-wavelength detector. The initial separations were run in a solvent of 27:73 acetonitrile and ammonium formate buffer 35 mM, pH 3.0. Fractions of 0.4 ml were collected in polypropylene plates on a Gilson FC80 collector. For further purification of the derivatives, the selected fractions were pooled, evaporated to dryness, redissolved in 200 μl of the running solvent, and reinjected. The combined N^α-Dnp-histidine and imidazolyl-Dnp-histidine were recycled with 8:92 acetonitrile and buffer, and the combined O-Dnp-tyrosine, ε-Dnp-lysine, and Dnp-alanine fractions were recycled in 25:75, acetonitrile and buffer. For radioactivity measurements, the fractions were mixed with 10 ml of Aquasol scintillation cocktail (New England Nuclear) and analyzed by a programmable LKB 1215 RackBeta scintillation counter with automatic quench correction and conversion to disintegrations per minute. It was necessary to correct the calculated ^{3}H:^{14}C ratios according to the relative proportions of the various Dnp derivatives in the singly-labeled [^{14}C]Dnp-glucagon.

RESULTS

The procedure for competitive labeling of glucagon outlined above is a variation from those previously described (5), the changes being designed to take into account the particular nature of glucagon, especially its N-terminal histidine residue (6) and the constraints of the HPLC separation. The main features are (a) the [^{3}H]Dnp-F labeling was carried out with native glucagon at 1 μM concentration, where glucagon is dissociated to the physiologically active monomer form; (b) the samples were not saturated with Dnp-F since this would have generated exclusively the di-Dnp derivative of histidine and the distinction between the two functional groups would have been more difficult to make; (c) taurine was added to remove excess Dnp-F reagent; and (d) the internal standard used was alanyl-alanine, generating in the hydrolysate Dnp-alanine which is more readily handled and separated from other Dnp derivatives by HPLC than the Dnp-imidazole lactate of previous experiments.

After hydrolysis, the samples contained the desired [^{3}H]- and [^{14}C]Dnp products N^α-Dnp-histidine, imidazolyl-Dnp-histidine, ε-Dnp-lysine, and O-Dnp-tyrosine from the protein and Dnp-alanine from the internal standard, along with the major byproducts Dnp-OH and Dnp-taurine. For purification of these Dnp derivatives by HPLC a reverse-phase C18 5-μm column was used at 62°C. The scale of the study was small enough for a normal analytical column to serve as a preparative one. The samples were first separated isocratically using 27:73, acetonitrile and 35 mM ammonium formate buffer, pH 3.0, mixture. In Fig. 1A the separation of standard Dnp derivatives by this procedure is shown. The various Dnp derivatives have quite different absorbance maxima and the wavelength chosen for monitoring the absorbance of the eluate, 320 nm, is a compromise one. Samples from the competitive labeling showed several additional components (Fig. 1B). The size of the peaks in this chromatogram is not a good guide to the radiochemical purity, however, as counting of eluant fractions (Fig. 1C) showed large peaks of ^{3}H radioactivity in regions of low spectroscopic absorbance. Since these peaks were not as evident in ^{14}C counting (data not shown) they may arise from radiolysis of the highly labeled [^{3}H]Dnp-F.

For further purification the fractions containing imidazolyl-Dnp-histidine and N^α-Dnp-histidine from the initial run were combined and rerun with isocratic elution using 8:92 proportions of acetonitrile and buffer. The separation of standard compounds in this system is shown in Fig. 2A. This chromatogram includes three Dnp compounds not relevant to the glucagon study but likely to be encountered in other competitive labeling work, S-Dnp-cysteine and Dnp-glutamic acid and Dnp-aspartic acid. The latter two frequently arise from the N-terminal residues in proteins such as immunoglobulins. In Fig. 2B, separation of a radiolabeled sample from the glucagon study shows, in addition to the two desired derivatives, a small peak (marked Pp) of a product that arises through photochemical degradation of imidazolyl-Dnp-histidine. The two other peaks in Fig. 2B were Dnp-taurine and an unidentified hydrolysis by-product from the Dnp-alanyl-alanine. Overall recovery of radioactivity in the various fractions was 82% in this step.

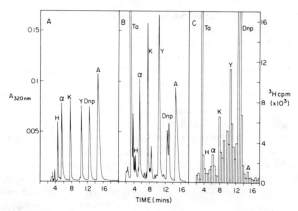

FIG. 1. Isocratic HPLC separations of Dnp amino acid derivatives at 62°C using 27:73 acetonitrile and 35 mM ammonium formate buffer, pH 3.0. The samples were a mixture of standard compounds (A) and a hydrolysate from the competitive-labeling study (B and C). The abbreviations are: H, imidazolyl-Dnp-histidine; α, N^α-Dnp-histidine; K, ϵ-Dnp-lysine; Y, O-Dnp-tyrosine; Dnp, Dnp-OH; A, Dnp-alanine; Ta, Dnp-taurine. In the standard run, the amounts of K, Y, and Dnp were 3.8, 2.8, and 4.8 nmol, respectively.

Fractions from the first chromatography containing ϵ-Dnp-lysine, O-Dnp-tyrosine, and Dnp-alanine were combined and rechromatographed using 25:75, acetonitrile:buffer elution. The recovery of radioactivity was 92% in this recycle. Despite the triple runs required for each competitive-labeling sample, the overall purifications and handling times were more satisfactory than those from a gradient procedure.

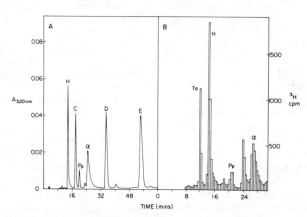

FIG. 2. Isocratic HPLC separations of hydrophilic Dnp amino acid derivatives using 8:92, acetonitrile and 35 mM ammonium formate buffer, pH 3.0. The samples were a mixture of standard compounds (A) and pooled imidazolyl-Dnp-histidine and N^α-Dnp-histidine fractions from a separation of the kind shown in Fig. 1 (B). The abbreviations are: H, imidazolyl-Dnp histidine; C, S-Dnp-cysteine; Pp, photochemical degradation product of imidazolyl-Dnp-histidine; α, N^α-Dnp-histidine; D, Dnp-aspartic acid; E, Dnp-glutamic acid; Ta, Dnp-taurine. In the standard run, the amounts of C, D, and E were 12 nmol each.

The $^3H/^{14}C$ ratios of the purified Dnp derivatives from the glucagon samples were converted to $\alpha_x r$ values, where α_x is the degree of ionization of the functional group and r is its pH-independent second-order rate constant relative to that of the internal standard, using the equation of Kaplan et al. (5),

$$\alpha_x r = \alpha_s (^3H/^{14}C)_x/(^3H/^{14}C)_s,$$

where α_s is the degree of ionization of the internal standard at the pH of the labeling (the pK of alanyl-alanine was taken to be 8.2) and $(^3H/^{14}C)_x$ and $(^3H/^{14}C)_s$ are the corresponding radioactivity ratios for the functional group and the internal standard.

The resulting pH-reactivity profiles are shown in Fig. 3, together with theoretical titration curves fitted by a nonlinear least-squares procedure to the data. The parameters of these curves, i.e., the pK and relative reactivity r, are summarized with their estimated errors in Table 1.

DISCUSSION

While a single isocratic HPLC procedure is sufficient for identification and spectro-

TABLE 1

CHEMICAL PROPERTIES OF THE FUNCTIONAL GROUPS OF GLUCAGON

Group	pK	Relative reactivity, r
Histidine imidazole	7.84 ± 0.13	0.31 ± 0.02
Histidine α-NH$_2$	7.98 ± 0.11	0.84 ± 0.04
Lysine	8.46 ± 0.10	7.68 ± 0.35
Tyrosine	10.22 ± 0.05	45 ± 2

photometric quantitation of most Dnp derivatives of amino acids, a more elaborate separation procedure was necessary to obtain fractions of sufficient purity for their ^3H and ^{14}C contents to be determined. This was largely due to the presence of byproducts with high ^3H activity but low absorption at 320 nm. The two-stage procedure used gave satisfactory purification, as judged by the $^3H:^{14}C$ ratios of fractions across the peaks, on a 25-cm analytical column.

The behavior of the imidazolyl-Dnp-histidine in HPLC is worth noting. The tautomerization of the neutral form of the imidazole moiety of histidine has been characterized by NMR (7,8) and Raman spectroscopy (9), the ratio of the N^1 and N^3-protonated forms of histidine being 8.8:1.2 when the α-amino group is protonated, and 4:1 when it is not. Since the uncharged form of the imidazole is the one which reacts with Dnp-F, in principle two Dnp derivatives might be formed. Earlier work had indicated only the N^1-Dnp product was being formed (10). Despite the greater resolving power of HPLC, we did not notice any secondary peaks or distortion of the imidazolyl-Dnp-histidine peak that might have indicated a second derivative. The photochemical lability of imidazolyl-Dnp-histidine is well-known and the photoproduct was present in our samples. The structure of this compound and the mechanism that apparently produces only one imidazolyl-Dnp-histidine derivative are worth further study.

Other difficulties encountered with the Dnp amino acid derivatives were the dis-

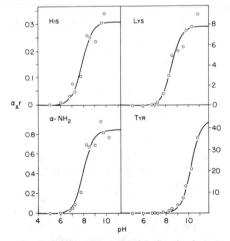

FIG. 3. pH-Reactivity profiles of the functional groups of glucagon. The lines are theoretical titration curves whose parameters are given in Table 1.

torted shape of the N^α-Dnp-histidine peaks, which is evident in the standard runs and is not due to impurities; the poor solubility of O-Dnp-tyrosine, which the double-label procedure circumvents since quantitative recoveries are not required; and the occurrence of an unidentified hyrolysis by-product from the Dnp-alanyl-alanine (the internal standard derivative), which eluted with N^α-Dnp-histidine in the initial runs. Also, as the column aged, the acetonitrile proportion in the solvent for the initial runs needed to be reduced by 1% to maintain resolution of the Dnp derivatives.

The peptide hormone glucagon was chosen for study by competitive labeling because of the interesting properties of its N-terminal histidine residue. Chemical modification of the α-amino or imidazole groups of the histidine alters glucagon's biological activity (11,12) and the residue is a feature glucagon shares with secretin (13) and porcine vasoactive intestinal peptide (14). The chemical properties of the histidine in the derivative S-methylglucagon have been studied by NMR methods (15) and a pK of 5.3 was found for the imidazole moiety, and a pK of 7.2 for the α-amino group. While the imidazole pK is unusual for protein histidine residues, it agreed well with studies of model compounds for N-terminal histidines. Reaction-rate studies with trinitrobenzenesulfonic acid indicated a pK of 7.9 for the α-amino group, and for the sole lysine residue a pK of 10.2 (12). The pK values of the two tyrosines have been determined spectrophotometrically to be 9.9 and 10.7 for the monomer form of glucagon (16).

The results, summarized in Table 1, obtained by the competitive-labeling approach should be regarded as preliminary, more data being desirable for the key region around pH 8.0. The pK values were in agreement with values obtained in experiments conducted in parallel which used electrophoretic and thin-layer chromatographic separation procedures as previously described (3,5).

It is evident that there are several unusual features in the properties of the glucagon residues. The reactivity of the imidazole groups is higher than usually found in proteins (3,17) and its pK is not only above normal values, but very much greater than that reported from NMR studies of S-methylglucagon (15). The α-amino group of the histidine has a pK similar to that determined previously by a chemical method (12) and to those of other N-terminal residues such as the dipeptide internal standard used here, and aspartic acid and glutamic acid N-termini in immunoglobulins (3), but a little above the value for S-methylglucagon determined by NMR (15). The reactivity of the α-amino group is also close to that of the internal standard, so this functional group appears normal in both respects.

An explanation that reconciles the different pK obtained for the imidazole group and its reactivity would be for the imidazole group to deprotonate around pH 5.3 as NMR (15) and titration curves (11) indicate, and for the imidazole group then to have its normally low reactivity suppressed further by the nearby charged α-amino group. When the α-amino group in turn deionizes at higher pH, the reactivity of the imidazole would be greatly enhanced, resulting in the dominant transition in the pH-reactivity profile. In effect, the imidazole would be acting as a "reporter group" for the α-amino group, and the similarity of the two pK values is consistent with this explanation. Similar interactive effects have been seen with the thiol and amino functions of cysteine (18).

The other functional groups of glucagon also show interesting properties. The tyrosine residues were not separately studied and the results represent the average properties of the two residues. The pK of 10.2 is a satisfactory average of the two values 9.9 and 10.7 obtained by curve-fitting of the spectrophotometric titration data (16). The reactivity toward Dnp-F is higher than the average values obtained from studies of proteins (3), probably due to the proteins having some tyrosine buried in the interior of the protein and hence of low reactivity. The overall average reactivity will thus be reduced from that for more accessible residues. The smaller size of glucagon would make steric screening a much lesser factor.

The single lysine residue shows a lower pK than usual for lysine residues in proteins. Its reactivity is several times greater than that of the N-terminal α-amino group as expected from previous studies (3). There is a suggestion in the scatter of the data about the derived pH curve (Fig. 3) that there may be two pH transitions occurring, one closer to the normal value. Additional data will be obtained to examine this further. Previous reaction rate studies indicated a pK of 10.36 for the lysine (12) but these experiments lacked sufficient data points in the critical region between pH 8 and 10.5. It is worth noting that studies of insulin by competitive labeling (19) showed that its sole lysine residue has the unusually low pK of 7.8.

In carrying out this study, we found that the laborious separation steps involved in competitive labeling could be greatly speeded up by the use of HPLC to separate the Dnp derivatives, though some redesign of the overall experiment was necessary. In addition, the high yields from HPLC make it possible to reduce the scale of the experiment, resulting in economies in the use of the radiolabeled reagents. It should be possible by the present approach to study much smaller protein samples. The experiments reported here used a total of 3 mg of glucagon, and were carried out at concentrations where glucagon is in its monomeric, and hence biologically relevant, form.

ACKNOWLEDGMENTS

We thank Mrs. JoAnne Johnson for excellent technical assistance. This work was supported in part by the Medical Research Council of Canada. NRCC publication No. 20372.

REFERENCES

1. Kaplan, H., Stevenson, K. J., and Hartley, B. S. (1971) *Biochem. J.* **124**, 289–299.
2. Bosshard, H. R. (1979) *Methods Biochem. Anal.* **25**, 273–301.
3. Kaplan, H., Long, B. G., and Young, N. M. (1980) *Biochemistry* **19**, 2821–2827.
4. Shaltiel, S. (1967) *Biochem. Biophys. Res. Commun.* **29**, 178–183.
5. Duggleby, R. G., and Kaplan, H. (1975) *Biochemistry* **14**, 5168–5175.
6. Bromer, W. W., Sinn, L. G., and Behrens, O. K. (1957) *J. Amer. Chem. Soc.* **79**, 2807–2810.
7. Reynolds, W. F., Peat, I. R., Freedman, M. H., and Lyerla, J. R. (1973) *J. Amer. Chem. Soc.* **95**, 328–331.
8. Blomberg, F., Mauer, W., and Ruterjans, H. (1977) *J. Amer. Chem. Soc.* **99**, 8149–8159.
9. Ashikawa, I., and Itoh, K. (1979) *Biopolymers* **18**, 1859–1876.
10. Henkart, P. (1971) *J. Biol. Chem.* **246**, 2711–2713.
11. Epand, R. M., Epand, R. F., and Grey, V. (1973) *Arch. Biochem. Biophys.* **154**, 132–136.
12. Epand, R. M., and Wheeler, G. E. (1975) *Biochim. Biophys. Acta* **393**, 236–246.
13. Mutt, V., Jorpes, J. E., and Magnusson, S. (1970) *Eur. J. Biochem.* **15**, 513–519.
14. Mutt, V., and Said, S. I. (1974) *Eur. J. Biochem.* **42**, 581–589.
15. Rothgeb, T. M., England, R. D., Jones, B. N., and Gurd, R. S. (1978) *Biochemistry* **17**, 4564–4571.
16. Gratzer, W. B., and Beaven, G. H. (1969) *J. Biol. Chem.* **244**, 6675–6679.
17. Cruickshank, W. H., and Kaplan, H. (1975) *Biochem. J.* **147**, 411–416.
18. Wallenfels, K., and Streffer, C. (1966) *Biochem. Z.* **346**, 119–132.
19. Sheffer, M. G., and Kaplan, H. (1979) *Canad. J. Biochem.* **57**, 489–496.

Purification by Reverse-Phase High-Performance Liquid Chromatography of an Epidermal Growth Factor-Dependent Transforming Growth Factor*

Mario A. Anzano, Anita B. Roberts, Joseph M. Smith, Lois C. Lamb, and Michael B. Sporn

Laboratory of Chemoprevention, National Cancer Institute, Bethesda, Maryland 20205

Transforming growth factors (TGF) are low molecular weight, acid-stable polypeptides that confer a malignant phenotype on nonneoplastic cells. TGF from murine sarcoma virus transformed 3T3 cells were isolated by acid/ethanol extraction, Bio-Gel P-30 chromatography, and reverse-phase high-performance liquid chromatography. Using a μBondapak C_{18} column with an acetonitrile gradient, TGF activity can be resolved into two peaks, one of which requires epidermal growth factor (EGF) to induce colony formation of indicator cells in soft agar. Subsequent rechromatography of the EGF-dependent TGF on a μBondapak CN column using an n-propanol gradient resulted in a 430-fold purification over the acid/ethanol extract and showed soft agar activity at a concentration of 4 ng/ml in the presence of 2 ng/ml EGF. Sodium dodecyl sulfate–polyacrylamide gel electrophoresis showed one main band with an apparent molecular weight of 13,000.

Transforming growth factors (TGF)[1] are low molecular weight (6000–20,000), acid-stable polypeptides that confer a malignant phenotype on nonneoplastic indicator cells and induce anchorage-independent growth in soft agar (1). This property has been correlated with tumorigenicity *in vivo* (2,3). Polypeptides belonging to this family of TGF have now been found in both neoplastic (1,4–7) and nonneoplastic (8) tissues. To ascertain whether these TGF are identical, a comparison of their biological and chemical properties in highly purified form is necessary. In this report, we describe the sequential application of two reverse-phase high-performance liquid chromatography

(RP-HPLC) systems as the final steps in the purification from murine sarcoma virus (MSV) transformed 3T3 cells of a particular class of TGF which is dependent on epidermal growth factor (EGF) to induce normal rat kidney fibroblasts to form colonies in soft agar.

MATERIALS AND METHODS

Chemicals. Bio-Gel P-30 (100–200 mesh), protein assay kit, and reagents used for polyacrylamide gel electrophoresis were purchased from Bio-Rad Laboratories (Richmond, Calif.). Low molecular weight protein standards for gel electrophoresis were obtained from Pharmacia (Piscataway, N. J.) and Bethesda Research Laboratories (Gaithersburg, Md.). Acetonitrile and n-propanol for HPLC were products of Burdick and Jackson Laboratories (Muskegon, Mich.). Sequanal grade trifluoroacetic acid (TFA) was obtained from Pierce Chemical Company (Rockford, Ill.). Deionized water was distilled in glass. Agar Noble was ob-

---

* This paper was presented at the International Symposium on HPLC of Proteins and Peptides, November 16–17, 1981, Washington, D. C.

[1] Abbreviations used: TGF, transforming growth factors; RP-HPLC, reverse-phase high-performance liquid chromatography; MSV, murine sarcoma virus; EGF, epidermal growth factor; TFA, trifluoroacetic acid; DMEM, Dulbecco's modified Eagle's medium; SDS, sodium dodecyl sulfate.

tained from Difco Laboratories (Detroit, Mich.), and tissue culture media and calf serum were obtained from Gibco Laboratories (Grand Island, N. Y.). Unless otherwise stated, all chemicals were of reagent grade quality.

Extraction of TGF. MSV-transformed 3T3 cells, line 3B 11-1C (1), were supplied by the Frederick Cancer Research Facility under Contract NO1-CO-75380, National Cancer Institute. Packed cells (1.5 kg) were extracted using an acid/ethanol procedure as described previously (4).

Bio-Gel P-30 chromatography. Gel-filtration chromatography of acid/ethanol extracts on Bio-Gel P-30 (100–200 mesh) was performed using a 21.5 × 100-cm column with downward flow at a rate of 250 ml/h. Fractions of 210 ml were collected, and aliquots were taken for soft agar and protein assay. Protein was determined using the dye-binding procedure (9) with bovine plasma globulin as standard. For reference, the elution volumes of ribonuclease and insulin were also determined.

Reverse-phase high-performance liquid chromatography. Chromatography was performed using an Altex high-performance liquid chromatography system. This consisted of two Model 100A dual-reciprocating pumps controlled by a Model 420 microprocessor solvent controller–programmer. The effluent was monitored at 280 nm using an Altex Model 153 analytical uv detector and at 210 nm with a variable wavelength Schoeffel Spectroflow monitor SF 770 with the outputs coupled to a Hewlett–Packard 7130A dual-pen recorder. Samples containing 40–50 mg protein in 2 ml of 0.1% TFA in water were centrifuged at 5000g to remove insoluble materials and injected by means of a Valco injector with 2-ml loop onto a Waters semipreparative μBondapak C$_{18}$ or CN column (0.78 × 30 cm) fitted with Bio-Rad ODS-10 microguard column. The mobile phase containing 0.1% TFA was either acetonitrile for μBondapak C$_{18}$ or n-propanol for μBondapak CN. All solvents were thoroughly de-

gassed by sonication. The flow rate was 0.8 ml/min and 1-min fractions were collected. Aliquots ranging from 20 to 30 μl were lyophilized for soft agar assay.

Soft agar assay. Anchorage-independent growth in soft agar was determined using normal rat kidney fibroblasts, clone 49F, as reported previously (4). Lyophilized samples were dissolved in 0.2 ml sterile 4 mM HCl containing 1 mg/ml bovine serum albumin. To the sample was added 0.6 ml of Dulbecco's modified Eagle's medium (DMEM) containing 0.5% agar, 10% calf serum, 100 units/ml penicillin, and 100 μg/ml streptomycin. Finally, 0.2 ml of normal rat kidney fibroblast cells containing 3 × 10^4 cells/ml in DMEM were added, mixed, and 0.7 ml of the resultant mixture was pipetted onto a 0.7-ml base layer of 0.5% agar in the supplemented DMEM using 35-mm Falcon petri dishes. Plates were incubated at 37°C in a humidified 5% CO$_2$/95% air atmosphere and stained after 7 days by incubation for an additional 24 h with 0.7 ml 2-(p-iodophenyl)-3-(p-nitrophenyl)-5-phenyl tetrazolium chloride at a concentration of 0.5 mg/ml in water. Colony formation was quantitated using a Bausch & Lomb Omnicon image analysis system programmed to count and size colonies on a logarithmic scale from 500 to 353,000 μm^2 as described (8). Specific activity of the TGF in the soft agar assay was calculated from the linear portion of the dose response curves by determining the protein concentration giving 400 colonies greater than 3100 μm^2/cm^2, under the described assay conditions. Based on this protein concentration, specific activity, expressed in colonies per microgram protein, was calculated as the total colonies per plate (9.6 cm^2) induced by the total micrograms protein per plate (1.4 ml).

Sodium dodecyl sulfate (SDS)-polyacrylamide gel electrophoresis. Protein samples at various steps of purification were analyzed on 1.5-mm slab gels with a polyacrylamide gradient of 15 to 30% and a discontinuous buffer system as described by

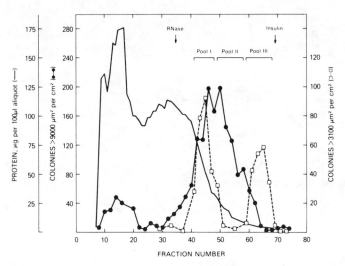

FIG. 1. Bio-Gel P-30 chromatography of TGF. Chromatography of 8.7 g protein of acid/ethanol extract from MSV-transformed 3T3 cells on a 21.5 × 100-cm column of Bio-Gel P-30. The sample was applied in 295 ml of 1 M acetic acid and eluted in 210-ml fractions at a flow rate of 250 ml/h. Protein and soft agar assays were determined on lyophilized 250-μl aliquots. Marker proteins indicated by arrows are RNase (13,800) and insulin (6000).

Laemmli (10). Proteins were fixed using formaldehyde (11) and stained with Coomassie brilliant blue. The molecular weight of TGF was determined by comparing its migration to that of protein standards electrophoresed in adjacent wells.

RESULTS

Acid/ethanol extraction and Bio-Gel P-30 chromatography. MSV-transformed 3T3 cells were extracted with acid/ethanol, precipitated with ethanol/ether, dialyzed, and lyophilized as described previously (4). The yield of lyophilized protein from 1.5 kg of packed cells was 9.2 g (6.1 mg protein/g of cells). A portion of the extracted protein (8.7 g) was then chromatographed on Bio-Gel P-30 in 1 M acetic acid as shown in Fig. 1. Total protein recovery was 94%. Analysis of the soft agar colony-forming activity assayed in the absence of EGF showed two peaks of activity designated as pool I (fractions 41–48) and pool III (fractions 59–68). In the

presence of 2 ng/ml EGF, a broad peak of soft agar activity was observed between fractions 40 and 60, corresponding to an apparent molecular weight between the ribonuclease (13,800) and insulin (6,000) markers. The soft agar colony-forming activity in fractions 49 to 58 (pool II) was strongly dependent on the presence of EGF. A dose response curve of Bio-Gel P-30 pools I, II, and III in the presence of 2 ng/ml EGF gave specific activities of 750, 2500, and 490 colonies/μg protein, respectively (Table 1). Pool II was chosen for further purification because it had the highest specific activity; however, as shown in Table 1, the total colony-forming units recovered in pool I (21%) were slightly greater than the total units in pool II (16%).

Reverse-phase high-performance liquid chromatography. Figure 2 shows a typical profile of 38 mg protein of Bio-Gel P-30 pool II applied to a semipreparative μBondapak C_{18} column using a gradient of acetonitrile

TABLE 1

PURIFICATION OF AN EPIDERMAL GROWTH FACTOR-DEPENDENT TRANSFORMING GROWTH FACTOR
FROM 1.5 kg OF MSV-TRANSFORMED 3T3 CELLS

Step	Protein recovered (mg)	Total activity[a] (colonies × 10^{-7})	Specific activity (colonies/µg protein)	Degree of purification	Activity recovery (%)
1. Acid/ethanol extraction	9200	430	460	1	100
2. Bio-Gel P-30 chromatography					
Pool I	1200	90	750	1.6	21
Pool II	270	68	2500	5.4	16
Pool III	25	1.2	490	1.1	0.28
Sequential RP-HPLC of Bio-Gel P-30 pool II					
3. µBondapak C_{18}	16	24	15,000	33	5.6
4. µBondapak CN	0.58	12	200,000	430	2.8

[a] Activity is calculated as described under Materials and Methods.

containing 0.1% TFA. The profile at 280 nm indicates good separation of proteins (Fig. 2A). Soft agar activity assayed in the absence of EGF (Fig. 2B) shows a sharp peak eluting at 31% acetonitrile (peak I) and a broader peak at 37% acetonitrile (peak II). Marker EGF is eluted at 35% acetonitrile. In the presence of 2 ng/ml EGF, the soft agar activity of peak II, but not of peak I, was stimulated approximately 800-fold, as measured by the number of colonies >9000 μm^2 (Fig. 2C). Protein recoveries of samples ranging from 350 µg to 50 mg ranged from 80 to 111%. Assayed in the presence of EGF, the soft agar activity recovered was 5.6% of the original acid/ethanol extract or 35% of the Bio-Gel P-30 pool II. The biological activity was stable at 4°C for 4 to 6 weeks in the HPLC solvent. Subsequent rechromatography of the µBondapak C_{18}-pooled EGF-dependent TGF activity (peak II, fractions 80–92) on a semipreparative µBondapak CN column using a n-propanol gradient containing 0.1% TFA showed preferential retention of the EGF-dependent TGF compared to contaminating proteins (Figs. 3A and B). Soft agar assay in the presence of 2 ng/ml EGF showed that the TGF activity corresponded to the small protein peak between fractions 70 and 82 (48% n-propanol). The specific activity of this pooled TGF region, as determined by a dose response curve of soft agar activity, was 200,000 colonies/µg protein, a 430-fold overall purification relative to the acid/ethanol extract (Table 1, Fig. 4).

Dose response curves of TGF activity assayed in the presence of 2 ng/ml EGF at various steps of purification are shown in Fig. 4, and a summary of the purification scheme together with the calculated total soft agar activity, specific activity, as well as recovery of soft agar activity and protein with TGF activity are shown in Table 1.

Molecular weight. To further assess the progress of the purification, various TGF preparations were subjected to electrophoresis on SDS–polyacrylamide gels under reducing conditions using a gradient of 15 to 30% acrylamide. Bio-Gel P-30 chromatography removed the high molecular weight impurities (Fig. 5, lane D). RP-HPLC on µBondapak C_{18} further removed 95% of the impurities (Table 1), and the most highly purified EGF-dependent TGF obtained after µBondapak CN showed only one main band with a molecular weight of approximately 13,000 (Fig. 5, lane F).

DISCUSSION

A sequence of two different RP-HPLC systems has been employed to purify TGF from MSV-transformed 3T3 cells. Following extraction with acid/ethanol and chro-

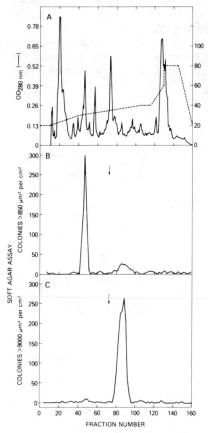

FIG. 2. RP-HPLC of TGF on a μBondapak C₁₈ column. Protein (38 mg) from Bio-Gel P-30 pool II was dissolved in 2 ml of 0.1% TFA, centrifuged at 5000g, and loaded onto a semipreparative μBondapak C₁₈ column (0.78 × 30 cm). The gradient consisted of 20 to 40% acetonitrile as indicated, at a flow rate of 0.8 ml/min. Aliquots (30 μl) of 0.8-ml fractions were used for soft agar assay. (A) Optical density profile at 280 nm; (B and C) soft agar activity profile in the absence and presence of 2 ng/ml EGF, respectively. Arrows indicate the elution of marker EGF.

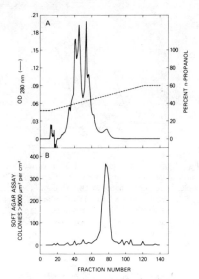

FIG. 3. RP-HPLC of TGF on a μBondapak CN column. Lyophilized peak II TGF (fractions 80–92, Fig. 2C) from the μBondapak C₁₈ column (1.85 mg protein) was dissolved in 2 ml of 0.1% TFA and loaded onto a semipreparative μBondapak CN column (0.78 × 30 cm). A linear gradient of 32 to 60% n-propanol containing 0.1% TFA over 112 min at a flow rate of 0.8 ml/min was used. Aliquots of 40 μl out of 0.8-ml fractions were used for soft agar assay. (A) Optical density profile at 280 nm; (B) soft agar activity in the presence of 2 ng/ml EGF.

matography on Bio-Gel P-30, these TGF, which are defined operationally by the ability to stimulate nonneoplastic anchorage-dependent indicator cells to form colonies in soft agar, can be further separated into two distinct classes using a μBondapak C₁₈ column and a linear gradient of acetonitrile containing 0.1% TFA as the counterion. TGF peaks I and II are eluted at 31 and 37% acetonitrile, respectively, while marker EGF is eluted at 35% acetonitrile. Peak I TGF resembles the sarcoma growth factor isolated from the conditioned media of these same cells as reported by De Larco and Todaro (1) because it competes with epidermal growth factor for membrane receptors and elutes earlier than the EGF marker on the

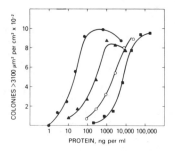

FIG. 4. Dose response curves of TGF at various steps of purification assayed in the presence of 2 ng/ml EGF. Samples were acid/ethanol extract (■); Bio-Gel P-30 pool II (○, fractions 49–58, Fig. 1); μBondapak C₁₈ peak II (▲, fractions 80–92, Fig. 2C), and μBondapak CN peak (●, fractions 70–82, Fig. 3B). Background of 60 colonies/cm² induced by EGF alone has been subtracted.

μBondapak C_{18} column using 0.1% TFA and an acetonitrile gradient (12). In addition, we observed that peak I TGF even at optimal concentrations induces relatively small colonies in soft agar (850–3100 μm^2) and is not potentiated by EGF (Figs. 2B and C). Peak II TGF represents a new class of TGF, which by itself has negligible activity but which, in the presence of 2 ng/ml EGF, induces the formation of large colonies in soft agar (>9000 μm^2). This new class of EGF-dependent TGF (Fig. 2C, peak II) does not compete for or enhance the binding of [^{125}I]EGF to normal rat kidney fibroblasts or mink lung epithelial cells (data not shown). Following rechromatography of this peak II TGF on a μBondapak CN column using a n-propanol gradient containing 0.1% TFA, soft agar colony formation could be observed at 4 ng/ml in the presence of 2 ng/ml EGF; however, no colonies showing two or more cell divisions could be observed when the soft agar assay was carried out in the absence of EGF, even at 3000-fold higher concentrations of the EGF-dependent TGF (6 μg/ml). It therefore appears that purified peak II TGF has no intrinsic activity, but requires EGF or possibly EGF-related polypeptides

that interact with EGF receptors to induce indicator cells to form colonies in soft agar. Preliminary results show that the soft agar activity of peak II TGF is potentiated not only by EGF, but also by peak I TGF, which competes for binding to the EGF receptors. Optimal concentrations of peak I TGF potentiate the soft agar colony-forming activity of peak II TGF to the same extent as native EGF, even though peak I is not immunologically reactive to anti-EGF antibodies (12). The precise nature of the interaction of these

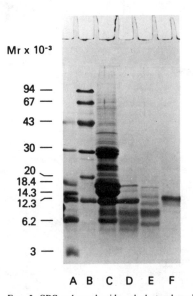

FIG. 5. SDS–polyacrylamide gel electrophoresis of marker proteins and TGF at different steps of purification. Samples were (A) 20 μg of BRL protein standard consisting of ovalbumin (43,000), α-chymotrypsinogen (25,700), β-lactalbumin (18,400), lysozyme (14,300), cytochrome c (12,300), bovine trypsin inhibitor (6,200), and insulin A and B chain (3,000); (B) 15 μg of Pharmacia protein standard consisting of phosphorylase b (94,000), bovine serum albumin (67,000), ovalbumin (43,000), carbonic anhydrase (30,000), soybean trypsin inhibitor (20,000), and α-lactalbumin (14,400); (C) 80 μg of acid/ethanol extract; (D) 12 μg of Bio-Gel P-30 pool II; (E) 7 μg of μBondapak C₁₈ pool; (F) 3 μg of μBondapak CN pool. Lanes D–F are as described in the legend to Fig. 4.

two classes of TGF with each other and their relationship to the secreted sarcoma growth factor (SGF) characterized by De Larco and Todaro (1) from the same MSV-transformed 3T3 cell line (3B 11-1C), are of major interest and are currently being investigated in our laboratory.

SDS–polyacrylamide gel electrophoresis of the EGF-dependent TGF peak after μBondapak CN chromatography showed one major band corresponding to a molecular weight of 13,000, higher than the molecular weight of approximately 9,000 obtained using gel filtration on Bio-Gel P-30. This slight discrepancy could be due to the possible hydrophobic nature of peak II TGF, as evidenced by its elution from the μBondapak column only at high organic solvent concentrations. This hydrophobicity might cause the molecular weight of TGF obtained by gel filtration to be low since other workers have observed that EGF is retarded on Bio-Gel columns presumably due to interaction of the aromatic, hydrophobic residues in the peptide with the polyacrylamide backbone (13). Analysis of the amino acid composition of the EGF-dependent TGF should indicate whether hydrophobic and aromatic residues are present.

RP-HPLC in combination with other protein-separation techniques offers a promising method to isolate cellular TGF and related polypeptides. Other workers have utilized RP-HPLC in isolating biologically active peptides such as interferon (14), insulin (15, 16), adrenocorticotropin (17,18), somatostatin (15), enkephalin (19), and neuropeptides (20). Good resolution has been shown to depend on the presence of an ionic modifier in the organic solvent (21). Based on the original observation of Bennett (22), we have successfully used 0.1% TFA as the counterion with both acetonitrile and n-propanol as the mobile phase. TGF are particularly suitable for separation on HPLC because of their stability at low pH and solubility in organic solvents, properties compatible with the derivatized silica column packing material used in HPLC. The volatile solvents employed reduce protein manipulation for bioassay. The HPLC gradient can be readily modified to increase resolution. Although the time of elution may vary slightly from day to day, the same chromatographic profile in terms of peak ratios is always obtained. The technique, as applied to the purification of TGF, has been shown to give high resolution, is rapid, and provides good recoveries of protein with retention of biological activity.

ACKNOWLEDGMENTS

The authors thank George Todaro and Joseph De Larco, Laboratory of Viral Carcinogenesis, National Cancer Institute, and Henry Hearn, Raymond Gilden, Charles Benton, and Michael Grimes, Frederick Cancer Research Facility, for their assistance in providing the cells for this project. We also thank Ms. Ellen Friedman for typing the manuscript.

REFERENCES

1. De Larco, J. E., and Todaro, G. J. (1978) *Proc. Nat. Acad. Sci. USA* **75**, 4001–4005.
2. Tucker, R. W., Sanford, K. K., Handleman, S. L., and Jones, G. M. (1977) *Cancer Res.* **37**, 1571–1579.
3. Cifone, M. A., and Fidler, I. J. (1980) *Proc. Nat. Acad. Sci. USA* **77**, 1039–1043.
4. Roberts, A. B., Lamb, L. C., Newton, D. L., Sporn, M. B., De Larco, J. E., and Todaro, G. J. (1980) *Proc. Nat. Acad. Sci. USA* **77**, 3494–3498.
5. Todaro, G. J., Fryling, C., and De Larco, J. E. (1980) *Proc. Nat. Acad. Sci. USA* **77**, 5258–5262.
6. Ozanne, B., Fulton, R. J., and Kaplan, P. L. (1980) *J. Cell Physiol.* **105**, 163–180.
7. Moses, H. L., Branum, E. L., Proper, J. A., and Robinson, R. A. (1981) *Cancer Res.* **41**, 2842–2848.
8. Roberts, A. B., Anzano, M. A., Lamb, L. C., Smith, J. M., and Sporn, M. B. (1981) *Proc. Nat. Acad. Sci. USA* **78**, 5339–5343.
9. Bradford, M. M. (1976) *Anal. Biochem.* **72**, 248–254.
10. Laemmli, U. K. (1970) *Nature (London)* **227**, 680–685.
11. Steck, G., Leuthard, P., and Burk, R. R. (1980) *Anal. Biochem.* **107**, 21–24.
12. Roberts, A. B., Anzano, M. A., Lamb, L. C., Smith,

J. M., Frolik, C., Todaro, G. J., Marquardt, H., and Sporn, M. B. (1982) *Nature (London)* **295**, 417–419.

13. Savage, R. C., and Cohen, S. (1972) *J. Biol. Chem.* **247**, 7609–7611.

14. Rubinstein, M., Levy, W. P., Moschera, J. A., Lai, C., Hershberg, R. D., Bartlett, R. T., and Pestka, S. (1981) *Arch. Biochem. Biophys.* **210**, 307–318.

15. Rivier, J. E. (1978) *J. Liquid Chromatogr.* **1**, 343–366.

16. Hearn, M. T. W., Hancock, W. S., Hurrell, J. G. R., Fleming, R. J., and Kemp, B. (1979) *J. Liquid Chromatogr.* **2**, 919–933.

17. Seidah, N. G., Routhier, R., Benjannet, S., Lariviere, N., Gossard, F., and Chretien, M. (1980) *J. Chromatogr.* **193**, 291–299.

18. Bennett, H. P. J., Browne, C. A., and Solomon, S. (1981) *Biochemistry* **20**, 4530–4538.

19. Lewis, R. V., Stern, A. S., Kimura, S., Stein, S., and Udenfriend, S. (1980) *Proc. Nat. Acad. Sci. USA* **77**, 5018–5020.

20. Schwandt, P., and Richter, W. O., (1980) *Biochim. Biophys. Acta* **626**, 376–382.

21. Hearn, M. T. W., and Hancock, W. M. (1979) *Trends Biochem. Sci.* **4**, N58–N62.

22. Bennett, H. P. J., Browne, C. A., and Solomon, S. (1980) *J. Liquid Chromatogr.* **3**, 1353–1365.

Preparative Size-Exclusion Chromatography of Human Serum Apolipoproteins Using an Analytical Liquid Chromatograph[1]

C. Timothy Wehr,* Robert L. Cunico,* Gary S. Ott,†[,2] and Virgie G. Shore†

*Varian Instrument Group, Walnut Creek Instrument Division, 2700 Mitchell Drive, Walnut Creek, California 94598; and †Lawrence Livermore National Laboratory, Biomedical Sciences Division, University of California, Livermore, California 94550

Apolipoproteins, extracted from human serum high-density lipoproteins, can be resolved and recovered with high yield from a preparative MicroPak TSK Type 3000SW size-exclusion column using Tris-buffered 6 M urea or 6 M guanidinium chloride mobile phases. Adequate resolution of some apolipoprotein pairs is only achieved at low flow velocities and low sample loads, necessitating repetitive injections of small amounts of material for preparative isolation. An analytical high-performance liquid chromatograph equipped with a simplified sample introduction scheme and low-pressure switching valves for fraction collection was used to isolate milligram quantities of HDL apolipoproteins.

Elucidation of the structure and function of human serum lipoproteins requires a convenient method for isolation of sufficient amounts of purified apolipoproteins for biochemical characterization. This has usually been accomplished using anion-exchange chromatography on low-pressure gels such as DEAE-cellulose (1,2) and, more recently, using high-performance anion-exchange supports (3). The latter method is rapid and provides excellent resolution of apolipoproteins, but has certain limitations: gradient elution is required, recovery is variable, and use of large sample loads can result in adsorption and gradual loss of column performance.

Recently, we have found that apolipoproteins can be resolved on an analytical high-performance steric-exclusion column, the MicroPak TSK 3000SW, with nearly quantitative (>80%) recovery (4). Although the separation was not as useful for analytical

purposes as that obtained by anion exchange, it seemed to be a promising preparative technique for obtaining high yields of individual apolipoproteins. We have investigated the performance of a preparative MicroPak TSK 3000SW column as a function of flow rate, sample load, and mobile phase composition. Based on these results, a simplified sample introduction and fraction collection scheme for isolation of apolipoproteins was devised.

MATERIALS AND METHODS

Materials. Protein standards (bovine serum albumin, Cohn Fraction II γ-globulins, and myoglobin) were obtained from Sigma Chemical Company (St. Louis, Mo.). Tris–HCl and urea were also obtained from Sigma. Ultrapure guanidinium chloride was obtained from Schwarz/Mann (Spring Valley, N. Y.). HPLC grade potassium phosphate and potassium chloride were obtained from Fisher Scientific Company (Fairlawn, N. J.). Mobile phases for HPLC were prepared with water obtained from a Hydroservices water purification system (Durham, N. C.).

[1] This paper was presented at the International Symposium on HPLC of Proteins and Peptides, November 16–17, 1981, Washington, D. C.

[2] Present address: Bio-Rad Laboratories, 2200 Wright Ave., Richmond, CA 94804.

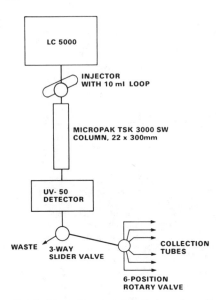

FIG. 1. Schematic presentation of chromatography system for preparative chromatography using repetitive injections.

Preparation of HDL apolipoproteins.
HDL[3] (ρ = 1.065–1.210 g/ml) was isolated from the fresh sera of normal males by the sequential flotation ultracentrifugation procedure of Lindgren *et al.* (5). VLDL and LDL were removed by centrifugation of a 1.063-g/ml solution for 18 h at 14°C at 40,000 rpm in a Beckman Type 40.3 rotor (Beckman Instruments, Palo Alto, Calif.). Total HDL was isolated by centrifugation of a ρ = 1.21-g/ml solution for 48 h under similar conditions. The lipoprotein composition of these fractions was analyzed by gradient polyacrylamide gel electrophoresis on 2.5 to 27% gradient polyacrylamide gels (Isolab, Akron, Ohio). Albumin and LDL contamination was minimal. Salt was removed by dialysis (2 × 100:1 v/v) against nitrogen-saturated double-distilled water.

---

[3] Abbreviations used: HDL, high-density lipoproteins; VLDL, very low-density lipoproteins; LDL, low-density lipoproteins; SDS, sodium dodecyl sulfate.

The HDL was delipidated with ethanol/diethyl ether as previously reported (6). The protein concentration of the resultant apo HDL preparation was determined by the method of Lowry *et al.* (7). Prior to chromatography, apo HDL was adjusted to a concentration of approximately 10 mg/ml, and 30 min prior to chromatography denaturant was added to the mobile phase concentration.

Chromatographic system. Preparative high-performance liquid chromatography was performed using a Varian Model 5060 liquid chromatograph equipped with a Valco Model AH60 automatic loop injector, MicroPak TSK 3000SW preparative size-exclusion column (22 × 300 mm), and Varian Model UV-50 variable wavelength detector (Varian Instrument Group, Walnut Creek, Calif.). For investigations on the effect of sample load and flow rate on chromatography of apolipoproteins, the injector was equipped with a 1.0-ml loop. For unattended repetitive injection and collection of protein standards or HDL apolipoproteins, the liquid chromatograph was modified as shown in Fig. 1. The air-actuated injector was equipped with a 10-ml loop. The outlet line from the detector was connected to an air-actuated Rheodyne Model 5301 three-way slider valve with a Model 5300 pneumatic actuator, which diverted the column effluent from the waste line to an air-actuated Rheodyne Model 5011A six-position rotary valve for fraction collection (Rheodyne, Cotati, Calif.). Actuation of both valves was controlled by a single solenoid operated by a time-programmable external event (powered relay) in the liquid chromatograph. At the beginning of a preparative run, the 10-ml loop was manually filled with sample. During the run, a series of collection cycles was carried out, with an aliquot of the loop volume injected at each cycle. At the initiation of each collection cycle, a sample aliquot was introduced onto the column by pumping at a reduced flow rate with the autoinjector in the "inject" position. In most cases, a 0.5-ml aliquot of the loop volume was introduced onto the column by pumping at 1.0 ml/min for 30 s; following this, the injector was returned to the "load" position and flow increased to the normal rate for the remainder of the collection cycle. During the separation, up to six fractions could be collected by diverting the eluant stream via the slider valve from waste to sequential ports of the rotary valve. At the end of a collection cycle, the liquid chromatograph program looped to start injection of the next sample aliquot. Using this protocol, up to twenty 0.5-ml aliquots of a 10-ml sample were injected over a 24-h period of unattended operation; for preparative isolation of apolipoproteins, fractions collected from each cycle were combined and held in an ice bath during the preparative run.

Mobile phases. Protein standards were separated using a mobile phase consisting of 0.1 M potassium phosphate + 0.1 M KCl (pH 6.8). Human serum HDL apolipoproteins were separated using 50 mM Tris–HCl (pH 7.0) + 6 M urea or 6 M guanidinium hydrochloride. Mobile phase solutions were filtered through a 0.45-μm filter (Type HA, Millipore, Bedford, Mass.) prior to use.

Gel electrophoresis. Fractions collected following chromatography on a MicroPak TSK 3000SW column were concentrated by vacuum dialysis and analyzed by sodium dodecyl sulfate–polyacrylamide gel electrophoresis on 10% gels using methods previously described by Shore *et al.* (8). These methods were modifications of those described by Weber and Osborne (9).

RESULTS AND DISCUSSION

Size-Exclusion Chromatography of HDL Apolipoproteins

Previous work with analytical size-exclusion columns (10) indicated that optimum efficiencies for small molecules could be achieved at flow velocities between 0.25 and 0.5 mm/s (0.5–1.0 ml/min for a 7.8 × 300-mm column), whereas column efficiencies were considerably lower for proteins due to the poor mass transfer kinetics displayed by macromolecules in size-exclusion chromatography. Efficiency is inversely related to molecular weight, and adequate resolution of proteins with molecular weights greater than 50,000 often requires operation at flow velocities less than 0.25 mm/s.

To maximize sample throughput in the isolation of HDL apolipoproteins, the effects of flow velocity and sample load were investigated using the 22 × 300-mm preparative column. Flow-rate dependency of apolipoprotein resolution is seen in Fig. 2. Baseline resolution of apo AI (M_r 28,000) and apo AII (M_r 17,400) was only achieved at flow velocities less than 0.13 mm/s (2.0 ml/min). The HETP vs flow rate plot for the apo CII + apo CIII peak did not reach a minimum at flow velocities as low as 0.06 mm/s (1.0 ml/min).

Maximum sample throughput is achieved at the highest sample loads without causing a serious degradation in resolution (4). Determination of the effect of sample load on column efficiency for ovalbumin (Fig. 3) indicated that a sample load of 10 mg only reduced the efficiency by 10% and suggested the 22 × 300-mm column would be useful for sample loads in the range of 10 to 50 mg. Comparable effects of sample load on efficiency were observed with the apo C proteins using a Tris–urea mobile phase (Fig. 4). However, the apo AI/apo AII separation degraded rapidly at sample loads above 10 mg/ml, possibly due to protein–protein interactions at the high sample concentrations required to achieve sample loads greater than 10 mg. Large volumes of dilute sample could not be injected due to loss of resolution with injection volumes greater than 1% of the total column volume.

These results indicate that adequate resolution of all HDL apolipoproteins by size-

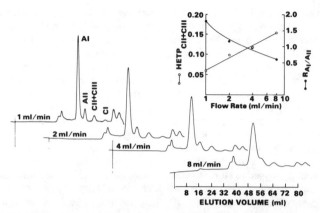

FIG. 2. Effect of flow rate on separation of HDL apolipoproteins. One-milliliter volumes of a 1-mg/ml solution of HDL apolipoproteins in 50 mM Tris–HCl (pH 7.0) + 6 M urea were injected onto a 22 × 300-mm MicroPak TSK Type 3000SW preparative column and eluted with 50 mM Tris–HCl (pH 7.0) + 6 M urea at the indicated flow rates. Proteins were detected by absorbance at 250 nm. Height equivalent of a theoretical plate (HETP) (○) as calculated from the apo CII + apo CIII peak and resolution of apo AI and apo AII (R) (●) are plotted as a function of flow rate.

exclusion chromatography requires operation at low flow velocities and injection of small volumes of dilute samples. To realistically use this technique for preparative isolation of lipoproteins, repetitive injections with unattended operation is necessary.

Preparative Chromatography Using Repetitive Injections

Preparative chromatography with overlay of fractions collected from repetitive injections requires high precision of retention

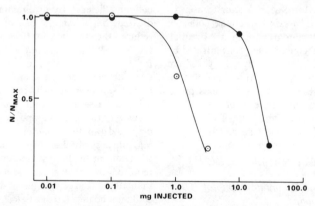

FIG. 3. Effect of sample load on column efficiency. Ovalbumin at concentrations of (○) 1 to 10 mg/ml (for injection on a MicroPak TSK Type 3000SW, 7.8 × 300-mm analytical column) or (●) 10 to 100 mg/ml (for injection on a MicroPak TSK Type 3000SW, 22 × 300-mm preparative column) was injected and eluted with 0.1 M potassium phosphate + 0.1 M KCl (pH 6.8). Flow rates and injection volumes were 1.0 ml/min and 0.1 ml for the analytical column and 8.0 ml/min and 1.0 ml for the preparative column.

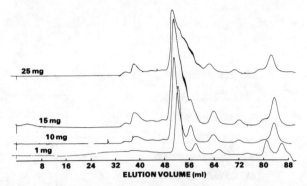

FIG. 4. Effect of sample load on separation of HDL apolipoproteins. One-milliliter volumes of HDL apolipoproteins at the indicated concentration were injected on a MicroPak TSK 3000SW 22 × 300-mm column and eluted with 50 mM Tris–HCl (pH 7.0) + 6 M urea at a flow rate of 1.0 ml/min. Proteins were detected at 250 nm.

times to minimize cross contamination of fractions. Performance of the preparative system described under Materials and Methods was evaluated by repetitively injecting a sample of bovine serum albumin in an overnight run. Injections were made by flow programming the chromatograph at 1.0 ml/min for 30 s to introduce 0.5-ml aliquots from the 10-ml loop during the injection period of each cycle. Precision data for peak areas and heights from 15 injections (Table 1)

demonstrate that the system can reliably deliver a constant sample volume from cycle to cycle. The retention-time precision is high

TABLE 1

PREPARATIVE SYSTEM PERFORMANCE: PRECISION OF REPETITIVE SAMPLING

	N	Mean	σ	σ Relative (%)
Retention time (min)	15	8.58	0.005	0.05
Peak height (mm)	20	131.8	1.82	1.38
Peak width at half height (s)	15	29.34	0.22	0.75
Peak area (μV s)	10	9,962,785	65,681	0.66

Note. Column, MicroPak TSK 3000SW, 22 × 30 mm; mobile phase, 0.1 M potassium phosphate + 0.1 M KCl, pH 6.6; flow rate, 8.0 ml/min; injection volume, 0.5 ml; sample, bovine serum albumin 10 mg/ml.

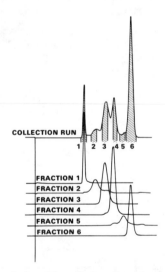

FIG. 5. Repetitive collection of protein standards. Aliquots (0.5 ml) of a mixture of protein standards (γ-globulin, bovine serum albumin, and myoglobin, each at 10 mg/ml prepared in the mobile phase) were injected by flow programming at 1.0 ml/min for 30 s, and then eluted at 8.0 ml/min with 0.1 M potassium phosphate + 0.1 M KCl (pH 6.6). Shaded areas indicate fractions collected with the slider and rotary valves.

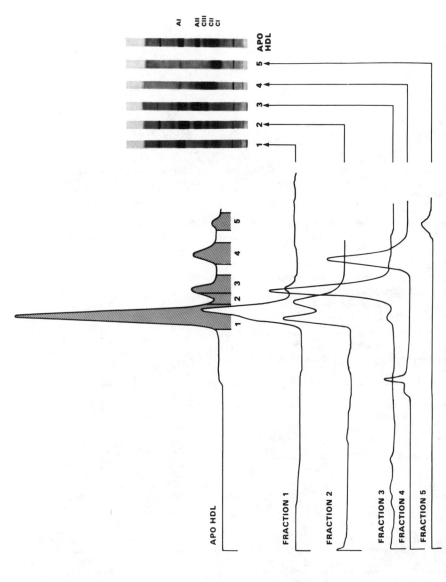

FIG. 6. Preparative isolation of HDL apolipoproteins using repetitive injections. Aliquots (0.5 ml) of a 10-mg/ml solution of HDL apolipoproteins in 50 mm Tris–HCl (pH 7.0) + 6 m urea were injected and eluted at 1.0 ml/min with 50 mm Tris–HCl (pH 7.0) + 6 m urea. Shaded areas indicate fractions collected for polyacrylamide gel electrophoresis analysis. Fractions were collected into containers held at 4°C.

enough to permit time-programmed fraction collection based on elution times.

To assess the performance of the entire preparative system, repetitive injections of a mixture of protein standards were made and six fractions were collected from each injection by time programming the slider and rotary valves (Fig. 5). The first three peaks in the chromatogram are components of γ-globulin, presumably representing aggregates, dimers, and monomers of IgG. After 20 injection cycles, the collected fractions were rerun individually on the same system fitted with a 0.5-ml injector loop. Chromatograms of the individual components demonstrate that repetitive runs with overlaid fraction collection permit isolation of fractions whose purity is determined only by chromatographic resolution. Fractions 1,

3, 4, and 6 from the four major peaks are greater than 90% pure, whereas fractions 2 and 5 from the poorly resolved minor peaks show contamination from neighboring components.

Preparative Chromatography of HDL Apolipoproteins

Evaluation of the preparative chromatography system with protein standards indicated that well-resolved proteins could be recovered in amounts of 100 mg or more with reasonable purity. Based on these results, the system was applied to the preparative isolation of HDL apolipoproteins (Fig. 6). To achieve maximum resolution, injection volume and protein concentration were kept at 0.5 ml and 10 mg/ml, while elution

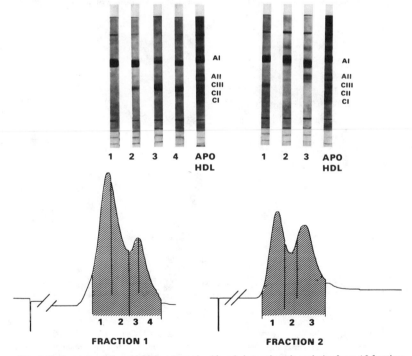

Fig. 7. Rechromatography and SDS–polyacrylamide gel electrophoresis analysis of apo AI fractions. Fractions 1 and 2 from Fig. 6 were concentrated and reinjected under the same conditions. Cuts were taken as shown and analyzed by SDS–polyacrylamide gel electrophoresis.

126 WEHR ET AL.

was carried out at a flow velocity of 0.06 mm/s (1.0 ml/min). Under these conditions, the apolipoproteins eluted as four baseline-resolved peaks: apo AI, apo AII, apo CII + apo CIII, and apo CI. The fraction-collection valves were programmed to collect these four peaks plus material eluting on the trailing edge of apo AI and the leading edge of apo AII. Fractions from 12 injection cycles were concentrated by vacuum dialysis and analyzed by SDS–polyacrylamide gel electrophoresis. Concentrated fractions were also rechromatographed using the same chromatographic system fitted with a 0.1-ml injection loop. As judged by rechromatography (Fig. 6), the apo AII, apo CII + apo CIII, and apo CI fractions were reasonably homogeneous, although the apo CII + apo CIII fraction showed evidence of high-molecular-weight material eluting near the void

volume and apo AII had about 5% contamination with apo AI. The SDS–gel electrophoretic patterns are in qualitative agreement with rechromatography results. (Note that staining characteristics of different apolipoproteins are dissimilar, so that relative band intensities need not correlate with protein concentration.)

Fractions 1 and 2 taken from the leading and tailing sections of the apo AI peak showed unexpected behavior upon rechromatography. Two components were observed in both fractions, one eluting as apo AI and a second eluting between apo AI and apo AII. The ratio of these two components appeared to vary depending on sample concentration, urea concentration, or sample handling. SDS–gel electrophoresis analysis of these fractions showed a broad apo AI band with evidence of some material moving

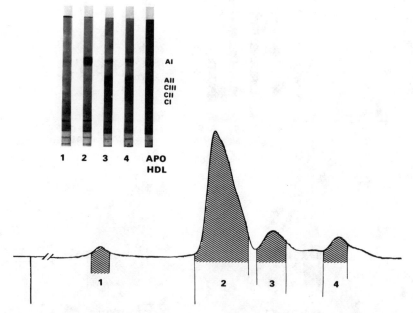

FIG. 8. Size-exclusion chromatography of HDL apolipoproteins with a Tris–guanidinium chloride mobile phase. A 0.5-ml injection of HDL apolipoproteins (10 mg/ml in 50 mM Tris–HCl, pH 7.0, + 6 M guanidinium chloride) was eluted at 1.0 ml/min with 50 mM Tris–HCl (pH 7.0) + 6 M guanidinium chloride.

slower than the main band and the presence of significant amounts of C proteins.

To determine the composition of the material eluting between apo AI and apo AII, fractions 1 and 2 were rechromatographed and a series of cuts were taken for gel electrophoresis analysis (Fig. 7). The gel patterns indicate that material eluting between apo AI and apo AII (cuts 2, 3, and 4 of fraction 1, and cut 1 of fraction 2) contain apo CIII, while cuts eluting near the apo AI and apo AII regions do not. It is unlikely that apo C proteins in the apo A regions arise from cross contamination in the fraction-collection system during repetitive collection, because this would result in contamination of the leading edge of the apo AI peak with apo CI, and contaminating free apo C proteins should not be expected to coelute with apo A proteins upon rechromatography. An alternative explanation is an association of apo A and apo C proteins under these conditions. Protein–protein interactions among apolipoproteins are known to occur in aqueous solution (see Ref. (11) for review); apo AI, apo AII, and apo CI all self-associate at concentrations as low as 1 mg/ml. Association of apo C with apo A proteins in the presence of urea has not been reported and was not observed using Tris–urea mobile phases in anion-exchange chromatography of HDL apolipoproteins; the association may depend upon the particular protein concentrations and chromatographic conditions. This effect was not seen when HDL apolipoproteins were chromatographed in a Tris–guanidinium chloride mobile phase (Fig. 8). Under these conditions, the presence of C proteins in the apo A fraction was minimal.

The system described here provides a suitable means of isolating up to several hundred milligrams of material using an analytical instrument. Application of the system to preparative isolation of HDL apolipoproteins showed that the purity of isolated components depends not only upon retention-time precision and chromatographic resolution, but also upon the behavior of protein species at high concentrations in the mobile phase.

ACKNOWLEDGMENTS

The authors thank Laura Correia for excellent technical assistance and Sally Bird for help in preparing the manuscript.

REFERENCES

1. Shore, B., and Shore V. G. (1969) *Biochemistry* **8,** 4510.
2. Shore, V. G., and Shore, B. (1973) *Biochemistry* **12,** 502.
3. Ott, G. S., and Shore, V. G. (1982) *J. Chromatogr.,* in press.
4. Klink, F. E., Wehr, C. T., and Ott, G. (1981) *Fed. Proc.* **40,** 1888.
5. Lindgren, F. T., Jensen, L. C., and Hatch, F. T. (1972) *in* Blood, Lipids, and Lipoproteins: Quantitation, Composition, and Metabolism (Nelson, G., ed.), p. 184, Wiley-Interscience, New York.
6. Shore, V., and Shore, B. (1968) *Biochemistry* **7,** 3396.
7. Lowry, O. H., Rosebrough, N. J., Farr, A. L., and Randall, R. J. (1951) *J. Biol. Chem.* **193,** 265.
8. Shore, V. G., Shore, B., and Lewis, S. B. (1978) *Biochemistry* **17,** 2174.
9. Weber, K., and Osborne, M. (1969) *J. Biol. Chem.* **244,** 4406.
10. Wehr, C. T., and Abbott, S. R. (1979) *J. Chromatogr.* **174,** 409.
11. Scanu, A. M., and Teng, T.-L. (1979) *in* The Biochemistry of Atherosclerosis (Sanu, A. M., ed.), pp. 107–121, Dekker, New York.

Use of Reverse-Phase High-Performance Liquid Chromatography in Structural Studies of Neurophysins, Photolabeled Derivatives, and Biosynthetic Precursors[1]

David M. Abercrombie,[*][†][2] Christopher J. Hough,[*] Jonathan R. Seeman,[*]
Michael J. Brownstein,[‡] Harold Gainer,[§] James T. Russell,[§]
and Irwin M. Chaiken[*]

Laboratory of Chemical Biology, National Institute of Arthritis, Diabetes, and Digestive and Kidney Diseases; †Department of Chemistry, University of Maryland, College Park, Maryland 20742; ‡Laboratory of Clinical Science, National Institute of Mental Health; and §Laboratory of Developmental Neurobiology, National Institute of Child Health and Human Development, National Institutes of Health, Bethesda, Maryland 20205

The neurophysins are a class of hypothalamo–neurohypophyseal proteins that function as carriers of the neuropeptide hormones oxytocin and vasopressin. Currently, we are using reverse-phase high-performance liquid chromatography for structural characterization of the neurophysins, their chemically modified derivatives, and biosynthetic precursors. A cyanopropylsilyl (Zorbax CN) matrix has been found to be efficient and convenient for separation of major tryptic peptides of performic acid, oxidized or reduced, and alkylated neurophysins. Using this peptide mapping system we have studied the site of modification of a photoaffinity-labeled derivative of bovine neurophysin II by separation and identification of covalently modified peptides. In addition, this system has been used for mapping subfemtomole amounts of radioactively labeled biosynthetic precursors of the neurophysins. This procedure has allowed identification of neurophysin sequences within both pre-pro-neurophysins produced by *in vitro* translation and rat pro-neurophysins produced by *in vivo* pulse labeling.

Liquid chromatography has long been a central theme in the chemical and biochemical study of proteins. Yet, while high-resolution separations of both analytical and preparative use are expected goals of many types of current liquid chromatography methodology, HPLC recently has provided significantly improved liquid chromatographic approaches for several types of separatory demands. In our ongoing structural studies of the neurophysins, their chemically modified derivatives, and biosynthetic precursors, reverse-phase HPLC has provided solutions to peptide mapping problems not easily addressed for this protein by other methods, chromatographic and otherwise. This has become evident, for example, in

efforts to obtain chemical evidence, with very small (subfemtomole) amounts of material, that neurophysin precursors obtained as translation products by *in vitro* synthesis from hypothalamic mRNA and as pulse-labeled products *in vivo* contain intact neurophysin sequences. Separation of large, subtly modified peptide fragments in photoaffinity-labeling studies has presented another demand. We have previously described the use of reverse-phase HPLC on cyanopropylsilyl columns to effect separation of the large tryptic fragments of performic acid-oxidized BNP I and II[3] (1). The present report de-

---

[1] This paper was presented at the International Symposium on HPLC of Proteins and Peptides, November 16–17, 1981, Washington, D.C.

[2] This paper is taken in part from a dissertation submitted to the Graduate School, University of Maryland, in partial fulfillment of the requirements for the Ph.D. degree in Biochemistry.

[3] Abbreviations used: BNP I and BNP II, bovine neurophysins I (oxytocin associated) and II (vasopressin as-

scribes the use of this HPLC method in studies of rat as well as bovine neurophysins, biosynthetic precursors of these molecules, and a hormone binding-site derivative.

MATERIALS AND METHODS

BNP I and II were extracted from acetone-dried or freeze-dried posterior pituitary tissue (2). Gel-filtration and ion-exchange chromatography of the crude neurohysins on Sephadex G-75 and DEAE-Sephadex A-50, respectively, followed established procedures (3). Mature bovine neurophysins used throughout this study were purified additionally by affinity chromatography on a Met-Tyr-Phe-diaminohexyl agarose matrix (4). The homogeneity of the neurophysins was assessed routinely by amino acid analysis and polyacrylamide gel electrophoresis.

Mature rat neurophysins were obtained as a mixture of RNP I and RNP II by a procedure analogous to that used for bovine neurophysins. This involved acid extraction of rat freeze-dried posterior pituitary tissue, Sephadex G-75 fractionation of the extract (in 0.1 M formic acid), and affinity chromatography of the resulting neurophysin fraction on lysine–vasopressin Sepharose (5). The presence in the rat neurophysin mixture of RNP I and II as the major components was assessed by one- and two-dimensional gel electrophoresis.

Radioactively labeled rat neurophysin precursors (pro-RNP I ("Pro-pressophysin") and pro-RNP II ("Pro-oxyphysin")) were prepared following established procedures (6). This involved bilateral injection of [^{35}S]-cysteine adjacent to the supraoptic nucleus

of normal (for pro-RNP I) and Brattleboro (for pro-RNP II) rats and dissection of the supraoptic nucleus from frozen sections of rat hypothalami. The supraoptic nuclei were homogenized in 0.1 M HCl and the extract applied to a Sephadex G-75 column equilibrated with 0.05 mM HCl containing 1 mg/ml bovine serum albumin. Gel-filtration fractions corresponding to the specific precursors were collected. The homogeneity of the [^{35}S]cysteine-labeled products was determined to be greater than 90% for both pro-RNP I (p*I* 6.1) and pro-RNP II (p*I* 5.4) as assessed by isoelectric focusing and two-dimensional gel electrophoresis.

In vitro translation products were obtained from a 300-μl rabbit reticulocyte lysate translation reaction in a manner analogous to that used previously for wheat germ extract translation (7,8). The reaction was performed by adding 21 μg of bovine hypothalamic poly (A)-containing RNA (in 100 μl of sterile water) to 200 μl of a master mixture made from a rabbit reticulocyte lysate translation kit (BRL). The master mixture contained 100 μl of reticulocyte lysate in 3.5 mM MgCl$_2$, 0.05 mM EDTA, 25 mM KCl, 0.5 mM dithiothreitol, 25 μM hemin, 50 μg/ml creatine kinase, 1 mM CaCl$_2$, 2 mM (ethylene glycol bis(β-aminoethyl ether))N,N'-tetraacetic acid (EGTA), and 70 mM NaCl; 30 μl of "reaction mixture" composed of 250 mM 4-(2-hydroxyethyl)-1-piperazineethanesulfonic acid (Hepes), 400 mM KCl, 100 mM creatine phosphate, and 500 μM of each of the common amino acids except cysteine; 50 μl of [^{35}S]cysteine (1 mCi/ml, 500–1100 Ci/mmol, New England Nuclear); 13 μl of 2 M potassium acetate; and 7 μl of sterile water. The translation mixture was incubated for 90 min at 30°C. For immunoprecipitation, this mixture was diluted to 3 ml with immunoprecipitation buffer (IPB). Affinity purified neurophysin antibody directed against [poly-DL-alanyl-poly-L-lysine]-conjugated BNP II (9) was added to a final concentration of 0.06 A_{280} units. After incubation at 4°C overnight, immunoprecipitation was effected by adding

---

sociated); RNP I and RNP II, rat neurophysins I (vasopressin associated) and II (oxytocin associated); IPB, immunoprecipitation buffer (containing 20 mM Tris-HCl, pH 7.15, 10 mM EDTA, 0.05 M NaCl, 0.01% Triton X-100, 20 mM cysteine, 100 Kallikrien units/ml aprotinin, and 1 mg/ml bovine serum albumin); TEAP, triethylammonium phosphate buffer, prepared by adjusting 0.25 N phosphoric acid to pH 3.0 with distilled triethylamine.

20 μl of Cowan I strain *Staphylococcus aureus* cells (Pansorbin, Calbiochem) which were suspended in IPB to 10% v/v. After a 30-min incubation at room temperature, the centrifuged pellets of Cowan cell immunoprecipitates were washed extensively to remove residual [^{35}S]cysteine. A two-thirds portion of this pellet, destined for reverse-phase HPLC peptide mapping, was supplemented with 0.5 mg of native BNP II and then treated as before with performic acid followed by trypsin (1). The supernatant from the anti-BNP II-mediated immunoprecipitation was immunoprecipitated with anti-BNP I and the resulting immunoprecipitate treated as above.

Hormone binding-site photoaffinity-labeled BNP II was prepared and purified essentially as described elsewhere (10,11). Briefly, a photolysis solution (7.4 ml total volume) was prepared by dissolving affinity-purified BNP II (0.9 mM) with the [^{3}H]Met-Tyr-*p*-azido-Phe amide photoaffinity reagent (0.14 μCi/μmol) in 0.4 M ammonium acetate, pH 5.7, at a fivefold molar excess of tripeptide over protein. A Corex glass reaction vessel containing the photolysis solution was immersed in a 5°C water bath and exposed for 90 min to light from a medium-pressure mercury lamp surrounded by a Corex glass filter. Following photolysis, the solution was adjusted to pH 3 with glacial acetic acid to dissolve precipitated protein. The photolabeled protein was isolated by Sephadex G-25 fractionation, to remove decomposed photoaffinity reagent, and then fractionated on a Met-Tyr-Phe-diaminohexyl agarose affinity matrix, to separate hormone binding-site photolabeled protein (unretarded peak) from nonhormone binding-site photolabeled protein (retarded peak).

Performic acid-oxidized or reduced–carboxymethylated protein samples were trypsin digested as before (1). Performic acid oxidation of protein samples was performed as described previously (1). Reduction–carboxymethylation of proteins was performed according to standard procedures (12) with

some modifications. Protein disulfide bonds were reduced by mixing 0.2 or 0.5 mg of protein, 0.48 g ultrapure urea, 250 μl of 0.2 M dithiothreitol, 40 μl of 0.15 M disodium EDTA, 300 μl of 1.5 M Tris–HCl, pH 8.6, and water to 1.0 ml. After the contents were stirred in a dark amber reaction vial at room temperature under nitrogen, the resulting sulfhydryl groups were carboxymethylated by adding 86 μl of freshly prepared 1.45 M iodoacetate (0.224 g of recrystallized iodoacetic acid dissolved in 1.0 ml of 1.0 M NaOH) to the reaction mixture. Fifteen minutes after the addition of iodoacetate, the reaction was applied to a Sephadex G-25 column (PD-10, Pharmacia) preequilibrated with water. Protein eluted at the void volume of the column was pooled and lyophilized.

Reverse-phase HPLC peptide maps of proteins, derivatives, and precursor preparations were obtained using the basic procedure described before (1), in most cases with a Varian Model 5000 high-performance liquid chromatograph and for the analysis of cell-free translation products with a Waters Associates gradient liquid chromatograph equipped with a Model 600 solvent programmer, Model U6K injector, and two Model M-6000 pumps. A particular performic acid-oxidized or reduced–carboxymethylated protein was trypsin digested and a 10- or 50-μl aliquot containing the sample injected onto a Zorbax CN HPLC column (0.46 × 25 cm, DuPont) previously equilibrated with triethylammonium phosphate (TEAP). Peptides were eluted by a linear gradient of acetonitrile mixed with TEAP starting at 0% acetonitrile at 0 min and going to 40% acetonitrile at 60 min (flow rate 0.8 ml/min). The gradient was adjusted to 0 to 25% acetonitrile in the 60-min period for mapping on the Zorax CN column when peptide retention became reduced after a prolonged period of use. Such column aging was observed to occur generally after 50 to 100 gradient runs. Elution of peptides from the reverse-phase column was followed by monitoring the absorbance of the column eluant at 215 nm.

In most cases, the eluant was collected in 0.5- or 1.0-min fractions for analysis of radioactivity or amino acid content, as indicated. Fractions prepared for amino acid analysis were dried *in vacuo* and then hydrolyzed exhaustively in 6 N HCl (110°C, 24 h, *in vacuo*). For peptide maps of cell-free translation products and pulse-labeled precursor preparations, 90–95% of the volume of each fraction collected was mixed with 5 ml of scintillation cocktail (Aquasol, New England Nuclear) for measurement of radioactivity using a Tracor Analytic Mark III scintillation counter. Based on both the contents of amino acid residues in acid hydrolysates of HPLC peaks (for native neurophysins) and the amounts of eluted radioactivity for ^{35}S-labeled precursors, the yield of HPLC-eluted tryptic peptides was generally in the range of 70 to 90%.

RESULTS

Separation of tryptic fragments of reduced–carboxymethylated and performic acid-oxidized bovine neurophysins. Previous experiments have shown that Zorbax CN elution can separate most of the tryptic fragments, including all of the large fragments, of performic acid-oxidized BNP I and II (1). This is shown for BNP II by the dashed profile in Fig. 1B. The amino acid sequences of the separated peptides, denoted OT-1 through OT-7, are given in Fig. 2. This elution includes OT-7, a sequence variant of OT-6 that had not been observed in our earlier reported study. The only peptides of BNP II that are not separated by the HPLC elution are OT-8 and OT-9, dipeptides which elute in the breakthrough fraction. Similar chromatography of BNP I digests separates its OT peptides 1 through 5 (1), which have the sequences specified in Fig. 2.

As shown by the dashed elution profile in Fig. 1A, the neurophysin tryptic fragments obtained from reduced–carboxymethylated protein also are separable with the cyano-propylsilyl reverse-phase system. Here, the relative elution positions of the S-carboxy-methylcysteine-containing peptides, CT-1 through CT-4, are the same as those of the cysteic acid-containing peptides OT-1 through OT-4 derived from the same sequences of neurophysin (Fig. 1A vs B). As with the profile for oxidized peptides, the identification of peptides in the profile for reduced-carboxymethylated protein were made by amino acid analysis of collected fractions.

Characterization of chemical properties of neurophysin photoaffinity labeled at the hormone binding-site. Photolytic reaction of neurophysins with L-Met-L-Tyr-p-azido-L-Phe amide, an analog of the neurophysin hormone ligands oxytocin and vasopressin, results in the inactivation of these proteins due to the covalent, 1:1 labeling of the hormone binding site (10,11). We have used the cyanopropylsilyl–HPLC mapping method to determine several properties of this derivative. It has been found that peptide label can be released from the derivative with specific effects on the covalent structure of the residual protein. Performic acid oxidation of photolabeled protein leads to a residual protein yielding the HPLC tryptic peptide map shown by the continuous profile in Fig. 1B. Here, all peptides elute as they do for native BNP II except for OT-4. The latter is missing from the derivative map, replaced by OT-4'. Amino acid analysis of OT-4' has shown that it contains the same sequence as that of OT-4, namely residues 44–66, but with Tyr 49 missing. On this basis, it has been concluded that Tyr 49 is the likely site of covalent photolytic modification and, thus, that this residue is in or close to the major hormone binding site of BNP II (and probably other neurophysins as well). The chemical events leading to release of the photoaffinity label and concomitant loss of Tyr 49 are not known at present but presumably involve some type of rearrangement of the linkage group between label and protein during performic acid oxidation.

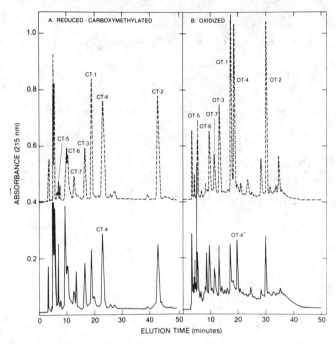

FIG. 1. Reverse-phase HPLC peptide maps of native and photoaffinity-labeled BNP II. (A) Native BNP II (0.5 mg, dashed profile) and photolabeled BNP II (0.2 mg, solid profile) were reduced with dithiothreitol, carboxymethylated, and subsequently trypsin digested as described under Materials and Methods. The lyophilized tryptic digests were dissolved in 12 μl of TEAP buffer and a 10-μl aliquot injected onto a Zorbax CN column (Dupont No. 2532) equilibrated with TEAP. A gradient of acetonitrile from 0 to 25% was started at sample injection and lasted for 60 min. Peptides eluting from the column were identified by amino acid analysis of fractions (0.5 min) corresponding to major peaks in the elution profile. (B) Native BNP II (0.25 mg, dashed profile) and photolabeled BNP II (0.1–0.2 mg, solid profile) were performic acid oxidized and trypsin digested as described under Materials and Methods. The tryptic digests were dissolved in 12 μl of TEAP buffer and a 10-μl aliquot injected onto the same Zorbax CN column as in (A). Fractions (1.0 min) were analyzed as in (A). Peptide numbering is based upon the system indicated in Fig. 2. OT, oxidized-tryptic fragment; CT, carboxymethylated-tryptic fragment. The shallow gradient of acetonitrile in (A) was used to offset a significant loss of peptide retention observed for column No. 2532 after approximately 9 months of use.

Interestingly, the HPLC map of reduced–carboxymethylated, photolabeled neurophysin is identical to that of unmodified native protein (Fig. 1A). Apparently, the reduction—alkylation treatment caused release of the peptide without modification of the amino acid to which it was attached covalently. The release of photolabel is probably due to reductive conditions, since it also can be effected by dithiothreitol treatment alone (data not shown).

Identification of pre-pro-neurophysins from in vitro translation of bovine hypothalamic mRNA. Much interest in the neurophysin–neuropeptide system has centered on the biosynthetic origin of the neurophysin–hormone complexes. Cell-free synthesis using hypothalamic mRNA has yielded high-mo-

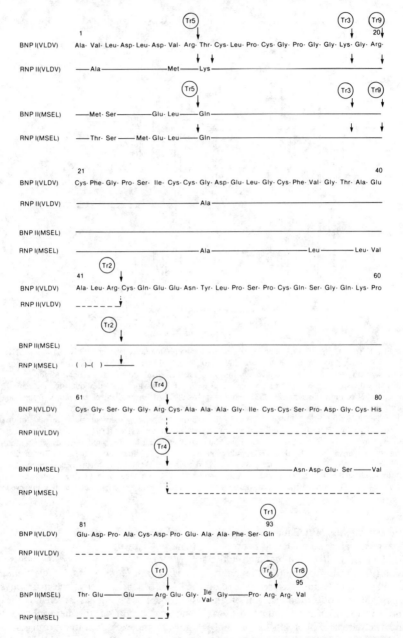

FIG. 2. Amino acid sequences of bovine and rat neurophysins. The sequences shown are based upon published data for BNP I (13), BNP II (14), and RNP I and II (15). The MSEL and VLDV classifications

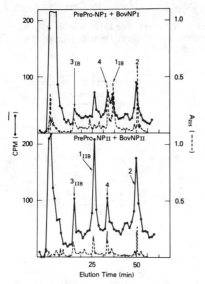

FIG. 3. Reverse-phase HPLC peptide maps of reticulocyte lysate translation products immunoprecipitated by anti-BNP I (top) and anti-BNP II (bottom). The lyophilized trypsin digests from a translation at 135 mM K⁺ (see Materials and Methods) were dissolved in TEAP and injected onto Zorbax CN (DuPont No. 1826). The column, equilibrated with TEAP, was eluted with a 0–40% acetonitrile gradient in 60 min at a flow rate of 0.8 ml/min. Eluant was monitored at 215 nm and collected in 1-min fractions. Fractions were counted in 5 ml of scintillation cocktail. The positions of the four cysteic acid-containing peptides of oxidized authentic BNP I or BNP II are indicated in each map and refer to the tryptic fragments of BNP I or BNP II as shown in Fig. 2. Peak number subscripts IB and IIB denote that the peptides are specific for BNP I and BNP II, respectively.

lecular-weight proteins (17–25,000) which are recognized by antibodies to the 10,000-dalton mature neurophysins (7,8). Unfortunately, the amounts of specific translation

products produced are quite small, in the subfemtomole range, thus obviating extensive structural characterization at the protein level. However, in the case of the bovine system, peptide mapping has yielded dependable evidence that the translation products in fact contain neurophysin sequences (1,17). This has been achieved most quantitatively by reverse-phase HPLC. The profiles of Fig. 3 show this for the ^{35}S-containing tryptic peptides obtained from performic acid-oxidized translation products. The four expected cysteic acid-containing peptides, OT-1 through OT-4, were observed in the maps for translation products I and II, those products precipitated using antineurophysin I and II, respectively. In the case of the anti-I-recognized product, a peak eluting in the position of the OT-1 of BNP II was observed, indicating that in this experiment antineurophysin I immunoprecipitated translation product II in addition to translation product I.

More striking than simply chromatographic overlap, the HPLC procedure has allowed quantitative estimates to be made of the amount of ^{35}S in each labeled peak. As shown in Table 1 for the profiles of Fig. 3, the contents of ^{35}S in the peaks assignable as OT-1 through OT-4 are consistent with the contents of cysteic acid expected for bovine neurophysins. Because the nonspecific background counts are difficult to determine in the profiles of Fig. 3, the estimated ^{35}S contents must be considered somewhat approximate. Even so, the calculated and expected values correspond well, especially when one considers that subfemtomole amounts of peptide were being analyzed and small selective losses of a particular tryptic peptide were possible. The ^{35}S content data provide a

(13,14) define the type of neurophysins based upon the specific sequence characteristics in positions 2, 3, 6, and 7. The MSEL proteins shown are associated *in vivo* with vasopressin, and the VLDV neurophysins with oxytocin. The circled symbols "Tr(number)" indicate the positions of tryptic cleavage. These symbols are based upon the peptide numbering system for BNP II (16) and refer to the fragment immediately preceding the cleavage point. The dashed lines and dashed arrows indicate, respectively, areas of proposed sequence homology and trypsin cleavage homology between bovine and rat neurophysins in as yet unsequenced parts of the rat neurophysins. These proposed added sequence aspects are based upon the HPLC maps shown in Fig. 4.

TABLE 1

CORRESPONDENCE BETWEEN ^{35}S-RADIOISOTOPE CON ENT IN HPLC PEAKS OF MAPS OF PERFORMIC ACID-OXIDIZED NEUROPHYSIN PRECURSORS AND [^{35}S]CYSTEIC ACID CONTENTS EXPECTED FROM NATIVE NEUROPHYSIN SEQUENCES

	Bovine *in vitro* translation products			
	Pre-Pro-NPII		Pre-Pro-NPI	
OT-peptide[a]	Calculated ^{35}S[b]	Expected $CySO_3H$[c]	Calculated ^{35}S[b]	Expected $CySO_3H$[c]
1_I	—	—		
1_{II}	5.2	5	4.8[d]	5[d]
2	5.0	4	3.8	4
3	1.5	2	2.7	2
4	2.4	3	3.7	3

	Rat *in vivo* pulse-labeled proteins			
	Pro-NPI		Pro-NPII	
OT-peptide[a]	Calculated ^{35}S[e]	Expected $CySO_3H$[c]	Calculated ^{35}S[e]	Expected $CySO_3H$[c]
1_I	4.9	5	—	—
1_{II}	—	—	5.1	5
2_I	4.2	4	—	—
2_{II}	—	—	3.7	4
3	1.8	2	2.2	2

[a] Tryptic peptides correspond to peaks in Fig. 3 for bovine translation products and Fig. 4 for rat pulse-labeled proteins. Subscripts I and II define a peptide as originating from neurophysins I and II, respectively.

[b] Number of [^{35}S]cysteic acid residues/peptide calculated by dividing the cpm in a particular peak by the average cpm/[^{35}S]cysteic acid residue. The latter is computed by dividing the total cpm in all [^{35}S]-containing OT-peptide peaks by the total number of [^{35}S]cysteic acid residues expected based upon the sequences of neurophysins I and II. The cpm value for each peak of Fig. 3 was not corrected for background due to the irregularity of the latter.

[c] Values represent the number of cysteic acid residues per peptide based on the amino acid sequences of BNP I and II.

[d] These values represent the sum of calculated ^{35}S and expected cysteic acid residues for peptides OT-1_I and 1_{II}, both present in the upper profile of Fig. 3.

[e] Number of [^{35}S]cysteic acid residues/peptide calculated as indicated in footnote b, except that the cpm value in each peak from Fig. 4 was corrected for the small and relatively constant background.

quantitative confirmation that the peptide maps in Fig. 3 indeed represent neurophysin sequences. Based on this and other arguments (17), translation products I and II have been identified as pre-pro-neurophysins, respectively.

Mapping of rat neurophysins and rat pro-neurophysins. The cyanopropylsilyl–HPLC peptide mapping method also has been useful in characterizing native rat neurophysins and microamounts of pulse-labeled rat proneurophysins. As shown in Fig. 4A, several of the tryptic peptides of a performic acid-oxidized rat neurophysin mixture have been separated. The tentative identification of these peptides as indicated in Fig. 4 was based

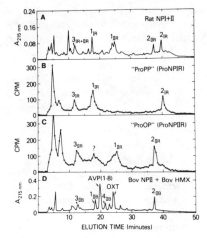

FIG. 4. Reverse-phase HPLC peptide maps of mature rat neurophysins and pro-neurophysins from rat hypothalamus. Performic acid-oxidized, trypsin-digested samples dissolved in TEAP were injected (in 50 μl) onto Zorbax CN (DuPont No. 3207). Peptides were eluted with a 0–40% acetonitrile gradient (over 60 min). The mature rat neurophysin mixture (A) was prepared as described under Materials and Methods and was performic acid oxidized without further treatment. The two rat precursor samples (B,C) were prepared as described. Immediately before performic acid oxidation, 0.1 mg of native BNP II was added to each sample as an internal standard. For the precursor maps, fractions were collected every 0.5 min and used for scintillation counting. The mature BNP II and bovine neurohypophyseal hormone mixture (D, presented for comparison) was prepared by mixing 25 μg of performic acid-oxidized, trypsin-digested BNP II and 50 μg of a mixture of the performic acid-oxidized, trypsin-digested bovine neurohypophyseal hormones, oxytocin and vasopressin. The numbers above the peaks in each map refer to the tryptic peptides of the neurophysins as indicated in Fig. 2. AVP (1–8) and OXT indicate, respectively, the positions of elution for the 1–8 fragment of oxidized vasopressin and the intact, oxidized nonapeptide, oxytocin. Peak number subscripts IR, IIR, IB, and IIB denote that peptides are specific for RNP I, RNP II, BNP I, and BNP II, respectively. The peak denoted "?" is as yet unidentifiable as neurophysin related.

on analogy of elution position with bovine neurophysin peptides, for OT-1_{IR}, OT-1_{IIR}, OT-2_{IIR}, and OT-3, and expected differences

in sequence between the two rat neurophysins, for OT-2_{IR}. The lack of OT-4 in the rat neurophysin maps cannot be explained at present, since sequence information in the 44–66 region (Fig. 2) is not yet available. Interestingly, the HPLC mapping data have allowed some deductions to be made about aspects of the sequences of the rat proteins which are as yet unsolved (15). These proposed additions are denoted in Fig. 2.

Using the native rat neurophysin maps as a basis, [^{35}S]cysteine-containing pulse-labeled pro-neurophysins (6) have been evaluated for their contents of neurophysin sequences, in a manner analogous to that used for the pre-pro-neurophysins. The HPLC maps are given in Figs. 4B and C for the tryptic peptides of the performic acid-oxidized forms of both rat pro-neurophysin I ("pro-pressophysin") and pro-neurophysin II ("pro-oxyphysin"). The correspondence of peptides OT-1, 2, and 3 between the pulse-labeled and native protein forms supports the conclusion that the pro forms contain intact neurophysin sequences. As for the maps of bovine neurophysin translation products (Fig. 3), the relative contents of ^{35}S in the HPLC peaks at pulse-labeled proteins are consistent with the contents of cysteic acid expected based on known or likely sequences of the rat neurophysins (Fig. 2). This agreement is shown by the data in Table 1.

The maps of pulse-labeled proteins in Fig. 4 also show that all of the major labeled peaks obtained, except for the two early peaks, correspond to native rat neurophysin peptides. No major peaks that correspond to the elution positions of performic acid-oxidized oxytocin or the 1–8 sequence of vasopressin (see Fig. 4D) were observed. Thus, while it has been proposed that neurophysins and their associated peptide hormones oxytocin and vasopressin are biosynthesized as parts of common precursors (18–20), the HPLC peptide mapping results have not been helpful in providing chemical information on this aspect of precursor structure. The same result has been noted previously for the *in vitro* translated pre-pro-neurophysins (17).

DISCUSSION

Reverse-phase HPLC has provided a useful method for high-sensitivity peptide mapping of neurophysins and neurophysin-containing proteins. The technique has been especially helpful as a micromethod for assessing the presence of neurophysin sequences in protein biosynthetic precursors that can be obtained only in small amounts. As is evident in Figs. 1 and 4, the mapping method allows neurophysin-specific tryptic fragments, including a difficult set of 20- to 27-residue peptides, to be separated and their amounts quantitated reproducibly. These analytical expectations overcome weaknesses that we had experienced previously in both separation and quantitation with two-dimensional paper electrophoresis–chromatography (7,8). The generally excellent resolving power and reproducibility of the cyanopropylsilyl reverse-phase peptide mapping method also is evident in our analyses of subtly modified peptides in photoaffinity-labeling studies (see Fig. 1).

The present results suggest that the HPLC method could prove helpful in future studies of the neurophysin system. It would seem, for example, that tryptic peptide mapping using cyanopropylsilyl columns can be an effective way to make sequence comparisons between neurophysins of different species. Areas of sequence homology can be deduced from areas of chromatographic overlap, while peptides of different retention time can be isolated and characterized to identify sequence differences. While the present comparative mapping of rat and bovine neurophysins (Figs. 1 and 4) was not carried to such fruition due to the limited availability of isolated rat proteins, even here some preliminary addition to available rat sequence data can be made (see Fig. 2). The method

also should be helpful for evaluating the proteolytically degraded neurophysins that often accompany the intact, mature proteins obtained from the neurohypophysis (15), as well as chemically modified derivatives and molecules that have been defined immunologically as neurophysin related (21). The present data have shown that the use of TEAP–acetonitrile gradients with cyanopropyl matrices, initially used by Rivier (22) for separation of relatively large polypeptides, can be conveniently applied to the above problems in studying the neurophysin system. Nonetheless, the above chromatographic system does not enable separation of dipeptides in neurophysin tryptic digests, may be of less value for separation of more complex mixtures of smaller peptides from other types of proteolytic digests, and demands subsequent steps if one wishes to remove the nonvolatile phosphate. Thus, there would be benefit in exploring other HPLC reverse-phase matrices and gradient systems to aid in studies similar to those described here for neurophysins.

REFERENCES

1. Chaiken, I. M., and Hough, C. J. (1980) *Anal. Biochem.* **107**, 1–16.
2. Hollenberg, M. D., and Hope, D. B. (1968) *Biochem. J.* **106**, 557–564.
3. Breslow, E., Aanning, H. L., Abrash, L., and Schmir, M. (1971) *J. Biol. Chem.* **246**, 5179–5188.
4. Chaiken, I. M. (1979) *Anal. Biochem.* **97**, 302–308.
5. Robinson, I. C. A. F., Edgar, D. H., and Walter, J. M. (1976) *Neuroscience* **1**, 35–39.
6. Russell, J. T., Brownstein, M. J., and Gainer, H. (1980) *Endocrinology* **107**, 1880–1891.
7. Giudice, L. C., and Chaiken, I. M. (1979) *Proc. Nat. Acad. Sci. USA* **76**, 3800–3804.
8. Giudice, L. C., and Chaiken, I. M. (1979) *J. Biol. Chem.* **254**, 11,767–11,770.
9. Fischer, E. H., Curd, J. G., and Chaiken, I. M. (1977) *Immunochemistry* **14**, 595–602.
10. Klausner, Y. S., McCormick, W. M., and Chaiken, I. M. (1978) *Int. J. Pept. Protein Res.* **11**, 82–90.
11. Abercrombie, D. M., McCormick, W. M., and Chaiken, I. M. (1982) *J. Biol. Chem.* **257**, 2274–2281.

12. Means, G. E., and Feeney, R. E. (1971) Chemical Modification of Proteins, pp. 218–219, Holden-Day, San Francisco.

13. Chauvet, M. T., Codogno, P., Chauvet, J., and Acher, R. (1979) *FEBS Lett.* **98**, 37–40.

14. Chauvet, M. T., Chauvet, J., and Acher, R. (1975) *FEBS Lett.* **58**, 234–237.

15. North, W. G., and Mitchell, T. I. (1981) *FEBS Lett.* **126**, 41–44.

16. Wuu, T. C., and Crumm, S. E. (1976) *Biochem. Biophys. Res. Commun.* **68**, 634–639.

17. Chaiken, I. M., Fischer, E. A., Giudice, L. C., and Hough, C. J. (1982) *in* Hormonally Active Brain Peptides: Structure and Function (McKerns, K., ed.), Plenum, in press.

18. Sachs, H., Fawsett, P., Takabatake, Y., and Portanova, R. (1969) *Recent Progr. Horm. Res.* **25**, 447–491.

19. Brownstein, M. J., Russell, J. T., and Gainer, H. (1980) *Science* **207**, 373–378.

20. Land, H., Schutz, G., Schmale, H., and Richter, D. (1982) *Nature (London)* **295**, 299–303.

21. Béguin, P., Nicolas, P., Boussetta, H., Fahy, C., and Cohen, P. (1981) *J. Biol. Chem.* **256**, 9289–9294.

22. Rivier, J. E. (1978) *J. Liq. Chromatogr.* **1**, 343–366.

Purification of Radiolabeled and Native Polypeptides by Gel Permeation High-Performance Liquid Chromatography[1]

C. Lazure,[2] M. Dennis, J. Rochemont, N. G. Seidah, and M. Chrétien

Clinical Research Institute of Montreal, 110 Pine Avenue West, Montréal H2W 1R7, Canada

Some results and observations concerning the use of protein columns are presented. The combined use of four protein columns having different fractionation ranges together with a volatile triethylamine formate buffer allowed the sieving of various polypeptides according to their molecular weights over a range of 500 to 150,000. The addition of 4 or 6 M guanidine-HCl permitted the reduction of aggregation with no sacrifice in resolution or linearity. With that denaturant, rapid separation, and molecular weight determination in the range 500–90,000 is easily accomplished. Moreover, sample recoveries as determined with radiolabeled proteins always exceeded 70% while radioimmunoassay techniques can be directly applied to the column eluate. Applications to quick identification of natural fragments of a serine protease, tonin, analysis of maturation products of pro-opiomelanocortin in an *in vitro* pulse experiment and finally quantitation by radioimmunoassays of pituitary peptides and elution of their ^{125}I-labeled derivatives are described.

The application of high-performance liquid chromatography (HPLC) to peptide and protein purification and characterization has enjoyed tremendous success and recognition in the past few years. This wide interest clearly arises from the speed, excellent resolving power, high sensitivity, and ease of operation inherent to HPLC technology. As reviewed by Rubinstein (1) and by Regnier and Gooding (2) separations of proteins, peptides, or amino acid derivatives can now be accomplished using ion-exchange, affinity, normal phase, reverse phase, and gel permeation methodologies. The last method can potentially circumvent many problems like time consumption, limited fractionation range, and difficult sample recoveries frequently encountered with classical methods such as ultracentrifugation, gel filtration on crosslinked dextrans or polyacrylamide and gel electrophoresis.

Recently, several laboratories have reported the use of gel permeation by HPLC which proves itself valuable in terms of resolution, recovery, and rapidity. Using commercially available supports such as TSK-gel PW or SW type (3–6), Synchropak GPC-100 (7), or protein columns (8–10) and nonvolatile buffers eluants of high ionic strength with or without denaturants such as guanidine-HCl (11,12) or sodium dodecyl sulfate (13), this technique has been used successfully for molecular weight determinations and for the separation of complex mixtures at an analytical or semipreparative scale.

In this report, we describe an HPLC gel permeation system which permits the rapid separation and molecular weight determination of proteins in the range of 500–90,000 using triethylamine phosphate (TEAP)[3] or

---

[1] This paper was presented at the International Symposium on HPLC of Proteins and Peptides, November 16–17, 1981, Washington, D. C.

[2] To whom all correspondence should be addressed.

[3] Abbreviations used: TEAP, triethylamine phosphate; TEAF, triethylamine formate; PBS, phosphate-buffered saline; MEM, minimum essential medium; POMC, pro-opiomelanocortin; LPH, lipotropin; SDS, sodium dodecyl sulfate; RIA, radioimmunoassay; MSH, melanotropin; END, endorphin; CLIP, corticotropinlike intermediate lobe peptide.

formate (TEAF) with or without 4 or 6 M guanidine-HCl as denaturant. Sample recoveries as assessed with radiolabeled proteins always exceeded 70% and radioimmunoassay procedures could be performed directly on the column eluate following appropriate dilution of the denaturant. Several applications of this procedure which demonstrate the utility of the system are presented.

MATERIALS AND METHODS

Unlabeled molecular weight markers for calibration curve determinations were bovine thyroglobulin, ovomucoid trypsin inhibitor (Sigma), bovine serum albumin, ovalbumin, ribonuclease A, chymotrypsinogen A (Pharmacia), cytochrome c, α-lactalbumin, lima bean trypsin inhibitor (Pierce), myoglobin, oxidized A and B chains of insulin (Schwarz/Mann), immunogobulin G, dinitrophenyl-glycine (Calbiochem), adrenocorticotropin (ACTH, this laboratory), and γ-melanotropin (synthesized by N. Ling, Salk Institute).

^{14}C-Labeled molecular weight markers for internal standards and recovery studies were obtained from Bethesda Research Laboratories. Iodinated [β-^{125}I]endorphin (synthesized by Serge St. Pierre, Sherbrooke, Que.), α-melanotropin (Ciba/Geigy), and ACTH were prepared by a minor modification of the method of Hunter and Greenwood (14) and purified on Sep-Pak C_{18} cartridges (Waters Assoc.) as described previously (15).

Formic acid was purchased from BDH, phosphoric acid and triethylamine (redistilled before use) from Fisher, and guanidine-HCl from Schwarz/Mann. All buffers used for HPLC gel permeation were filtered on 0.45 and 0.22-μm filters (Millipore) and degassed before use.

Protein analysis columns consisting of one I-60, two I-125's, and one I-250 (0.78 × 30 cm) connected in series were purchased from Waters Scientific. All runs were made isocratically on a Beckman liquid chromatograph Model 421 coupled to an Hitachi variable wavelength spectrophotometer Model 100-40 and Waters Model 730 data module.

All samples were dissolved in HPLC buffer (triethylamine formate or phosphate, 0.2 M, pH 3.0, with or without 4 or 6 M guanidine-HCl) applied to the column in a total volume of 100 μl. The columns were eluted at room temperature at a flow rate of 1.0 ml/min and where necessary, 0.5-ml fractions collected.

Rat submaxillary gland tonin (16) was reduced and alkylated with iodoacetic acid under denaturing conditions according to standard procedures. The desalted, lyophilized material was redissolved in HPLC buffer and analyzed.

Preparation and analysis of [^{3}H]Phe-labeled pituitary peptides. Neurointermediate pituitaries were dissected from 250-g male Sprague–Dawley rats (Charles River Labs), washed with phosphate-buffered saline (PBS), and preincubated in minimum essential medium lacking phenylalanine (Phe-free MEM; 0.1 ml/pituitary) for 1 h at 37°C in 95% O_2/5% CO_2. The tissue was then transferred to Phe-free MEM containing 1 mCi/ml [2,3,4,5,6-^{3}H]phenylalanine (sp act 134 Ci/mmol, Amersham Corp.) (0.1 mCi/pituitary) and incubated for 1 h as described above. Following the incubation, the tissue was washed with PBS/1 mM phenylalanine, heated for 5 min at 90°C in 5 M acetic acid containing 0.3 mg/ml iodoacetamide and phenylmethylsulfonylfluoride and 1 mM phenylalanine, and then sonicated for 30 s. Insoluble material was removed by centrifugation for 1 h at 40,000g at 4°C. The supernatant was applied to a column (0.9 × 100 cm) of Sephadex G-75 superfine equilibrated and eluted with 1 M acetic acid and precoated with 0.5 ml 25% human serum albumin. Fractions of 1.0 ml were collected, 10-μl aliquots were counted in 3.0 ml aquasol, and appropriate fractions were pooled.

Analysis of POMC immunoreactivity in

pituitary extracts. Separated anterior and neurointermediate pituitary lobes were dissected from 250-g male Sprague–Dawley rats and extracted as described above for [^{3}H]Phe-labeled proteins. Aliquots of the supernatant corresponding to 0.05 lobes were diluted with 20 vol of HPLC buffer and subjected to HPLC gel permeation. Fractions of 0.5 ml were mixed with equal volumes of radioimmunoassay (RIA) buffer (0.5 M phosphate, 0.5% human serum albumin, 0.01 M EDTA, 0.2% Triton X-100) and stored at −40°C until assayed.

RIAs were performed using antibodies raised against human β-endorphin (supplied by N. Ling, Salk Institute), α-melanotropin (supplied by H. Vaudry, France), and ovine ACTH (produced in this laboratory).

For the assays, 50 μl of fractions diluted as described above or of standard peptides in RIA buffer/HPLC buffer (1/1) were mixed with 150 μl of RIA buffer and 100 μl appropriately diluted antiserum (final concentration of guanidine hydrochloride = 0.3 M) and incubated at 4°C for 48 h. At this time, 100 μl of iodinated peptide (15,000–20,000 cpm) diluted in RIA buffer was added and a further 5-h incubation was performed. Following separation of bound and free hormone by standard second antibody procedures, radioactivity in the pellet was determined in a LKB Rackgamma II counter.

RESULTS AND DISCUSSION

I. Calibration Curves

In trying to avoid problems associated with nonvolatility, pH, and high salt content found with already proposed buffer systems, we decided to investigate the use of triethylammonium formate (TEAF) or phosphate (TEAP) buffers as proposed recently by Rivier (10). Although we were able to duplicate Rivier's results using two I-125 protein columns connected in series with and without the addition of an organic modifier, acetonitrile, it was still clear that the cali-

bration curve was concave downward rather than linear when using the equation

$$K_{av} = \frac{V_e - V_0}{V_t - V_0} = A_1 M + A_0, \quad [1]$$

where V_e, V_0, and V_t represent the elution, void, and salt volumes, respectively, and M the molecular weight. This observation appears to be typical of gel permeation by HPLC and Himmel and Squire (6) suggested instead the use of $F(v)$ versus $M^{1/3}$ defined as:

$$F(v) = \frac{(V_e)^{1/3} - (V_0)^{1/3}}{(V_t)^{1/3} - (V_0)^{1/3}}$$

$$= A_1 M^{1/3} + A_0. \quad [2]$$

Using this approach, we were able to obtain a linear relationship between $F(v)$ and $M^{1/3}$ throughout the fractionation range of I-125 columns (500–40,000 daltons) using either TEAF or TEAP buffers with and without acetonitrile (results not shown). Extension of these results to a series of four columns of varying fractionation ranges is illustrated in Fig. 1. In this application, a protein column I-60 (1000–20,000 daltons) was connected to two I-125 (2000–50,000 daltons) followed by one I-250 (10,000–500,000 daltons) keeping the length and bore sizes of the connecting tubing to a minimum. The eluting buffer was 0.2 M TEAF, pH 3.0, chosen because of its volatility, its transparency in the uv range, and its good ability for dissolving peptides and proteins. At lower TEAF concentrations some adsorption of peptides on the columns was observed, especially on the I-60 column. In Figs. 1A and D a calibration curve using Eq. [1] was constructed at a flow rate of 1 ml/min (higher flow rates up to 2 ml/min or lower flow rates down to 0.5 ml/min were not deleterious to the separation pattern). As seen, this system allows a very useful fractionation range of 500–150,000 daltons in less than 50 min. Moreover, since the 0.2 M TEAF buffer permits monitoring of the separation at 230 nm, a very high limit of sensitivity can thus be obtained. Using Eq. [2], it is clearly seen that a straight line (Fig. 1D) can be obtained throughout the fractionation range thus enabling accurate determination of molecular weights. However, it was noted that the use of TEAF buffer alone leads to aggregation (mainly dimerisation) of various peptides and proteins. To circumvent that problem, the same columns were run using 4 and 6 M guanidine-HCl in 0.2 M TEAP, pH 3.0; the results obtained are shown in Figs. 1B, C, E, and F. Here again, a linear calibration curve together with increased sharpening of the peaks and no aggregation was obtained even though reduction of the fractionation range to 500–90,000 daltons was noted: this effect can probably be explained by the action of the denaturant upon the pore sizes of the column. Using both methods, molecular weights of various compounds were de-

termined. Among those, the values for native tonin (16) ($M = 29,839$) a glycosylated serine protease, for the glycosylated human and porcine N-terminal part of pro-opiomelanocortin (17) ($M_r \simeq 15,000$) and for bovine *Escherichia coli* enterotoxin (Lallier, B., *et al.*, submitted) ($M_r = 2900$) correlated well with values obtained with classical methods. It is worth noting that glycosylated proteins or peptides did not exhibit, with that system, discrepancies in elution behavior as compared to unglycosylated polypeptides; this behavior was also noticed by Ui (12). It should be noted that it is essential for accurate computation according to Eq. [2] to include in the sample an internal marker for V_t and V_0 even though the reproducibility of the system is such that those values do not change very much, thus allowing a good approximation of molecular weight. It is also

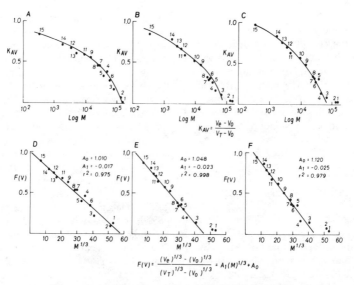

$$F(v) = \frac{(V_e)^{1/3} - (V_0)^{1/3}}{(V_T)^{1/3} - (V_0)^{1/3}} = A_1(M)^{1/3} + A_0$$

FIG. 1. Relationship between apparent molecular weight (M) and elution volume on protein columns. Data plotted according to K_{av} versus log M (A, B, and C) and according to $F(v)$ versus $M^{1/3}$ (D, E, and F) using 0.2 M TEAF (A, D), 0.2 M TEAP–4 M guanidine-HCl (B, E), and 0.2 M TEAP–6 M guanidine-HCl (C, F) at a pH of 3.0. Proteins used were immunoglobulin G (1), bovine serum albumin dimer (2) and monomer (3), ovalbumin (4), myoglobin dimer (5), ovomucoid trypsin inhibitor (6), chymotrypsinogen A (7), cytochrome c dimer (8), α-lactalbumin (9), lima bean trypsin inhibitor (10), ACTH (11), oxidized insulin B (12) and A (13) chains, γ- or α-MSH (14), and sucrose (15). V_0 and V_t were determined using thyroglobulin and DNP-Gly or sodium azide, respectively.

possible to generate a calibration curve while running the sample, as we have done, by adding trace amounts of radiolabeled proteins; this can be easily done by using a mixture of commercially available ^{14}C-labeled proteins (BRL) or by labeling selected proteins by iodination or by reductive alkylation using [^{14}C]formaldehyde. Indeed, recoveries of the marker proteins were very good if they are run with unlabeled carrier proteins (>90% as measured peak-wise) or even without carriers (>70%). This facilitates greatly the obtainment of results since only the counting of aliquots is involved and also ensures reproducibility of the results while comparing multiple runs. It was also found that a linear relationship can be obtained while running at different pH values since comparable results were obtained (A_0 = 1.031, A_1 = −0.022, r^2 = 0.979; compared to Fig. 1) at neutral pH. As shown in Figs. 1C and F, the use of 6 M guanidine only led to a steeper calibration curve as compared to Fig. 1E with no loss in other parameters. In spite of cost, relatively long equilibration time and care of pumps (especially the piston and its seal), the use of guanidine is definitely advantageous; on the other hand, there is no reason why the same results could not hold true with 8 M urea even though this eluant was not investigated in this study.

II. Applications

A. Studies concerning a serine protease, tonin.

It is known from sequence data (16) that tonin preparations obtained from rat submaxillary glands are always contaminated by the presence of truncated chains arising from partial autolysis in a manner similar to kallikrein (18) and the γ-subunit of nerve growth factor (19). The presence of those fragments can only be seen when the intramolecular disulfide bridges are broken apart, but so far no methods at an analytical scale have been able to separate conclusively the postulated shorter chains. Indeed as can be seen in Fig. 2A, no evidence

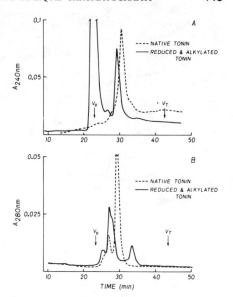

FIG. 2. Chromatogram of native tonin and reduced and alkylated tonin on protein columns using 0.2 M TEAF, pH 3.0 (A) and 0.2 M TEAP-4 M guanidine-HCl, pH 3.0 (B). V_0 and V_t were determined as in Fig. 1.

of such a fragmentation was obtained using the 0.2 M TEAF buffer alone. On the other hand, using the 0.2 M TEAP–4 M guanidine-HCl, it was clearly possible to resolve the suspected fragments; as can be seen in the chromatogram, a shoulder eluting at the position of reduced and alkylated tonin together with a small fragment eluting around 33 min ($M_r \simeq 9000$) could be seen. Confirmation of those results were obtained by amino acid analysis and also by sequencing of both fragments isolated using classical procedures on a Sephadex G75 in 1 M propionic acid and 8 M urea. The larger fragment corresponded to a chain starting around residue 65 of tonin while the smaller one clearly corresponded to the 1-65 segment (results not shown). In this application, it is clear that the use of the protein columns led in a very short time (50 min) to important information before embarking on a more time-consuming procedure devised to obtain

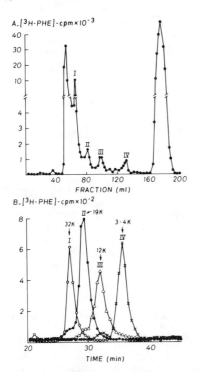

FIG. 3 (A) Chromatogram of [³H]Phe-labeled proteins obtained from in vitro incubation of neurointermediate lobe on Sephadex G-75 superfine (0.9 × 100 cm) using 1 M acetic acid as eluant. Fractions of 1 ml were collected and radioactivity was measured on aliquots by liquid scintillation counting. Peaks I, II, III, and IV were pooled for further characterization by HPLC gel permeation. (B) Chromatogram of ³H-labeled POMC precursor and related peptides obtained after separation on G-75 (A). Pooled fractions (I–IV) were separately analyzed on protein columns using 0.2 M TEAP–6 M guanidine-HCl, pH 3.0. Molecular weights indicated in the figure were computed according to the calibration curve presented in Fig. 1F. Internal standards (10–15 µg) included in the sample were thyroglobulin, ovalbumin, chymotrypsinogen A, oxidized insulin B chain, and DNP-glycine.

those fragments in a sizeable amount suitable for structure analysis. Moreover, with those columns, enough material can be accumulated so that direct amino acid analysis (or sequencing using radioactively labeled material) can be done as in the case of the

smaller fragment. Indeed, it was found that quick desalting of the column eluate can be accomplished using Sep-Pak C₁₈ cartridges (Waters) with minimal loss or that further purification can be done following the separation by directly injecting the pooled eluate onto a reverse-phase (HPLC) column without desalting, thus facilitating greatly the handling, analysis, and purification of the isolated fractions.

B. Separation and recovery of ³H-labeled POMC biosynthetic products. Pro-opiomelanocortin (POMC), the common glycoprotein precursor to adrenocorticotropin (ACTH) and β-lipotropin (β-LPH) is the most abundant protein synthesized in rat neurointermediate lobes. Since it represents 30% of the total of radioactive proteins after 1-h pulse incubation with [³H]Phe and since it can be further processed into various biologically active peptides such as ACTH, β-LPH, β-endorphin, etc., differing by their molecular weights, very low amounts (< 1 nmol) of material obtained after incubation were examined using the gel permeation system to investigate its potential use in pulse-labeling studies.

Neurointermediate pituitary peptides radiolabeled in vitro with [³H]Phe and chromatographed on Sephadex G-75 superfine produced the elution pattern shown in Fig. 3A. Four peaks (I–IV) known from previous studies (20) to be related to proopiomelanocortin were pooled and aliquots reanalyzed separately by gel permeation HPLC in 0.2 M TEAP–6 M guanidine-HCl, pH 3.0. As shown in Fig. 3B, the four peaks were clearly resolved in this system and had apparent molecular weights of 32,000 (peak I), 25,000 (peak II), 12,000 (peak III), and 3500 (peak IV) daltons, corresponding to POMC, ACTH/N-terminal fragment, β-lipotropin, and β-endorphin, the principal peptides synthesized in short-pulse incorporations of the neurointermediate lobe (20).

Analysis of duplicate aliquots of the G-75 pools by SDS–polyacrylamide gel electrophoresis (data not shown) gave apparent molecular weights which are in good agreement with those obtained by HPLC gel permeation. SDS–polyacrylamide gel electrophoresis, however, further resolved peaks I and II into two peaks differing in molecular weight by approximately 2000 daltons. The resolution of the HPLC technique thus appears limited in the conditions used in this study, though minor modifications such as decreasing the flow rate, lowering the concentration of guanidine-HCl, and/or separating the peaks into their ascending and descending components may permit partial separation of peptides differing only slightly in molecular weight (21).

Recoveries of the tritiated peptides was greater than 70% when applied without carrier proteins and exceeded 90% when molecular weight markers (10 μg each) covering the fractionation range of the columns were included in the sample. This HPLC gel permeation system thus provides a very rapid method for the separation and molecular weight determination of *in vitro* labeled peptides and their near-quantitative recovery for further analysis. Preliminary studies (data not shown) indicate that eluted material may be desalted on Sep-Pak C_{18} cartridges or applied directly without desalting to reverse phase HPLC columns for further purification and characterization. This last aspect is, in our opinion, of particular importance since, from past experience in this laboratory, it is known that separation of the various peptides derived from the POMC molecule is, even when attempted by reverse-phase chromatography alone, difficult and a lengthy procedure (3–4 h) whereas prior separation using the protein columns leads to easier and faster analysis.

C. Analysis of POMC immunoreactivity in pituitary extracts. It is known that the pro-opiomelanocortin molecule can be processed differently according to the localization of the maturation process. Indeed, ACTH and β-LPH together with the N-terminus part of the POMC are major end products in the pars distalis (anterior lobe) of the pituitary gland whereas β-endorphin, α-melanotropin, CLIP, and γ-LPH are more abundant in the pars intermedia (intermediate lobe). It was thus of interest to find out whether the gel permeation system is able to reveal differences in maturation pattern occurring in both lobes.

In an effort to further extend the usefulness of the HPLC gel permeation system, we investigated the possibility of coupling the method to radioimmunoassay procedures directly on the column eluate, a method commonly employed for partial characterization of peptides separated by gel filtration on Sephadex columns. The results of such studies on immunoreactive fragments of POMC in extracts of separated pituitary lobes are shown in Figs. 4A and B. The elution patterns of β-endorphin (β-END), ACTH, and α-MSH immunoreactivities are compared with the elution positions of radiolabeled standards run separately (Fig. 4A).

Neurointermediate lobe extracts reveal two minor peaks of β-END immunoreactivity at positions of [^{3}H]POMC and [β-^{125}I]lipotropin (β-LPH) and a major peak at the position of [β-^{125}I]endorphin. Immunoreactivity eluting after β-END may represent products of further cleavage of endorphin (22). ACTH immunoreactivity was observed at the positions of [^{3}H]POMC, [^{3}H]ACTH/N-terminal and ACTH with, in addition, two peaks of ACTH-like material eluted near the position of α-[^{125}I]MSH probably as a result of the demonstrated cross-reaction of α-MSH and related fragments in the ACTH-RIA (data not shown). The α-MSH antibody also reacted with material eluting at these positions; however, due to the specificity of this RIA for the N-acetylated N-terminal portion of α-MSH, no larger peaks of α-MSH were observed. Comparison of the amounts of immunoreactivity in the peaks clearly indicates that β-endorphin and α-MSH are the major POMC-related fragments in this tissue, in agreement with biosynthetic studies (20).

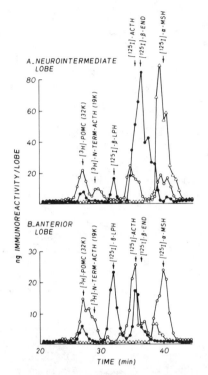

FIG. 4. Profile of POMC-related immunoreactive peptides extracted from neurointermediate (A) and anterior (B) pituitary lobes following separation on protein columns using 0.2 M TEAP–6 M guanidine-HCl, pH 3.0. Aliquots (50 μl) were assayed in duplicate for β-endorphin (●), ACTH (○), and α-MSH (Δ) as described in the text. The positions of radiolabeled markers as shown were determined in separate runs.

The elution positions of POMC immunoreactivity from anterior lobe were nearly identical with those observed from the neurointermediate lobe, but the relative amounts of the different fragments was strikingly different. In this tissue, ACTH and β-LPH were major components with only minor amounts of β-endorphin and α-MSH. The large amount of ACTH immunoreactive material eluting at the position of α-MSH may reflect the presence of nonacetylated ACTH fragments; such peptides have been

previously described in anterior lobe extracts (23). As observed in neurointermediate lobe, minor immunoreactive peaks eluting as shoulders slightly later than major peaks were observed; their identities and significance are not known.

CONCLUSIONS

The use of gel permeation by HPLC in many aspects of protein and peptides studies was illustrated in this report by various applications. It was shown that it could successfully replace classical methods for screening of molecular heterogeneity in enzyme studies since it is so rapid and requires microgram quantities of material that can be recovered and further analyzed. Its use also revealed its very high sensitivity since subnanomole amounts of proteins resulting from pulse-chase studies and picomole levels resulting from tissue extraction can be successfully analyzed using radiolabel monitoring or radioimmunoassay. In both cases, recoveries after gel permeation HPLC were always satisfactory thus allowing further characterization to be done on the sample. Moreover, this method can be applied with confidence to molecular weight determinations of glycosylated polypeptides such as tonin, NH₂-terminal part of POMC, etc. Since this system can be used in different applications as shown in this report, it is felt that such a technique could be of great utility in many fields such as metabolism, catabolism, enzymatic processing of precursor molecules, and protein purification. Therefore, it can be concluded that, because of its speed, reproducibility, high sensitivity, and versatility, gel permeation by HPLC definitely offers a good alternative to classical methods.

ACKNOWLEDGMENTS

This work was supported by a grant from the Medical Research Council of Canada (PG-2). The authors thank Mrs. D. Marcil for her help in typing the manuscript.

REFERENCES

1. Rubinstein, M. (1979) *Anal. Biochem.* **98**, 1–7.
2. Regnier, F. E., and Gooding, K. M. (1980) *Anal. Biochem.* **103**, 1–25.
3. Hashimoto, T., Sasaki, H., Aiura, M., and Kato, Y. (1978) *J. Chromatogr.* **160**, 301–305.
4. Rokushika, S., Ohkana, T., and Hatano, H. (1979) *J. Chromatogr.* **176**, 456–461.
5. Kato, Y., Komiya, K., Sasaki, H., and Hashimoto, T. (1980) *J. Chromatogr.* **190**, 297–303.
6. Himmel, M. E., and Squire, P. G. (1981) *Int. J. Pept. Prot. Res.* **17**, 365–373.
7. Gruber, K. A., Whitaker, J. M., and Morris, M. (1979) *Anal. Biochem.* **97**, 176–183.
8. Mole, J. E., and Niemann, M. A. (1980) *J. Biol. Chem.* **255**, 8472–8476.
9. Jenik, R. A., and Porter, J. W. (1981) *Anal. Biochem.* **111**, 184–188.
10. Rivier, J. A. (1980) *J. Chromatogr.* **202**, 211–222.
11. Ui, N. (1979) *Anal. Biochem.* **97**, 65–71.
12. Ui, N. (1981) *J. Chromatogr.* **215**, 289–294.
13. Imamura, T., Konishi, K., Yokoyama, M., and Konishi, K. (1979) *J. Biochem.* **86**, 639–642.
14. Hunter, W. M., and Greenwood, F. G. (1962) *Nature (London)* **194**, 495–496.
15. Seidah, N. G., Dennis, M., Corvol, P., Rochemont, J., and Chrétien, M. (1980) *Anal. Biochem.* **109**, 185–191.
16. Lazure, C., Seidah, N. G., Thibault, G., Genest, J., and Chrétien, M. (1981) Proceedings, VIIth American Peptide Symposium (Rich, D. H., and Gross, E., eds.), pp. 517–520.
17. Seidah, N. G., Rochemont, J., Hamelin, J., Lis, M., and Chrétien, M. (1981) *J. Biol. Chem.* **256**, 7977–7984.
18. Fiedler, F., and Hirschauer, C. (1981) *Hoppe Seyler's Z. Physiol. Chem.* **362**, 1209–1218.
19. Thomas, K. A., Silverman, R. E., Jeng, I., Baglan, N. C., and Bradshaw, R. A. (1981) *J. Biol. Chem.* **256**, 9147–9155.
20. Crine, P., Gossard, F., Seidah, N. G., Blanchette, L., Lis, M., and Chrétien, M. (1979) *Proc. Nat. Acad. Sci. USA* **76**, 5085–5089.
21. Seidah, N. G., Routhier, R., Benjannet, S., Larivière, N., Gossard, F., and Chrétien, M. (1980) *J. Chromatogr.* **193**, 291–299.
22. Mains, R. E., and Eipper, B. A. (1981) *J. Biol. Chem.* **256**, 5683–5688.
23. Lissitsky, J. C., Morin, O., Dupont, A., Labrie, F., Seidah, N. G., Chrétien, M., Lis, M., and Coy, D. H. (1978) *Life Sci.* **22**, 1715–1722.

Nonideal Size-Exclusion Chromatography of Proteins: Effects of pH at Low Ionic Strength[1,2]

W. KOPACIEWICZ AND F. E. REGNIER

Department of Biochemistry, Purdue University, West Lafayette, Indiana, 47907

Ideal size-exclusion chromatography separates molecules primarily on the basis of hydrodynamic volume. This is achieved only when the chromatographic support is neutral and the polarity nearly equal to that of the mobile phase. When this is not the case, the support surface may begin to play a role in the separation process. As the magnitude of surface contributions becomes larger, the deviation from the ideal increases. Because the separation mechanism is different than that of ideal size-exclusion chromatography, selectivity could be increased in nonideal size-exclusion chromatography. This paper explores the use of size-exclusion chromatography columns with mobile phases that cause proteins to exhibit slight deviations from the ideal size-exclusion mechanism. Although there are many ways to initiate nonideal size-exclusion behavior, the specific variable examined in this study is the influence of pH at low ionic strength. Individual proteins were chromatographed on SynChrom GPC-100, TSK-G2000SW, and TSK-G3000SW columns at low ionic strength. It was found that a protein could be selectively adsorbed, ion excluded, or chromatographed in an ideal size-exclusion mode by varying mobile-phase pH relative to the isoelectric point of the protein. In extreme cases, molecules could be induced either to elute in the void volume or beyond the volume of total permeation. It is postulated that these effects are the result of electrostatic interactions between proteins and surface silanols on the support surface. Optimization of size-exclusion separations relative to protein isoelectric points is discussed.

Size-exclusion chromatography (SEC)[3] is that chromatographic process whereby molecules are separated on the basis of their hydrodynamic volume. In the absence of solute interaction with the support surface, the log of solute molecular weight is inversely proportional to the fractional pore volume available to a molecule between the limits of total exclusion and total permeation. It is well known that under certain conditions proteins may deviate from this pure size-exclusion retention mechanism (1–5). When mobile-phase ionic strength is low, electrostatic in-

teractions may occur between charged solutes and charged SEC packing materials. At the other extreme of mobile-phase ionic strength, the salt concentration is sufficiently high to induce a solvophobic effect between the mobile phase and some hydrophobic proteins. Under these conditions the protein will adsorb to the surface of even slightly hydrophobic SEC supports.

Because virtually all inorganic liquid chromatographic media are porous and weakly ionic, macromolecular solute retention is a function of two different phenomena: size exclusion and sorption components. The size-exclusion component depends only on the physical dimensions of the SEC packing (pore diameter, pore volume, and pore shape), whereas surface interactions are a function of the chemical nature of the packing (hydrophobicity, charge, and bioaffinity).

[1] This is Journal Paper No. 9150 from the Purdue University Agricultural Experiment Station.
[2] This paper was presented at the International Symposium on HPLC of Proteins and Peptides, November 16–17, 1981, Washington, D. C.
[3] Abbreviations used: SEC, size-exclusion chromatography; nSEC, nonideal size-exclusion chromatography.

Pfannkoch (1) has shown that these surface interactions are independent and contribute differently under different conditions. At low ionic strength, electrostatic effects dominate, whereas size exclusion predominates at intermediate ionic strengths. High ionic strength induces hydrophobic interactions. The equation

$$k'_t = k'_s + \sum k'_i \qquad [1]$$

expresses this relationship in terms of a total capacity factor (k'_t). It will be recalled that $V_e = t_0 F (1 + k')$, where V_e is solute elution volume, t_0 is the retention time of a nonretained solute, and F is the volumetric flow rate in milliliters per minute. Equation [1] indicates that k'_t is equal to the size-exclusion contribution (k'_s) plus the sum of all sorption (k'_i) contributions. When there is no surface interaction, retention will be accomplished exclusively by an SEC mechanism. As the sorption component increases, the relative contribution of SEC decreases until it has minimal contribution in total retention. Those intermediate cases where both SEC and sorption mechanisms are contributing simultaneously are designated nonideal size-exclusion chromatography (nSEC) in this paper.

We examined protein retention on high-performance SEC columns at low ionic strength. Under these conditions, electrostatic interactions are maximized. Pfannkoch (1) has shown that all of the commercially available high-performance SEC columns have at least some anionic character. In the case of polymethylmethacrylate-based supports, this charge is the result of residual carboxyl groups on that support. With the silica based support, the residual negative charge is the result of underivatized surface silanols. Silanol groups act as weak acids with a pK of 3.5 to 4.0. Thus, at neutral conditions the silanols will be ionized, whereas under acidic conditions they are protonated and the support is neutral.

The central goal of most of the SEC work to date has been to identify nonideal size-exclusion effects and eliminate them so that columns may be used in the pure size-exclusion mode (1–5). In actuality, the resolving power of SEC is rather low. A good high-performance SEC column will only resolve a mixture into seven peaks with approximately a twofold difference in molecular weight between peaks. It would be useful to extend the fractionating power of SEC columns. This paper will demonstrate that this is possible by exploiting nonideal size-exclusion effects.

Exploitation of weak ionic interactions is the primary mechanism investigated in this study. The proposed model for accomplishing this is to operate the silica-based column at low ionic strength, such that surface charge or zeta potential would be maximized and charge shielding would be minimized. Silica, exhibiting a net negative charge, would therefore be expected to exclude anionic solutes from support pores and induce early elution, whereas cationic species would absorb electrostatically to the surface and increase retention. Because proteins are amphoteric, they characteristically demonstrate both behaviors. Proteins will be positively charged when the mobile-phase pH is below the isoelectric point and should be adsorbed. The only exception would be where the pH is too low for the support to be ionized. At pH values above the isoelectric point, the protein will have a net negative charge and be ion excluded. Manipulation of these interactions can be controlled by varying mobile-phase pH with respect to the isoelectric point and by small changes in ionic strength. When these factors are optimized, column selectivity is greatly enhanced such that it is possible to separate two proteins of equal molecular weight differing in pI.

MATERIALS

The pumping system was an Altex 110A reciprocating pump purchased from AN-SPEC (Ann Arbor, Mich.). A Perkin–Elmer LC 55 spectrophotometer (Norwalk, Conn.)

fitted with a flow cell served as our detector. The three size-exclusion columns used were the SynChropak GPC-100 (25 × .41 cm) (SynChrom, Linden, Ind.) and the TSK-G2000SW and TSK-G3000SW column (30 × .75 cm) (Toya Soda Corp., Japan).

Reagents. All the proteins used in this study were purchased from Sigma Chemical Corporation (St. Louis, Miss.). Reagent grade inorganic compounds were obtained from M.C.B. (Norwood, Ohio) and ICN Pharmaceuticals (Cleveland, Ohio).

EXPERIMENTAL

Protein samples were prepared by making filtered 1% (w/v) solutions in the appropriate buffer at neutral pH and an ionic strength of 0.05. In subsequent experiments involving a series of different pH buffers, little difference was seen in the retention time of samples prepared at neutral or the corresponding pH. Calculations for individual buffer solutions took into account the percentage of ionization and multiprotic equilibria so that the ionic strengths were corrected accordingly.

Between chromatographic runs, each support was equilibrated with at least 3 column volumes of the next buffer. Sample injections were repeated until reproducibility was obtained. In general, precision was within 3%. The resulting retention times were then used to calculate the distribution coefficient (K_d), where

$$K_d = \frac{V_e - V_o}{V_i}.$$

The terms V_e, V_o, and V_i represent solute elution volume, interstitial volume of the column, and the pore volume, respectively. The V_o was determined with DNA (whole calf thymus) and V_i with glycyl-L-tyrosine or ADP according to Pfannkoch (1).

RESULTS

To test the hypothesis that the extent and type of ionically induced nonideal size-exclusion behavior may be experimentally ma-

nipulated requires the study of (i) the influence of ionic strength on K_d at constant pH; (ii) the influence of pH of K_d at constant low ionic strength; and (iii) changes in selectivity (K_{d_2}/K_{d_1}) as a function of the first two variables. Experiments dealing with the manipulation of column selectivity will be presented along with the data on ionic strength and pH effects.

A. The influence of ionic strength on solute K_d at constant pH. It has been reported by numerous investigators that electrostatic interactions may occur between charged solutes and charged supports at ionic strengths below 0.1. As the ionic strength was reduced from 0.1 to 0.01, the distribution coefficient of myoglobin (horse heart) changed on a SynChropak GPC-100 column (Fig. 1). At pH 5.2, the K_d increased beyond the total permeation volume of the column ($K_d = 1.0$), indicating protein sorption. Changing the pH to 6.9 and repeating the experiment indicated little effect of ionic strength on K_d, while increasing the pH still more to 8.2 resulted in ion exclusion at low ionic strength as evidenced by a decrease in K_d. Because the pI of myoglobin is 7.3, it would have a net positive charge at pH 5.2, a net negative charge

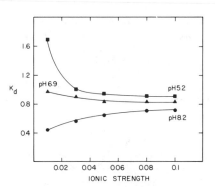

FIG. 1. Nonideal size-exclusion behavior of myoglobin: the influence of pH at low ionic strength. A 40-μl sample of myoglobin (horse heart) was chromatographed on a SynChropak GPC-100 eluted at 0.3 ml/min with phosphate buffer. The greatest selectivity was obtained at an ionic strength of 0.01.

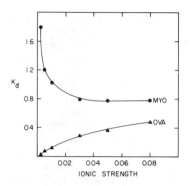

FIG. 2. Nonideal size-exclusion behavior of ovalbumin and myoglobin: the influence of ionic strength. Individual myoglobin (sperm whale) and ovalbumin (egg white) samples were chromatographed on a TSK-G3000SW column as a function of ionic strength at pH 6.3. The separation factor increased over fivefold at low ionic strength. Phosphate buffer at various ionic strengths was used to elute 40-μl samples at a flow rate of 0.8 cm^3/min.

at 8.2, and would approach neutrality at 6.9. At all of these pH values the support would still be negatively charged. The change in K_d as a function of ionic strength suggested that three separate retention mechanisms were involved: almost ideal size-exclusion at pH 6.9, a combination of size exclusion and sorption at pH 5.2, and a combination of size exclusion and ion exclusion at pH 8.2. The fact that the curves did not merge at 0.1 M ionic strength suggested that there was still some residual electrostatic effect or that the hydrodynamic volume of the protein changed with pH. In either case, the ability to manipulate protein retention and therefore column selectivity as a function of pH and ionic strength should be useful.

To test the hypothesis that selectivity and resolution may be increased by manipulation of nSEC effects, a pH was chosen between the isoelectric points by myoglobin and ovalbumin. Under these conditions, myoglobin would have a net positive charge and ovalbumin a net negative. When the K_d of these proteins was examined as a function of de-

creasing ionic strength at pH 6.3 on a TSK-G3000SW column, myoglobin exhibited sorption and ovalbumin became ion excluded (Fig. 2). The separation factor ($K_{d_{myo}}/K_{d_{ova}}$) was increased over fivefold by this simple manipulation of ionic strength. Although there was some decrease in the K_d of ovalbumin as a result of ion exclusion, the bulk of the increase in separation factor was contributed by myoglobin. Because the net charge on both the proteins and support is constant at any given pH, this increase in electrostatic interaction apparently results from charge deshielding at low ionic strength.

Because protein-support interactions are observed at low ionic strength, the possibility that protein–protein interactions occur between oppositely charged proteins was also examined. This was accomplished by repeating the experiment shown in Fig. 2 with the proteins chromatographed individually and together (Table 1). There was no significant difference in K_d whether the proteins were chromatographed individually or together. It may be concluded with this solute pair that surface interactions dominated intermolecular interactions.

Peak dispersion was also dependent on the retention mechanism. Size-excluded and ion-excluded molecules characteristically exhibited sharper peaks than sorbed species. Relative band-spreading data from the experiment described in Fig. 2 are presented in Table 1. The term σ is a measure of peak dispersion in a Gaussian peak and is defined as

$$\sigma = \frac{\text{peak area}}{\text{peak height} \times (2\pi)^{1/2}}.$$

Because σ is only valid for Gaussian peaks, it is used here as an approximation. Larger values of σ indicate increased band spreading and loss of column efficiency. Band spreading almost doubled when the K_d of myoglobin went from 0.78 to 1.70. It was concluded that the extensive peak dispersion seen when $K_d > 2$ precludes the use of longer retention times to increase resolution.

TABLE 1

EFFECT OF IONIC STRENGTH ON DISTRIBUTION COEFFICIENTS AND PEAK DISTORTION

Ionic strength	K_d				σ^c	
	OVA[a]	OVA[b]	MYO[a]	MYO[b]	OVA	MYO
0.0025	ND[d]	0.01	ND	1.7	14	ND
0.005	0.08	0.08	1.2	1.13	15	33
0.01	0.12	0.11	1.08	1.01	17	27
0.03	0.29	0.29	0.79	0.78	17	24
0.05	0.36	0.35	0.77	0.74	16	21
0.08	0.48	0.47	0.78	0.78	13	18

[a] Independent samples.
[b] Mixed samples.
[c] Approximation of peak dispersion in relative units.
[d] Not determined.

B. *The influence of pH on K_d at constant low ionic strength.* It was recognized early in these studies that although the horse heart myoglobin sample gave a single peak in the pure size-exclusion mode, it could be shown to be heterogeneous in the nSEC mode. The separation between components in this sample could be increased by decreasing the mobile phase pH from 8.4 to 6.2 at constant ionic strength (Fig. 3). Again, the increase in resolution was primarily the result of a sorptive process. The chromatograms obtained in this experiment on a TSK-G2000SW column are illustrated in Fig. 4A. A normal SEC chromatogram could be obtained by operating the column at 0.1 ionic strength (Fig. 4A). Decreasing mobile-phase pH at low ionic strength increased resolution in all cases (Figs. 4B–F). It will be noted again that the best resolution obtained between major components was obtained at pH 6.2 (Fig. 4B); however the best fractionation between all sample components was seen in Fig. 4D. This sample was fractionated into four peaks on an anion-exchange column (G. Vanecek, personal communication). Above pH 7, the components became mutually ion excluded and resolution was lost. Resolution (R_s), K_d, and band-spreading calculations are shown in Table 2. Resolution between the major components increased over threefold as the

pH decreased from 7.8 to 6.2; however at pH 6.2, R_s was compromised by increased band spreading. Again, increased retention due to ionic sorption was the major factor involved in superior resolution. As evidenced in Fig. 4D, resolution of components was quite good and peaks were still sharp. Decreasing the pH still further to pH 6.6 and 6.2 increased resolution slightly but was accompanied by ad-

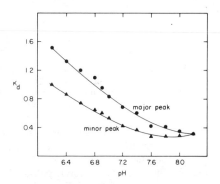

FIG. 3. Nonideal size-exclusion behavior of myoglobin: the influence of pH at low ionic strength. The upper and lower graphs represent the distribution coefficient of the major and minor components of myoglobin (horse heart) chromatographed on a TSK-G2000SW column as a function of pH at constant low ionic strength. Samples (40 μl) were eluted from the column at a flow rate of 0.7 cm³/min with 0.01 μ Tris–HCl buffer.

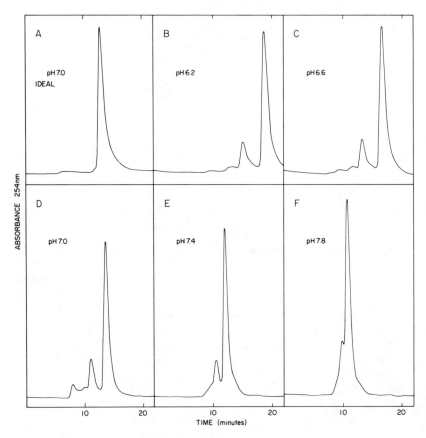

FIG. 4. Myoglobin chromatograms at different pH and low ionic strength. Myoglobin (horse heart) was chromatographed under varying nSEC conditions on a TSK-G2000SW column. The mobile phase for chromatogram A was 0.1 M phosphate. Tris–HCl of an ionic strength of 0.01 eluted samples B–F (40 μl). The flow rate (0.7 cm³/min), chart speed (0.5 cm/min), and AUFS (0.2) were constant throughout.

ditional band spreading. Resolution calculations between the major myoglobin components indicated that there was over a threefold increase when mobile-phase pH was changed from 7.8 to 6.6 (Table 2).

To further explore the nSEC phenomenon, a series of nSEC analyses were performed on several other proteins (Table 3). Distribution coefficients were determined over a range of pH values from approximately 2 to 8 at an ionic strength of 0.01 (Fig. 5). The horizontal dashed lines in each

octant of Fig. 5 indicate the protein distribution coefficient in the ideal size-exclusion mode, and the vertical line designates the isoelectric point. Referring to the pI reference, data to the right resulted from negatively charged species, while that to the left from positively charged species. The general trend was the occurrence of an ion-exclusion mechanism above the pI with a transition to increase retention due to ionic sorption below the pI. The K_d increased with decreasing pH to a maximum between pH 4 and 5.

Below pH 4 the distribution coefficient decreased independent of the properties of proteins. Although there are minor qualitative and quantitative differences between proteins, the data were consistent with the model proposed in the introduction. Ion exclusion occurred when columns were operated at low ionic strength above the pI. This gave way to a size-exclusion-dominated mechanism near the pI that in turn shifted to a mixed size-exclusion–ionic sorption mechanism below the pI. As the pH decreased, ionization of the protein increased concomitantly with increased retention. Superimposed on the increasing ionization of proteins under more acidic conditions was the collapse of support charge. Under increasingly acidic conditions, ionization of surface silanols was repressed and, therefore, the net charge at the support surface decreased. Because the ionization of

TABLE 2

VARIATION OF DISTRIBUTION COEFFICIENTS, RESOLUTION, AND BAND SPREADING OF MYOGLOBIN[a] AS A FUNCTION OF pH

pH	$K_d{}^b$	$R_s{}^c$	σ^b
6.2	1.51	1.64	15.0
6.6	1.21	1.57	12.4
7.0	0.83	1.44	11.2
7.4	0.60	0.94	11.5
7.8	0.41	0.45	11.2

[a] From horse heart.

[b] Calculated from major peak, pI 7.3; σ, approximation of peak dispersion in relative units.

[c] Calculated from major and minor peaks as shown by Pfannkoch (1).

proteins and loss of charge on supports were inversely related, retention went through a maximum below the pI and then declined.

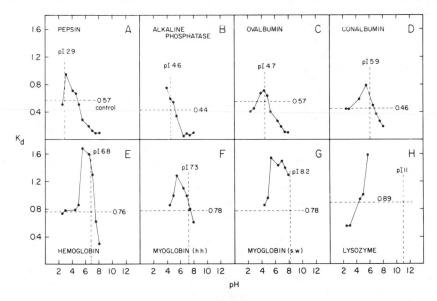

FIG. 5. Influence of pH on K_d of a variety of protein probes at low ionic strength. The distribution coefficient of a series of protein probes (Table 3) were calculated over a broad pH range to determine the extent of nonideal selectivity of a TSK-G3000SW column. Phosphate and sodium acetate solutions were used for the mobile-phase buffering over a pH range consistent with their buffering capacities. An ionic stength of 0.01 and a flow rate of 0.7 ml/min were maintained. Forty microliters of a 1% sample solution was introduced at each pH, and the proteins were detected by absorbance at 254 nm.

TABLE 3

Physical Parameters of Protein Probes

Protein	Origin	M_T		pI	$K_d{}^a$
Pepsin	Porcine stomach	33,000		2.9	.57
Alkaline phosphatase	Human placenta	116,000		4.6	.44
Ovalbumin	Egg white	43,500		4.7	.57
Conalbumin	Egg white	77,000		5.9	.46
Hemoglobin	Bovine blood	64,500		6.8	.76
Myoglobin	Horse heart	17,500	minor	6.9	.78
			major	7.3	
Myoglobin	Sperm whale	17,500	minor	7.7	.78
			major	8.2	
Lysozyme	Egg white	13,930		11^b	.89

[a] Calculated from DNA and ATP data in 0.1 M phosphate, pH 7.
[b] Isoelectric point of lysozyme in human sera. Data from Refs. (5,6).

Two observations of particular interest were that the transition from ion exclusion to adsorption occurred above the pI and the K_d obtained at the pI rarely equaled the K_d obtained under ideal SEC conditions. Both of these facts suggested that localized surface charges may play a significant role in the sorption process in addition to the net protein charge. In the sorption mode, as the pH approached the isoelectric point, these surface directed interactions could have occurred when the net negative charge of the protein was small enough to allow electrostatic interaction within the pore. At higher pH, the critical contact distance was never achieved because of exclusion.

As noted earlier, ionization of surface silanols was repressed below pH 4 and silica-based SEC supports became neutral. This should have resulted in pure SEC retention, because electrostatic interactions between solutes and the support were no longer possible. In the cases of conalbumin, hemoglobin, and myoglobin this seemed to be true. The distribution coefficients of these proteins in acidic mobile phases approached those observed in the pure SEC mode at neutral pH and 0.1 ionic strength. However, in the cases of lysozyme and ovalbumin the K_d became much smaller than that observed in the pure SEC mode. This phenomenon could

have resulted from either ion exclusion or a change in molecular structure. Because the support was neutral below pH 4, one would assume that this effect was the result of structural changes induced by acidic conditions. These changes would have had to have been sufficiently profound to have caused size-exclusion characteristics to be altered.

CONCLUSIONS

From the data presented here it may be concluded that nSEC can be a useful technique for the separation and purification of proteins. It was shown that nSEC conditions were unique for each sample. Because the region of maximum exploitation for electrostatic effects falls within 1 pH unit of the pI, one must have some prior knowledge of the sample or take a trial and error approach to mobile-phase selection. With rather large differences occurring in the elution profile over a 0.5 pH unit change, a trial and error search would require that a series of mobile phases varying by one-half pH unit or less be used.

Minimal interaction with support matrices occurred at a pH slightly above the pI of a protein. Maximum enhancement of selectivity and resolution was obtained in the sorptive mode below the pI. With the upper limit of support stability being pH 8 and the collapse of support ionization at pH 4, the operating region for electrostatic nSEC is from pH 5 to 8. In the case of proteins having higher pI values, such as lysozyme and cytochrome c, interaction is so strong that adsorption occurs, while anionic proteins become mutually ion excluded. As a general rule, exploitable electrostatic nSEC for proteins of a pI between 5 and 8 is achieved at ionic strengths between 0.005 and 0.025. Lower ionic strengths cause adsorption and band spreading, severely compromising the resolution. Although protein solubility problems are sometimes encountered at low ionic strength, this was not seen with analytical sample loads. Higher salt concentrations cause charge shielding and decrease ionic interactions. The delicate balance between protein charge, support charge, and mobile-phase ionic strength cannot be emphasized enough, since successful manipulation of this mode is only possible over a very finite range of conditions.

As mentioned earlier, prior knowledge of solute behavior is paramount. However, such knowledge will not only allow one to exploit nSEC, but may also aid in choosing conditions for ideal separations. In the long run, experience with nSEC should give the researcher a better understanding of retention mechanisms in addition to extending the general usefulness of the average SEC column.

ACKNOWLEDGMENT

This work was supported by NIH Grant OM 25431.

REFERENCES

1. Pfannkoch, E., Lu, K. C., Regnier, F. E., and Barth, H. G. (1980) *J. Chromatogr. Sci.* **18**, 430–441.
2. Engelhardt, H., Mathes, D. (1981) *Chromatographia* **14**, 325–332.
3. Schmidt, D. E., Jr., Giese, R. W., Conron, D., Karger, B. L. (1980) *Anal. Chem.* **52**, 177–182.
4. Engelhardt, H., Hearn, N. T. W. (1981) *J. Liquid Chromatogr.* **4**, 1261–1368.
5. Hearn, M. T. W., Grego, B., Bishop, C. H., and Hancock, W. S. (1980) *J. Liquid Chromatogr.* **3**, 1549–1560.
6. Righetti, P. G., Cardvaggio, T. J. (1976) *J. Chromatogr.* **127**, 1–28.
7. Sober, N. D. (ed.) (1973) Handbook of Biochemistry, CRC Press, Cleveland, Ohio.

Factors Influencing Chromatography of Proteins on Short Alkylsilane-Bonded Large Pore-Size Silicas[1]

M. J. O'HARE,* M. W. CAPP,* E. C. NICE,* N. H. C. COOKE,† AND B. G. ARCHER†[2]

Ludwig Institute for Cancer Research (London Branch), Royal Marsden Hospital, Sutton, Surrey SM2 5PX, Great Britain; and †Altex Scientific, Subsidiary of Beckman Instruments, 1780 Fourth Street, Berkeley, California 94710

The separation of proteins by high-performance gradient chromatography on short alkyl-chain bonded silica was studied with respect to pore size, mobile-phase composition, and temperature. Selectivity could be increased in particular cases by varying temperature or eluate compositon. Recovery of late-eluting, hydrophobic proteins was found to increase with flow rate and gradient slope. Recovery was also shown to be dependent on eluate composition—decreasing with added salt. The applicability of reverse-phase chromatography to proteins as large as 150 kD was demonstrated by the separation of monoclonal immunoglobulin.

The use of what may be broadly termed RP[3]-HPLC methods for the separation of polypeptides (<10 kD) has been a remarkable growth area in separation science, and procedures have now been well documented. Current popular systems, most of which involve gradient elution, have been optimized primarily by varying solvent composition (buffer and modifier) and, to a lesser extent, "ion-pairing" additives, pH, temperature, and specific bonded-phase chemistry. It has also been apparent for some time that a variety of proteins (>10 kD) can also be chromatographed with these methods (1). Comparatively few studies, however, have yet been directed specifically at the optimization of protein, as opposed to polypeptide, separations, notably in the quest for RP-based methods suited to larger materials (>60 kD). In particular, the issue of recovery has not

been systematically studied. Thus, although there have been some recent exceptions (2–6), many protein separations (>10 kD) have been carried out with 6- to 10-nm pore-diameter silica-based RP supports. Some have used C18 bonded phases (6–14), some C8 materials (2,3,15–17), and others alkylphenyl- (18), cyanopropyl- (2,13,14,19,20), or diphenyl-type surface chemistry (2).

Recently, we have shown that short-chain-type alkylsilane-bonded supports (C2–C6) offer specific advantages for protein chromatography, with enhanced recoveries and a greater range of compounds eluted, compared with C8 or C18 materials (21). The original investigation of these bonded phases was carried out with various small (8–10 nm) pore-size silicas and was limited to relatively small proteins. We document here the optimization of conditions for protein separations on large pore-size (30 nm) short alkyl-chain packings (C2–C4) and, in particular, demonstrate a dependence of recovery on separation conditions.

We also report the successful separation of immunoglobulin (156 kD) in the form of mouse monoclonal antibody.

[1] This paper was presented at the International Symposium on HPLC of Proteins and Peptides, November 16–17, 1981, Washington, D. C.

[2] To whom correspondence should be addressed.

[3] Abbreviations used: RP, reverse phase; BSA, bovine serum albumin; IgG, immunoglobulin G; TFA, trifluoroacetic acid; HFBA, heptafluorobutyric acid; ODS, octadecylsilane.

TABLE 1

DETAILS OF COLUMN PACKINGS USED[a]

Carbon chain length	Particle size (μm)	Pore size (nm)	Pore volume (cm³/g)	Specific area (m²/g)	End capping[b]
C2	10 ± 2[c]	30[d]	0.65	90	+
C3	4–6	8	0.57	200	+
C3	10 ± 2[c]	30[d]	0.65	90	+
C4	10 ± 2[c]	30[d]	0.65	90	−
C4	10 ± 2[c]	30[d]	0.65	90	+

[a] Parameters of spherical silica refer to underivatized material.
[b] All RP packings prepared by reaction with dimethylalkylchlorosilanes and capping, where appropriate, with trimethylchlorosilane (42). Reaction conditions were such as to result in maximal coverage. Typical values for a C3 material are 2.5% carbon and 4.7 μmol/m².
[c] 90% limits.
[d] 95% limits of 24 to 43 nm.

MATERIALS AND METHODS

Protein standards used included ovalbumin (44 kD), bovine serum albumin (BSA) (69 kD), phosphorylase b (92 kD), and myosin (200 kD) obtained as crystalline preparations of the highest available level of purity from Sigma Chemical Company (Poole, Dorset, Great Britain), except for myosin, which was a glycerol extract of rat skeletal muscle. [^{14}C]Methylated analogs of these proteins, prepared by reductive alkylation (22), were from Amersham International (Amersham, Great Britain). Preparations of immunoglobulin (IgG, 156 kD) tested included polyclonal rabbit IgG, prepared by ammonium sulfate precipitation, affinity-purified goat polyclonal F(ab′)$_2$ (antihuman immunoglobulin, 100 kD, Sera Lab, Crawley, Great Britain), and a mouse monoclonal IgG$_1$ as both crude ascites fluid containing 10–20 mg/ml protein from a mouse bearing a hybrid myeloma cell line (M8) and as a partially purified preparation of the latter obtained by ammonium sulfate precipitation. Sources of other peptides and proteins were as described previously (21). Polyclonal ^{125}I-labeled IgG was kindly supplied by Dr. J. Goding (WEHI, Melbourne, Australia).

Separations were carried out with 7.5 cm × 4.6-mm (i.d.) stainless steel columns packed with one of the short alkyl-chain bonded silicas detailed in Table 1. Unless otherwise specified, proteins were chromatographed at 30 or 45°C at a constant solvent flow rate of 1 ml/min using an Altex Model 324-40 or Spectra-Physics SP 8000. They were eluted with a continuous linear gradient between an aqueous primary solvent and a secondary organic modifier (acetonitrile or propan-1-ol). All solvents were HPLC grade. Trifluoroacetic (TFA) and heptafluorobutyric acids (HFBA) were obtained from Pierce Chemical Company (Rockford, Ill.). All proteins were injected in the aqueous primary solvent, typically 5–50 μg in 5 to 20 μl.

Eluted compounds were detected by uv absorbance (215 or 280 nm depending on organic modifier used) with Altex 155 or 160 or with Spectromonitor III (Laboratory Data Control) variable-wavelength spectrophotometers and, where appropriate, endogenous tryptophan fluorescence (225/340 or 280/370 nm) with a Schoeffel FS 970. Proof of the identity of peaks was obtained by size-exclusion HPLC (Beckman Spherogel/TSK SW3000 and 2000), polyacrylamide gel electrophoresis, and comparison of retention times with their slightly more hydrophobic [^{14}C]methylated analogs, where available. In the case of immunoglobulins, identity was checked by Ouchterlony gel diffusion against appropriate antisera after lyophilization of eluted peaks.

Protein recoveries were checked in three ways. The integrated areas of chromatographed protein peaks were compared with integrated areas of a tryptophan internal standard, and in some cases with the total uv-absorbing material in a corresponding aliquot of protein solution injected into an empty capillary column. The sizes of "ghost" peaks (if any) were determined as described previously (21). Good agreement ($\pm 10\%$) was obtained between these three methods in identifying proteins whose recoveries were significantly reduced under a given set of chromatographic conditions.

RESULTS

Effects of pore diameter and particle size on protein RP chromatography. Studies with polypeptide hormones (<10 kD) have shown (10) that optimum efficiencies and most effective resolution of complex mixtures are obtained with ODS-type packings when a primary solvent of relatively high molarity (>0.1 M) and low pH (<3) is used in conjunction with a low-viscosity organic modifier such as acetonitrile. For recovery of bioactive materials and their subsequent analysis by, for example, radioimmunoassay or bioassay, we have found it advantageous to use an unbuffered isotonic (0.155 M) NaCl solution adjusted to pH 2.1 with HCl (11). This system gave virtually identical results with all polypeptides tested in terms of efficiency, selectivity, and recovery compared with the 0.2 M phosphate, pH 2.1, buffer of Molnar and Horvath (23), and both systems were equally effective in separating smaller proteins, including RNase, lysozyme, cytochrome *c*, and myoglobin (10).

Larger proteins, such as bovine serum albumin (69 kD), however, were eluted from a high-coverage ODS packing on 8- to 10-nm pore-diameter silica as substantially broader peaks than the smaller proteins, and the more hydrophobic ovalbumin (44 kD) could not be eluted at all (<2% recovery). Short alkyl-chain-type packings (C2–C6) gave significant improvements in protein chromatography, particularly when used with propanol (15) as the secondary solvent; ovalbumin recoveries rose to 80% (C3) compared with 22% on C8 and 8% on C18-type supports (21). Compared with the acid chloride system, 0.01 M TFA (pH 2), preferred by some on account of its volatility, gave much poorer protein peak shapes on the small pore-diameter packings (Fig. 1).

To test the selectivities of the large pore-diameter RP packings prepared for this study (Table 1), each was used to chromatograph the mixture of polypeptides and smaller proteins that had been used previously to characterize the short alkyl-chain 8-nm pore-size packings (21).

The results showed virtually identical selectivities of all packings listed in Table 1 when 0.155 M NaCl, pH 2.1, with acetonitrile gradient elution was used (see Fig. 2), irrespective of whether they were capped or uncapped. Peak shapes of the larger proteins, such as BSA, were, furthermore, markedly improved on 30-nm materials compared with 8-nm-type supports (compare Figs. 1 and 4). However, the protein loadings accommodated by the large pore-size packings before significant deterioration of peak efficiency took place were somewhat lower (<500 µg) than with the smaller pore-size supports in columns of identical dimensions, on account of the lower surface area of the former (Table 1). All packings tested showed consistent chromatographic properties for extended periods of time (greater than 200 h).

Recoveries of ovalbumin from the large pore packings were, surprisingly, much reduced compared with 8-nm packings, when a 0.155 M NaCl, pH 2.1/propan-1-ol gradient elution system was used, falling from approximately 80% to virtually zero at loadings of 50 µg; recoveries of BSA were reduced from 100 to about 80% at this level. This

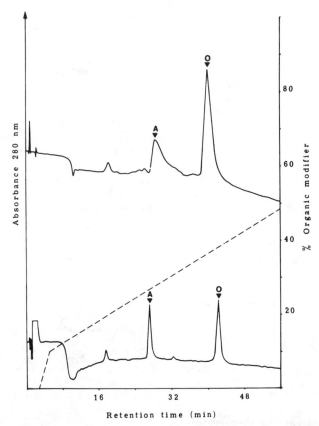

FIG. 1. The effect of primary solvent composition on the efficiency of protein standards on RP-HPLC. A C3 5-μm particle size (8-nm pore-diameter) packing (75 × 4.6-mm i.d. column) was used. The lower uv-absorbance trace ——— shows the profiles obtained with 0.155 M NaCl (pH 2.1)/propan-1-ol gradient elution, and that above with a 0.01 M TFA/propan-1-ol:TFA solvent system run under identical conditions of 1 ml/min at 45°C. The dotted line shows the continuous linear gradient profile. Note the considerable improvement in efficiencies with the high-salt primary solvent, which was capable of eluting ovalbumin in high yields (80%) from these small pore-size packings; A, bovine serum albumin (25 μg); O, ovalbumin (50 μg).

difference in protein recoveries on large and small pore-diameter silica-based packings was noted with all short alkyl-chain length materials tested (C2, C3, and C4) both as freshly packed and extensively used columns, and with an unrelated C8-type support on 30-nm pore-size silica (SynChropak RP-P). This inconvenient effect prompted us to reexamine the effects of temperature, solvent composition, flow rates, and gradient profiles so that optimum conditions for chromatography of larger and/or more hydrophobic proteins on the larger pore-size packings could be established.

Effect of temperature on protein RP chromatography. The effect of varying the temperature from 4 to 70°C on a series of peptide and protein standards was examined on a C3,

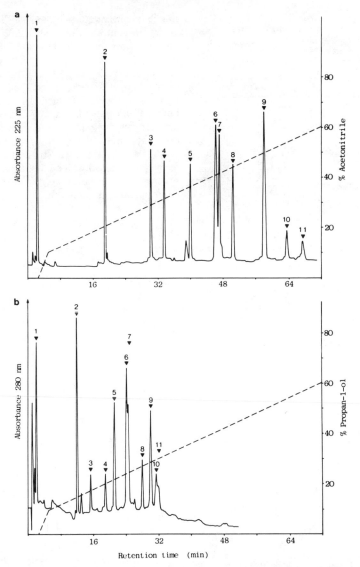

FIG. 2. Separation of peptides and proteins by RP-HPLC on large pore-size short alkyl-chain packing. A mixture of standards was injected onto a 75 × 4.6-mm (i.d.) column packed with a C3 capped 10-μm particle-size (30-nm pore-size) support, and eluted with acetonitrile and a primary solvent of 0.155 M NaCl/0.01 M HCl (a) and propan-1-ol and a primary solvent of 0.01 M TFA (b) using a gradient of the form illustrated by the dotted line. Standards used were 1, tryptophan; 2, ACTH$_{1-24}$; 3, RNase; 4, calcitonin (human); 5, lysozyme; 6, BSA; 7, α-lactalbumin (bovine); 8, lactoglobulin A; 9, prolactin (ovine); 10, prolactin (rat); and 11, growth hormone (rat). Both separations were carried out at 45°C and a flow rate of 1 ml/min.

O'HARE ET AL.

TABLE 2

EFFECT OF TEMPERATURE ON THE RETENTION TIMES
OF POLYPEPTIDES AND PROTEINS ON LARGE PORE-
SIZE, SHORT ALKYL-CHAIN PACKING[a]

	Retention times (min)					
Material	4°C	18°C	25°C	35°C	45°C	60°C
Tryptophan	10.8	8.0	6.0	4.5	3.4	2.6
ACTH 1-24	23.8	22.4	22.0	20.9	20.0	18.8
RNase	31.6	31.4	31.4	31.4	30.2	29.2
Lysozyme	39.8	40.4	40.8	40.6	40.2	39.2
BSA	41.6	42.8	44.0	44.2	44.4	44.4
Prolactin (rat)	51.0	55.6	57.6	59.6	60.4	59.8

[a] Compounds were eluted from a 7.5 cm × 4.6-m (i.d.) column packed
with the C3 10-μm, 30-nm pore-diameter support listed in Table 1, using
a linear gradient between 0.15 M NaCl (pH 2.1) and acetonitrile, of the
form illustrated in Fig. 1, at a constant flow rate of 1 ml/min.

30-nm pore-diameter packing (Table 2). The
most hydrophobic protein tested, rat prolactin, was in fact, eluted at a lower organic

modifier concentration at the lower temperatures. In contrast, the smaller, less hydrophobic compounds all showed lengthened
retention at low temperature, an effect shown
in several other studies (24,25). Differences
in the temperature-dependent behavior of
proteins can be used to facilitate their resolution, as illustrated in Fig. 2. Reduced temperature can also improve recoveries to some
extent; that of ovalbumin rose to about 20%
at 4°C (see Fig. 3) but was still far below the
recoveries obtained with an identical solvent
system (pH 2.1 acid saline/propan-1-ol) on
a corresponding 8-nm packing at 45°C (80%).

*Effect of solvent composition on protein
chromatography by RP-HPLC.* Although low-
molarity organic-based acid solvent systems
clearly give inferior efficiencies to high-mo-

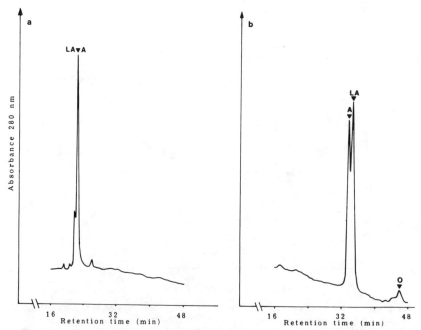

FIG. 3. The effect of temperature on selectivity, efficiency, and recovery of protein standards on RP-
HPLC. These chromatograms were obtained with a 75 × 4.6-mm (i.d.) column packed with a C3 capped
10-μm particle-size (30-nm pore-diameter) packing. In both cases proteins were eluted with a 0.155 M
NaCl (pH 2.1)/propan-1-ol gradient of the form illustrated in Fig. 1, at a flow rate of 1 ml/min. (a) The
temperature was maintained at 70°C when BSA (A) and bovine α-lactalbumin (LA) coeluted and no
peak corresponding to ovalbumin was seen. (b) At 4°C an injection of an identical protein mixture
resulted in partial separation of BSA and α-lactalbumin, and a detectable but small ovalbumin (O) peak.

larity salt-based systems at comparable pH (Fig. 1), they have been favored in some studies following the introduction of TFA in this role by Bennett *et al.* (26). Substitution of these low-molarity systems gave much improved recoveries of hydrophobic proteins such as ovalbumin from the large pore-size (30 nm) packings (Fig. 4). Losses of efficiencies compared to the high-molarity conditions (compare Figs. 1 and 4) were tolerable and permitted reasonable resolution of closely eluted materials, although some separations were lost (Fig. 4). Tests with standards showed that phosphorylase *B* was eluted as a sharp peak close to ovalbumin; myosin, however, gave no clear peak as either native material or its [^{14}C]methylated analog. Substitution of alternative salt-based buffers of higher molarity, such as the 0.2 M phosphate at pH 2.1 (23) for 0.155 M NaCl, decreased protein recoveries (Table 3). On the other hand, substitution of propan-1-ol for acetonitrile (15) improved hydrophobic protein recoveries in most cases (Table 3). Low-molarity systems based on strong acids (e.g., 0.01 M HCl/propan-1-ol) were equally effective as organic acids in the recovery of more hydrophobic proteins. Addition of perfluoroalkanoic acids to the secondary solvent used in conjunction with salt-based primary solvents did not, however, improve protein recoveries (e.g., ovalbumin). The effects of these and other solvent systems on recoveries of typical protein standards of varying hydrophobicities are detailed in Table 3.

As predicted (21), HFBA lengthened the retention times of most protein standards, an effect which might reduce the recovery of hydrophobic proteins. Nevertheless, it is apparent from Fig. 4 that some marked selective effects with proteins as well as peptides (27–29) could be obtained with this hydrophobic "ion-pairing" reagent with reversals in retention orders of, for example, α-lactalbumin and bovine serum albumin.

Effects of flow rates and gradient profiles on protein recoveries. Proteins diffuse relatively slowly compared to small peptides, and it has been shown (3) that low flow rates (0.1

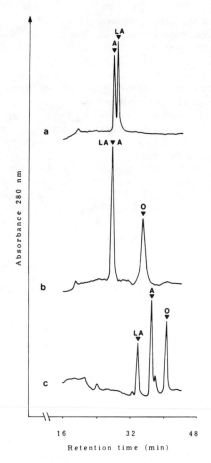

FIG. 4. The effect of primary solvent composition on selectivity, efficiency, and recovery of protein standards on RP-HPLC. Standard mixture of BSA (A), α-lactalbumin (LA), and ovalbumin (O) (25, 25, and 50 μg, respectively) was injected onto a 75 × 4.6-mm (i.d.) column packed with a C3 capped 10-μm particle-size (30-nm pore-diameter) support. Gradient elution with propan-1-ol was carried out at a flow rate of 1 ml/min at 45°C (see Fig. 1 for gradient profile). The primary solvent was (a) 0.155 M NaCl (pH 2.1), (b) 0.01 M TFA, and (c) 0.01 M HFBA; in the latter two the appropriate fluoroalkanoic acid was also added to the secondary solvent. Note that with isotonic acid saline no ovalbumin peak is seen at this loading, in contrast both to results with an 8-nm pore-diameter packing under identical conditions (see Fig. 1) and to the low-molarity fluoroalkanoic acids on the 30-nm pore-diameter packing. Note also lengthened retention times and changes in selectivity with HFBA.

O'HARE ET AL.

ml/min) can compensate for this effect and significantly improve efficiencies under isocratic conditions. When, however, effects of flow rates and gradient profiles on protein recoveries were examined with short alkyl-chain-type (C4) packings (Table 4), it was apparent that reducing flow rates markedly reduced recoveries of more hydrophobic proteins. Reducing gradient rate (expressed as change in composition per milliliter) also reduced recoveries.

At the lowest flow rates and lowest rates of gradient change overall protein recoveries were also at their lowest, with significant losses incurred even with less hydrophobic proteins such as BSA when acetonitrile was used as secondary solvent. The same trends were seen when a C3 packing was used and when 0.155 M NaCl (pH 2.1) and 0.01 M TFA were employed as the primary solvent and when propan-1-ol was used as secondary

solvent, although absolute recoveries of the less hydrophobic proteins were somewhat greater when the latter solvent was used. Recoveries of RNase, the least hydrophobic (and smallest) protein tested in this manner, were the least influenced by flow rate and rate of gradient change.

Effects of loading on protein recoveries. The effects of protein loadings on recovery were examined using conditions optimized for recovery of hydrophobic proteins (0.01 M TFA/propan-1-ol, 0.01 M TFA). The results are shown in Table 5. It is evident from this data that the recovery of BSA was reduced above loadings of about 50 μg and that of ovalbumin at even lower levels, whereas RNase recoveries were not significantly impaired at loadings up to and including 250 μg.

Separation of immunoglobulin by RP-HPLC. Conditions optimized for recoveries

TABLE 3

EFFECT OF ELUTING SOLVENT COMPOSITION ON PROTEIN RECOVERIES FROM LARGE PORE-SIZE (30 nm) RP PACKINGS[a]

Solvent system[b]		% Recovery[c]		
Primary	Secondary	RNase	BSA	Ovalbumin
NaCl[d]	Acetonitrile	110	88	<1
NaCl[d]	Propan-1-ol	87	85	23
Phosphate (pH 2.1)[e]	Acetonitrile	ND[g]	64	<1
Phosphate (pH 2.1)[e]	Propan-1-ol	ND	90	3
TFA[f]	Acetonitrile:TFA	100	99	86
TFA[f]	Propan-1-ol:TFA	97	98	96
HCl[f]	Acetonitrile:TFA	100	100	82
HCl[f]	Propan-1-ol:HCl	90	106	82
H₃PO₄[f]	Acetonitrile:H₃PO₄	97	97	61
H₃PO₄[f]	Propan-1-ol:H₃PO₄	98	92	86
TFA/NaCl[h]	Acetonitrile:TFA	95	88	2
TFA/NaCl[h]	Propan-1-ol:TFA	90	81	19

[a] Unless otherwise specified a 75 × 4.6-mm (i.d.) column with a C3 capped 10-μm (30-nm pore-diameter) packing was used.

[b] All recoveries were determined with a continuous linear gradient at 30°C and 1 ml/min. The rate of gradient change was 1.0%/ml with acetonitrile and 0.6%/ml with propan-1-ol.

[c] 10 μg each protein injected; duplicates agreed to better than ±10%.

[d] At 0.155 M (pH 2.1).

[e] Figures quoted for C4 capped packing. Concentration 0.2 M.

[f] At 0.01 M.

[g] Not determined.

[h] Concentrations 0.01 M/0.155 M.

TABLE 4

EFFECT OF FLOW RATE AND GRADIENT PROFILE ON
PROTEIN RECOVERIES FROM LARGE
PORE-SIZE RP PACKING[a]

Flow rate (ml/min)	Gradient change[b] (%/ml)	% Recovery		
		RNase	BSA	Ovalbumin
1.0	1.0	94	87	78
0.5	1.0	79	80	46
0.5	0.5	83	82	41
0.5	0.25	83	63	11
0.25	1.0	79	70	4

[a] C4 capped 10-μm particle size, 30-nm pore-diameter packing in 75 × 4.6-mm (i.d.) column.

[b] Linear gradient elution between aqueous 0.01 M TFA and acetonitrile with 0.01 M TFA at 30°C; 10 μg of each protein was injected.

of more hydrophobic proteins on large pore-size (30 nm) short-chain-type packings, as described above, were tested for their capability to resolve and recover naturally occurring mixtures of large relatively hydrophobic soluble protein molecules, as represented by IgG (156 kD). Results are illustrated in Fig. 5.

RP-HPLC of polyclonal IgG gave a broad peak (5–8 min) eluting 10 min after a BSA standard close to the phosphorylase *b* standard (Fig. 5c). A similar profile was seen when monoclonal IgG partially purified by ammonium sulfase precipitation was tested (Fig. 5b). Immunodiffusion studies confirmed that the broad peak contains monoclonal IgG as did the chromatography of ^{125}I-labeled IgG. Recovery of the IgG was greater than 53% as determined with ^{125}I-labeled IgG.

A practical application of these RP-HPLC methods can be seen in the separation of crude ascites fluid containing monoclonal antibody, illustrated in Fig. 5a. An early-eluting peak was obtained corresponding to serum albumin in retention time and molecular weight (~70,000) as determined by size-exclusion HPLC. A series of later eluting peaks >100 kD in size were also noted and two major peaks bracketed the area of the

chromatogram (33–36 min) where material which reacts with anti-IgG antiserum elutes.

DISCUSSION

In this and a previous (21) study of protein chromatography by RP-HPLC we have examined practical methods for optimizing the separation and recovery of compounds ranging from 10 to 200 kD in molecular mass. Specifically, the effects of different alkyl-chain lengths, pore diameters, and eluting solvent systems have been compared. Taking the current literature as a guide it seems that few, if any, problems of recovery are encountered when polypeptides of <10 kD are chromatographed using a variety of RP-HPLC packings based on 8- to 10-nm pore silica. Solvent systems of low pH and relatively high molarity do, however, result in high efficiencies and recoveries. Using these systems it is, for example, possible to chromatograph a hormone such as parathyrin (84 residues), recover it in high yield, and resolve it from a variety of other components and cogeners such as hormone fragments and oxidized and deamidated species in biological preparations, using a C18-type packing (30). For efficient chromatography and high recovery of larger proteins, however, alternative bonded phases are preferable (21). Nevertheless, the literature (1) reveals many

TABLE 5

EFFECT OF LOADING ON PROTEIN RECOVERIES[a]

Load (μg)	% Recovery		
	RNase	BSA	Ovalbumin
2.5	81	89	75
10	83	81	79
50	84	85	54
250	82	57	20

[a] Proteins were injected in aqueous solvent onto a 7.5 cm × 4.6-mm (i.d.) column with C3 capped 10-μm (30-nm pore-diameter) packing and eluted with 0.01 M TFA/propan-1-ol:TFA at 30°C and 1 ml/min with a gradient of 1%/ml. Recoveries were calculated by comparison with a tryptophan internal standard.

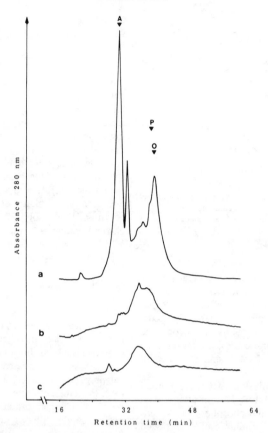

FIG. 5. Separation of immunoglobulins by RP-HPLC. The traces show uv-absorbance profiles of (a) 100 μl of ascites fluid containing 1 mg total protein from a mouse carrying a monoclonal antibody-producing hydridoma cell line (M8, IgG) compared with (b) a preparation of M8 partially purified by ammonium sulfate precipitation and (c) a preparation of polyclonal serum IgG. The elution positions of bovine serum albumin (A), phosphorylase *b* (P), and ovalbumin (O) are also shown. All separations were carried out on a 75 × 4.6-mm (i.d.) column packed with a C3 capped 10-μm (30-nm pore-diameter) support using gradient elution (see Fig. 1 for profile) between aqueous 0.01 M TFA and propan-1-ol with added TFA, at a flow rate of 1 ml/min and 45°C.

separations effected with small pore (6–10 nm) packings. Recent examples include β_2-microglobulin (12 kD) (31), calmodulin (16.5 kD) (18), human leukocyte interferon (18 kD) (16), prolactin and growth hormone (22 kD) (21), A-apolipoprotein (26 kD) (9), pro-opiocortin (30 kD) (17), and large (17–19 kD) pro-opiocortin fragments (13). Using 30- to 50-nm pore-diameter packings, RP separations have recently been extended to

tyrosinase (128 kD), cyanogen bromide collagen peptides (<60 kD) (6), and intact collagen (types I, II, and III 120 kD), with a variety of bonded phases (C18, C8, CN, and diphenyl).

Using ultrashort (C2–C4) alkylsilane-bonded silicas it is possible to chromatograph several proteins >100 kD, including γ-globulin (immunoglobulin) (Fig. 5). However, some differences have become evident in comparisons of small (8 nm) and large (30 nm) pore-diameter RP silicas with identical bonded RP surface chemistry (Fig. 1 and 4, Tables 2–5). Notably, the use of high-molarity (0.1–0.2 M) salt-based eluants resulted in marked losses with some proteins (e.g., ovalbumin) using the larger pore material. Increasing the ionic strength of the eluant is known to increase the strength of hydrophobic bonding (32). It may therefore seem self-evident that losses of more hydrophobic proteins will be incurred with the use of primary solvents such as 0.155 M (isotonic) NaCl. However, why such "salting-out" effects should occur to a much greater degree with large than with small pore-size packings is not immediately obvious (compare Figs. 1 and 4). Although simple low-molarity primary solvents (e.g., 0.01 M TFA or 0.01 M HCl) were more successful, it does not necessarily follow that such systems will be optimal for all larger proteins. The present study, however, does illustrate the differing behavior of various protein standards, notably in respect of questions of recoveries, a facet of protein RP chromatography that is often overlooked in purely chromatographic studies. The retention of a proportion of an injected protein and its partial elution on subsequent chromatograms as a "ghost" peak, which can occur under some circumstances, can, if unrecognized, potentially vitiate biological studies. In this context we have noted that low flow rates, while they may increase protein efficiencies (3), markedly reduce recoveries of larger and/or more hydrophobic materials (Table 4). Recovery is also decreased when shallow gradients are used.

The unexpected dependence of recovery on flow rate, gradient slope, sample mass, and temperature is at this time without a demonstrated explanation. Relevant information which permits reasonable speculation does, however, exist. Velicelebi and Sturtevant (35) have shown that the thermal denaturation of lysozyme in the presence of alcohols occurs below the temperatures used in this study. Also pertinent is the observation of Melander et al. (36) of irregular retention behavior of oligomeric polyethylene oxides which have different preferred conformations depending on solvent composition. If during an increasing gradient of organic modifier, the conformation of a protein is altered and consequently, its affinity for the stationary phase increased, ghost peaks and the anomalous dependence of recovery on flow rate, etc., might result. Solvent induced aggregation could be similarly involved.

Despite the successes achieved in this and other studies with RP chromatography of large proteins, it is clear that in its present form it will supplement rather than supplant separations by size-exclusion or ion-exchange chromatography. It would be of value, however, to be able to predict the behavior of proteins on this type of system. Considerable success has been achieved in predicting the retention of small peptides (<20 residues) (10,23,33,34) where conformation and sequence have only minor effects (37). Calculations of the net hydrophobicity of proteins (38) are of less predictive value (21) on account, presumably, of conformational effects and the presence of multiple hydrophobic domains, particularly in relatively inflexible proteins such as immunoglobulins (39). Calculations are often at variance with observations, even when "conventional" hydrophobic interaction systems based on alkyl- or phenyl-bonded Sepharose are used (40). Solubility in potential eluting solvent mixtures obviously dictates, to some degree, the extent to which methods of RP-HPLC can be applied to larger proteins. There are, however, also potentially complex "trade-offs" between efficiency as related to the composition of primary solvent and solubility in various secondary organic modifiers of different solvent strengths (34,41) and viscosities.

ACKNOWLEDGMENTS

We are very grateful to Dr. C. S. Foster for supplying us with mouse monoclonal antibody and for testing eluted immunoglobulin peaks by Ouchterlony gel diffusion.

REFERENCES

1. Regnier, F. E., and Gooding, K. M. (1980) *Anal. Biochem.* **103**, 1–25.
2. Lewis, R. V., Fallon, A., Stein, S., Gibson, K. D., and Udenfriend, S. (1980) *Anal. Biochem.* **104**, 153–159.
3. Jones, B. N., Lewis, R. V., Paabo, S., Kojima, K., Kimura, S., and Stein, S. (1980) *J. Liq. Chromatogr.* **3**, 1273–1383.
4. Fallon, A., Lewis, R. V., and Gibson, K. D. (1981) *Anal. Biochem.* **110**, 318–322.
5. Vale, W., Speiss, J., Rivier, C., and Rivier, J. (1981) *Anal. Biochem.* **113**, 1394–1397.
6. Van der Rest, M., Bennett, H. P. J., Solomon, S., and Glorieux, F. H. (1980) *Biochem. J.* **191**, 253–256.
7. Glasel, S. A. (1978) *J. Chromatogr.* **145**, 469–472.
8. Monch, W., and Dehnen, W. (1978) *J. Chromatogr.* **147**, 415–418.
9. Hancock, W. S., Bishop, C. A., Battersby, J. E., Harding, D. R. K., and Hearn, M. W. T. (1979) *J. Chromatogr.* **168**, 377–384.
10. O'Hare, M. J., and Nice, E. C. (1979) *J. Chromatogr.* **171**, 209–226.
11. Nice, E. C., Capp, M., and O'Hare, M. J. (1979) *J. Chromatogr.* **185**, 413–427.
12. Congote, L. F., Bennett, H. P. J., and Solomon, S. (1979) *Biochem. Biophys. Res. Commun.* **89**, 851–858.
13. Seidah, N. G., Routhier, R., Benjannet, S., Laviviere, N., Gossard, F., and Chretien, M. (1980) *J. Chromatogr.* **193**, 291–299.
14. Hancock, W. S., Capra, J. D., Bradley, W. A., and Sparrow, J. T. (1981) *J. Chromatogr.* **206**, 59–70.
15. Rubinstein, M. (1979) *Anal. Biochem.* **98**, 1–7.
16. Rubinstein, M., Rubinstein, S., Familletti, P. C., Miller, R. S., Waldman, A. A., and Pestka, S. (1979) *Proc. Nat. Acad. Sci. USA* **76**, 640–644.
17. Kimura, S., Lewis, R. V., Gerber, L. D., Brink, L., Rubinstein, M., Stein, S., and Udenfriend, S. (1979) *Proc. Nat. Acad. Sci. USA* **76**, 1756–1759.
18. Klee, C. B., Oldewurtel, M. D., Williams, J. F., and Lee, J. W. (1981) *Biochem. Int.* **2**, 485–493.
19. Rivier, J. (1978) *J. Liq. Chromatogr.* **1**, 343–366.
20. Asakawa, N., Tsuno, M., Hattori, T., Ueyama, M., Shindoa, A., and Miyake, Y. (1981) *J. Pharm. Soc. Japan* **101**, 279–282.
21. Nice, E. C., Capp, M. W., Cooke, N., and O'Hara, M. J. (1981) *J. Chromatogr.* **218**, 569–580.
22. Dottavio-Martin, D., and Ravel, J. M. (1978) *Anal. Biochem.* **87**, 562–565.
23. Molnar, I., and Horvath, C. (1977) *J. Chromatogr.* **142**, 623–640.
24. Hearn, M. T. W. (1980) *J. Liq. Chromatogr.* **3**, 1255–1276.
25. Barford, R. A., Sliwinski, B. J., and Rothbart, H. L. (1982) *J. Chromatogr.* **235**, 281–288.
26. Bennett, H. J. P., Hudson, A. M., McMartin, C., and Purdon, G. E. (1977) *Biochem. J.* **168**, 9–13.
27. Starratt, A. N., and Stevens, M. E. (1980) *J. Chromatogr.* **194**, 421–423.
28. Bennett, H. J. P., Browne, C. A., and Solomon, S. (1980) *J. Liq. Chromatogr.* **3**, 1353–1365.
29. Schaaper, W. M. M., Voskamp, D., and Olieman, C. (1980) *J. Chromatogr.* **195**, 181–186.
30. Zanelli, J. M., O'Hare, M. J., Nice, E. C., and Corran, P. H. (1981) *J. Chromatogr.* **223**, 59–67.
31. Alvarez, V. L., Roitsch, C. A., and Henriksen, O. (1981) in Immunological Methods (Lefkovits, I., and Pernis, B., eds.), Vol. 2, pp. 83–103, Academic Press, New York/London.
32. Hjerten, S. (1973) *J. Chromatogr.* **87**, 325–331.
33. Meek, J. L. (1980) *Proc. Nat. Acad. Sci. USA* **77**, 1632–1636.
34. Wilson, K. J., Honegger, A., Stotzel, R. P., and Hughes, G. J. (1981) *Biochem. J.* **199**, 31–41.
35. Velicelebi, G., and Sturtevant, J. M. (1979) *Biochemistry* **18**, 1180–1186.
36. Melander, W. R., Nahum, A., and Horvath, C. (1979) *J. Chromatogr.* **185**, 129–152.
37. Meek, J. L., and Rossetti, Z. L. (1981) *J. Chromatogr.* **211**, 15–28.
38. Bigelow, C. C. (1967) *J. Theor. Biol.* **16**, 187–211.
39. Williams, R. J. P. (1979) *Biol. Rev.* **54**, 389–437.
40. Creamer, L. K., and Matheson, A. R. (1981) *J. Chromatogr.* **210**, 105–111.
41. Mahoney, W. C., and Hermodson, M. H. (1980) *J. Biol. Chem.* **255**, 11,199–11,203.
42. Cooke, N. H. C., and Olsen, K. (1980) *J. Chromatogr. Sci.* **18**, 512–524.

High-Performance Liquid Chromatography Analysis in the Synthesis, Characterization, and Reactions of Neoglycopeptides

HARVARD MOREHEAD, PATRICK MCKAY, AND RONALD WETZEL[1]

Protein Biochemistry Department, Genentech, Inc., South San Francisco, California 94080

Reverse-phase HPLC was utilized to study the synthesis, properties, and reactions of the neoglycopeptides formed by reductive lactosylation of the lysine amino groups of a derivative of the immunostimulatory thymic polypeptide thymosin α_1. During the reaction of N^α-formyldesacetylthymosin α_1, a 28-amino acid polypeptide which contains four lysines, with lactose and sodium cyanoborohydride, over 40 intermediates and products were separated by HPLC and 14 of these partially characterized. Increasing levels of lactosylation reduced the elution time of the modified thymosin α_1 derivatives in several HPLC buffer systems investigated. Furthermore, comparisons of mobilities of intact glucosyl- or lactosyl-N^α-formyldesacetylthymosin α_1 showed that the main determinant of elution time was the total number of monosaccharide units per peptide molecule. HPLC analysis was also shown to be a useful tool for studying the susceptibility of neoglycopeptides to proteolytic degradation and will prove an aid to elucidation of the relative reaction rates of the individual thymosin α_1 lysine residues. It is expected that the trends in peptide mobility described will generally apply to the behavior of naturally occurring glycopeptides and glycoproteins in reverse-phase HPLC.

Despite the rapid development of polypeptide HPLC technology in the past several years and the rapidly expanding literature in the field, there have been very few reports on the HPLC behavior of glycosylated polypeptides. To our knowledge no one has addressed the question of the effect on peptide mobility of glycosylation, despite the obvious importance of such data to glycopeptide and glycoprotein analysis. One difficulty in this area is that of obtaining the same polypeptide in modified and unmodified forms. An initial solution to this problem is the study of chemically synthesized glycopeptides, or neoglycopeptides. Most natural enzymatic glycosylation of polypeptides occurs by modification of serine or asparagine residues (1). While chemically analogous reactions have not been developed, we felt that any reaction

which introduced a saccharide unit without introducing additional chemical moieties or altering peptide net charge would be a reasonable approximation of the natural chemistry.

Reductive alkylation is a commonly utilized method of introducing a stable covalent carbohydrate moiety into a polypeptide, by *in situ* reduction of the Schiff base formed by reaction of a reducing sugar with a lysine ε-amino group (2). Sodium cyanoborohydride is the reducing agent of choice due to its relatively wide range of pH stability and specificity for Schiff base reduction (3,4). In the past, reaction products have been characterized, if at all, in terms of mole carbohydrate per mole protein (5,6). Before applying this chemistry to proteins, we wanted to investigate in detail the course of such a reaction on a relatively simple substrate by a high-resolution technique such as HPLC. We chose as a substrate the N^α-formyl derivative of the thymic polypeptide desacetyl-

[1] To whom all correspondence should be addressed: Protein Biochemistry Department, Genentech, Inc., 460 Point San Bruno Blvd., South San Francisco, Calif. 94080

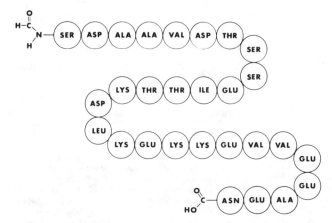

FIG. 1. Amino acid sequence of N^α-formyldesacetylthymosin α-1. Adapted from Goldstein *et al.* (14).

thymosin α_1 (Fig. 1), recently made available by recombinant DNA technology (7). Since the alpha amino group is formylated, only the four lysines of this polypeptide are susceptible to reaction.

In this paper we demonstrate the power of reverse-phase HPLC to separate the array of intermediates, products, and decomposition derivatives in the reaction using the disaccharide lactose. The effects of modification of a polypeptide with varying amounts of lactose are quantified using the algorithm of Meek and Rossetti (8). Finally, we show how HPLC can be used to study the proteolytic degradation of purified neoglycopeptides or glycosylation reaction mixtures, which in turn can be applied to simplifying the complex reaction kinetics of reductive alkylation of polypeptides.

MATERIALS AND METHODS

Materials

Desacetylthymosin α_1 was prepared by recombinant DNA techniques, and purified to chromatographic homogeneity (7). The formyl derivative was obtained as a side product of the cyanogen bromide reaction used to cleave the product from a precursor "fusion protein" (9). Sodium cyanoborohydride (re-

crystallized, Lot 32878) was from Alfa-Ventron (Danvers, Mass.). Fluorescamine (Lot 21F-0443) and D-lactose were from Sigma (St. Louis, Mo.). Spectrapor-3 dialysis tubing (M_r cutoff ~3500) was from Spectrum Medical Industries, Inc. (Los Angeles, Calif.).

The chromatography system consisted of two M-6000 pumps, an M-660 pump controller, a WISP 710B autoinjector (all from Waters Associates, Milford, Mass.), an LC-75 Variable Wavelength uv detector set at 210 nm (Perkin–Elmer Corp., Norwalk, Conn.), and a Kipp and Zonen (Delft, Holland) Model BD40 recorder. Chromatography was done on a 25-cm × 4.6-mm Altex Ultrasphere Octyl column with a 5-cm × 4.6-mm slurry-packed precolumn of the same material (Rainin Instrument Co., Woburn, Mass.).

Water for chromatography was double distilled in glass and further purified by passage over a Lobar C-8 column (E. Merck, Darmstadt, W. Germany). The uv grade acetonitrile was obtained from Burdick and Jackson Laboratories (Muskegon, Mich.).

In trifluoroacetic acid elution systems, both weak (10% acetonitrile) and strong (30% acetonitrile) solvents contained 0.1% trifluoroacetic acid (Sequanal Grade, Pierce Chemical Co., St. Louis, Mo.) and were titrated to an apparent pH 2.50 ± 0.05 with Ultrex grade ammonium hydroxide (J. T. Baker Chemical Co., Phillipsburg, N. J.).

In the phosphate system, the weak solvent consisted of 0.1 M monosodium phosphate (ACS Reagent grade, Mallinckrodt, Inc., St. Louis, Mo.) with 0.2% phosphoric acid (Purissimum grade, $\geq 98\%$, Fluka A. G., Buchs, Switzerland). The strong solvent was 0.1% phosphoric acid in acetonitrile.

The weak solvent in the perchlorate system was 0.1 M sodium perchlorate (ACS reagent grade, Mallinckrodt, Inc.) with 0.1% phosphoric acid. The strong solvent was 0.1% phosphoric acid in acetonitrile.

Methods

Reductive alkylation. For 10 days, 10 mg of N^α-formyldesacetylthymosin α_1, 40 mg of lactose, and 25 mg of sodium cyanoborohydride were incubated in 1.0 ml of 0.2 M potassium phosphate, pH 8.0, at 37°C. Then 100-μl aliquots were withdrawn at 24-h intervals to monitor the reaction. The samples were diluted to 1.0 ml with distilled water and dialyzed twice in Spectrapor-3 dialysis tubing at 4°C for 12 h against 1 liter of distilled water.

Determination of extent of lactosylation. The amount of lactose bound to the peptide was determined by a modification of the phenol/sulfuric acid method of Dubois *et al.* (10). Since the glucose moiety of lactose is reductively aminated in the product peptide as an acid-stable secondary amine, only galactose is detectible in the assay.

The number of unmodified lysines was determined by amino acid analysis, using an LKB Model 4400 Amino Acid Analyzer, with a Hewlett–Packard (Palo Alto, Calif.) Model 3390A integrator. Samples were hydrolyzed in evacuated, sealed vials for 24 h in 6 N hydrochloric acid at 110°C. The difference between the number thus obtained, and the four lysine residues present in the peptide, was taken as the number of modified lysines.

Unmodified lysines were also monitored by the fluorescamine reaction (11). Aliquots were added to 0.2 M potassium phosphate buffer, pH 8.0, with 750 μg of fluorescamine. Fluorescence was determined with a Perkin–Elmer Model 650-105 fluorescence spectrophotometer. Samples were excited at 390 nm, and emission monitored at 480 nm.

High-performance liquid chromatography. Reaction mixtures (Fig. 2) were chromatographed with trifluoroacetic acid buffers, using gradient curve 5 (convex) on the M660 controller, going from 5 to 25% acetonitrile in 40 min. All other analyses were performed using a linear 0 to 30% acetonitrile gradient in 40 min.

Products listed in Table 1 were collected from various HPLC separations such as those shown in Fig. 2. On reinjection, the products eluted at their characteristic retention times.

The peak eluting at 23.5 min was arbitrarily assigned the identification number L-1; later eluting major peaks were assigned progressively higher numbers. The prefix letter indicates the carbohydrate moiety (L, lactose; G, glucose).

RESULTS

The course of the reaction of N^α-formyldesacetylthymosin α_1 with lactose and sodium cyanoborohydride is shown in Figs. 2 and 3. Figure 2 shows that the starting material (expressed as percentage of total eluting peptide) is significantly reduced after 1 day and is essentially gone after 2 more days. In its place rise a series of peaks, all of which elute earlier in the acetonitrile gradient. After dramatic changes in the first few days the reaction slows down, although changes still occur in the HPLC profile at Day 10. The decline in rate at which the HPLC profile changes was not due to any significant decrease in concentration of cyanoborohydride. This was shown at Day 10 both by iodometric titration of reducing capacity (13) and by observing the rapid reaction of freshly added starting material (data not shown).

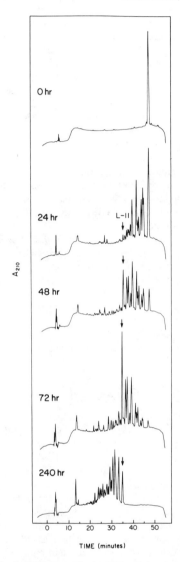

FIG. 2. Time course of lactosylation. Dialyzed aliquots (10–100 μl) of the reaction of N^α-formyldesacetylthymosin α_1 (10 mg/ml) with lactose (40 mg/ml) and sodium cyanoborohydride (25 mg/ml) at 37°C in 20 mM KH$_2$PO$_4$, adjusted to pH 8.0 with KOH were applied to an Altex Ultrasphere Octyl column and eluted with a convex gradient of 5 to 25% acetonitrile in 0.1% aqueous trifluoroacetic acid, pH 2.5. The arrow indicates the elution position of tetralactosylated product L-11.

Figure 3A shows the reaction course as measured by a number of assays conducted on the reaction mixture. One observes a decline in the number of available lysines measured either by fluorescamine assay or by amino acid analysis, and a parallel rise in the number of galactose (the nonreduced half of the disaccharide lactose) residues covalently attached to the polypeptide. Figure 3B shows the rapid decrease in starting material and the generation (as measured by HPLC) of the major reaction product, Peak L-11, tetralactosyl-N^α-formyldesacetylthymosin α_1. Figure 3B indicates that product formation is maximized at Day 6, after which further reactions give rise to the earlier-eluting peptides shown in Fig. 2. Amino acid analysis of peaks collected from various HPLC runs show that these late-appearing products (peaks L-1 through L-6) arise through cleavage of the polypeptide chain between residues 6 and 7.

Figure 4 shows the behavior of a mixture of some of the reaction products, described in Table 1, in several different buffer systems on the same HPLC column and with the same acetonitrile gradient. The trifluoroac-

TABLE 1

CHARACTERIZATION OF LACTOSYL PEPTIDES

Peptide[c]	Retention time (min)[a]	Modified lysines
L-1[b]	23.5	4
L-2[b]	26.0	3
L-3[b]	28.0	3
L-4[b]	30.0	3
L-5[b]	31.0	3
L-6[b]	31.5	4
L-7	34.5	3
L-8	35.0	4
L-9	36.0	4
L-10	37.0	4
L-11	39.0	4
L-12	41.0	3
L-13	42.0	2
L-14	44.0	2

[a] Retention times refer to Fig. 2.
[b] Fragmentation product (see text).
[c] For nomenclature, see Materials and Methods.

etic acid system (A) was used throughout this study because of the ease with which prep-collected samples can be desalted. The sodium perchlorate system (B) provides the best peak separation of the three systems tested, while acid phosphate (C) gives generally the least separation between neoglycopeptides in the same acetonitrile gradient, although it is capable of resolving the doublet of peptide L-14 which the trifluoroacetic acid system fails to resolve. In addition, peptides in the phosphate system are observed to elute earlier than in the trifluoroacetic acid and perchlorate systems.

The elution times of the lactosylated thymosin α_1 derivatives in Fig. 4 were used to calculate retention shifts for lactosyl lysine with respect to lysine, analogous to the re-

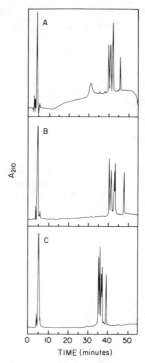

FIG. 4. Elution pattern of a mixture of N^α-formyl-desacetylthymosin α_1 derivatives in different chromatography systems. A mixture of tetra-, tri-, and dilactosylated derivatives (L-14, L-12, and L-11, respectively) and unmodified peptide were eluted with a linear gradient of 0 to 30% acetonitrile in 40 min on an Altex Ultrasphere Octyl column using an aqueous phase of (A) 0.1% trifluoroacetic acid adjusted to pH 2.5 with NH$_4$OH, (B) 0.1 M sodium perchlorate, 0.1% phosphoric acid, (C) 0.1 M monopotassium phosphate, 0.2% phosphoric acid. The broad band eluting at 31 min in (A) is an artifact. In each case the order of elution is: L-11, L-12, L-14, starting material. In systems B and C L-14 is resolved into a doublet (see Table 2).

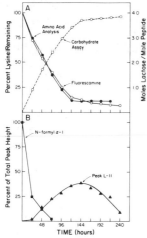

FIG. 3. Time course of the reaction described in Fig. 1 by following various parameters. (A) Loss of lysines was determined by amino acid analysis (Materials and Methods), as well as by measuring residual reactive amino groups using fluorescamine (11). Sugar incorporation was determined by a microanalysis for the galactose moiety of the covalently attached lactose (10). Before these determinations, each reaction time point was exhaustively dialyzed (Spectropor 3). (B) Loss of starting material and formation of the main product, peptide L-11, was determined by measuring relative peak heights in HPLC chromatograms like those shown in Fig. 2.

tention coefficient values in the algorithm of Meek and Rossetti (8). These values are shown in Table 2.

Figure 5 shows the relative contributions to peptide mobility of saccharide content and modified lysine content. In Fig. 5A, a sample of peptide L-13 (two lactosylated lysines) was co-injected with a sample of G-4 (four glucosylated lysines; synthesis and characteriza-

TABLE 2

VARIATION OF PEPTIDE RETENTION WITH DEGREE OF LACTOSYLATION

Peptide	Modified lysines	Retention shift per modified lysine (min)		
		Trifluoroacetate	Perchlorate	Phosphate
L-14	2	−1.8	−2.2, −2.4	−1.0, −1.1
L-12	3	−1.5	−2.2	−1.0
L-11	4	−1.5	−2.0	−1.0
Average shift		−1.6	−2.2	−1.0

tion are not described); the two peptides co-eluted. In Fig. 5B, a sample of peptide L-11 (four lactosylated lysines) was co-injected with a sample of G-4.

Figure 6 shows the effects of thermolysin digestion on N^α-formyldesacetylthymosin α_1 (a) and its tetralactosyl derivative (b).

DISCUSSION

HPLC analysis (Fig. 2) shows that the modification of N^α-formyldesacetylthymosin α_1 with lactose and cyanoborohydride is much more complex than indicated by analysis of the reaction mixture by less powerful methods (Fig. 3A). HPLC resolves several forms of a totally lactosylated product (peptides L-8, L-9, L-10, and L-11); the molecular differences between them have not been characterized. In addition, products L-1 through L-6 are derived from a specific polypeptide cleavage of N^α-formyldesacetylthymosin α_1 between residues 6 and 7. This cleavage was also observed in the absence of cyanoborohydride (McKay and Wetzel, unpublished results). HPLC analysis shows that product (L-11) formation is maximized at Day 6 (Fig. 3B).

Of the three acid/acetonitrile systems investigated, the perchlorate system (Fig. 4B) gives the best separation between thymosin α_1 molecules containing various amounts of lactose. Using the values of Meek and Rossetti (8), an elution time of 28 min was predicted for N^α-formyldesacetylthymosin α_1 compared to the experimental value of 48

min. Chromatography of several peptides analyzed by Meek and Rossetti indicated that most of the discrepancy between the determined and predicted mobilities of the thymosin α_1 derivative was due to the difference in the HPLC column used. A comparison of the retention times of the major di-, tri-, and tetralactosyl thymosins with the parent molecule revealed that, in all three mobile phases studied, the effect of increasing lactosylation was a reduction in retention time (Table 2, Figure 4). This effect is in accord with the model developed by Meek and Rossetti (8).

Not surprisingly, the effect on peptide mobility of the glycosylation of lysine de-

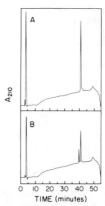

FIG. 5. Relative mobilities of several glycosylated N^α-formyldesacetylthymosin α_1 derivatives in the chromatographic system described in Fig. 4A. (A) Co-injection of peptide L-13 and peptide G-4, (B) co-injection of peptide L-11 and peptide G-4. See Results.

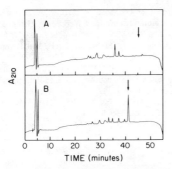

FIG. 6. Products of digestion of peptides with thermolysin at a 1:50 enzyme to substrate ratio for 16 h at RT in 1% ammonium bicarbonate, pH 8.3, buffer. Products were analyzed using the system described in Fig. 4A. (A) N^α-Formyldesacetylthymosin α_1, (B) the tetralactosylated derivative L-11. The arrows indicate the elution positions of the undigested peptide substrates.

pends on the size of the saccharide unit. The total number of monosaccharide units may be more important than their distribution in the neoglycopeptide. This is true for the thymosin α_1 derivatives in Fig. 5, which shows that two peptides containing the same total number of monosaccharide units are chromatographically indistinguishable, although one (L-13) contains two disaccharides and the other (G-4) four monosaccharides. By comparison, a thymosin α_1 with all four lysines modified with a disaccharide (L-11) elutes significantly ahead of the same peptide in which all four lysines are modified with a monosaccharide (G-4).

HPLC analysis showed that a peptide whose lysines are irreversibly derivatized with a disaccharide is resistant to thermolysin digestion compared to unmodified material (Fig. 6). Peptide L-11 was similarly shown to be resistant to trypsin digestion but of approximately unchanged susceptibility to *S. aureus* V8 (data not shown). Resistance of neoglycoproteins has been demonstrated previously by the monitoring of sensitivity of enzyme activity of glycosylated enzyme to proteolytic activity (5). HPLC systems such as those described here can be used to study

the effects of glycosylation on resistance of polypeptides to proteases at the molecular level. In addition, one should be able to elucidate the relative reactivities of the individual lysines of N^α-formyldesacetylthymosin α_1, and thus simplify the complex kinetic picture shown in Fig. 2, by analyzing proteolytic digests of reaction mixtures for relative amounts of proteolytic fragments with and without lactose using a proteolysis/HPLC mapping system such as that shown in Fig. 6. Proteolytic mapping may also help to characterize the various chromatographically distinct forms of tetralactosylated peptide isolated from the lactosylation reaction mixture (Table 1).

Reverse-phase HPLC has been shown to be a powerful tool for the analysis of reaction mixtures and isolated products in the chemical glycosylation of polypeptides. Neoglycopeptides formed as described here are similar in structure to glycoproteins formed by nonenzymatic glycosylation of proteins in the serum (12). The neoglycopeptide behavior described here should also be useful in HPLC analysis of enzymatically produced naturally occurring glycopeptides (1).

ACKNOWLEDGMENTS

The authors wish to express their gratitude to Maurice Woods for preparation of buffers and assistance with HPLC, Rodney Keck for amino acid analysis, Alane Gray and Mary McKay for artwork, and Jeanne Arch for preparation of this manuscript.

REFERENCES

1. Aplin, J. D., and Wriston, Jr., J. C. (1981) *CRC Crit. Rev. Biochem.* **10**, 259–306.
2. Stowell, C. P., and Lee, Y. C. (1980) *Advan. Carbohydr. Chem. Biochem.* **37**, 225–281.
3. Gray, G. R. (1974) *Arch. Biochem. Biophys.* **163**, 426–428.
4. Borch, R. F., Bernstein, M. D., and Durst, H. D. (1971) *J. Amer. Chem. Soc.* **93**, 2897–2904.
5. Marsh, J. W., Denis, J., and Wriston, J. C. (1977) *J. Biol. Chem.* **252**, 7678–7684.
6. Wilson, G. (1978) *J. Biol. Chem.* **253**, 2070–2072.

7. Wetzel, R., Heyneker, H. L., Goeddel, D. V., Jhurani, P., Shapiro, J., Crea, R., Low, T. L. K., McClure, J. E., Thurman, G. B., and Goldstein, A. L. (1980) *Biochemistry* **19,** 6096–6104.

8. Meek, J. L., and Rossetti, Z. L. (1981) *J. Chromatogr.* **211,** 15–28.

9. Wetzel, R., Ross, M. J., Levy, M. J., and Shively, J. E. submitted for publication.

10. Dubois, M., Gilles, K. A., Hamilton, J. K., Rebers, P. A., and Smith, F. (1956) *Anal. Chem.* **28,** 350–356.

11. Udenfriend, S., Stein, S., Boehlen, P., Dairman, W., Leimgruber, W., and Weigele, M. (1972) *Science* **178,** 871–872.

12. Koenig, R. J., Blobstein, S. H., and Cerami, A. (1977) *J. Biol. Chem.* **252,** 2992–2997.

13. Fischer, R. B. (1961) Quantitative Chemical Analysis, pp. 368–369, Saunders, Philadelphia.

14. Goldstein, A. L., Low, T. L. K., Thurman, G. B., Zaiz, M. M., Hall, N., Chen J., Hu, S.-K., Naylor, P. B., and McClure, J. E. (1981) *Rec. Prog. Hormone Res.* **37,** 369–412.

Use of Size-Exclusion High-Performance Liquid Chromatography for the Analysis of the Activation of Prothrombin[1,2]

DAVID P. KOSOW, SAM MORRIS, CAROLYN L. ORTHNER, AND MOO-JHONG RHEE

Plasma Derivatives Laboratory, American Red Cross Blood Services Laboratories,
9312 Old Georgetown Road, Bethesda, Maryland 20814

Prothrombin is a single-chain protein (M_r 72,000) which can be converted to the two-chain coagulation enzyme thrombin (M_r 37,000) by a protein present in the venom of *Echis carinatus*. During the course of this reaction, several intermediates and products are produced. We have found that size-exclusion high-performance liquid chromatography is a useful method for identifying the various intermediates as well as for determining the rates at which they are produced. The intermediates were identified by comparison with either authentic proteins or sodium dodecyl sulfate–polyacrylamide gel electrophoresis of duplicate samples. The separation of prothrombin, thrombin, and the other products was accomplished in less than 18 min using a TSK Type SW 3000 column equilibrated with 0.1 M Tris buffer, pH 7.2, containing 0.15 M NaCl. Estimates of the molecular weights of the products determined from calibration curves did not necessarily agree with actual values. Meizothrombin, the two-chain active form of prothrombin, has a longer elution time than prothrombin, although both molecules have the same molecular weight. Ca(II) increased the elution time of prothrombin and Fragment 1. The rate of disappearance of prothrombin as determined by peak-height analysis agreed with the rate of formation of active enzyme as determined by active site titration. Thus, high-performance liquid chromatography is a rapid analytical tool for the study of the proteolytic modification of proteins.

Prothrombin is a single-chain zymogen (M_r 72,000) which can be converted to the two-chain coagulation enzyme thrombin (M_r 37,000) by proteolysis. Our laboratory has been studying the mechanism of activation of human prothrombin by the venom coagulant enzyme of *Echis carinatus* (ECV-P).[3] During the course of this reaction several intermediates and products are formed. Pre-vious investigations have used SDS–polyacrylamide gel electrophoresis as the major means of identifying prothrombin, its activation fragments, intermediates, and products (1–4). We have found that high-performance gel-permeation chromatography (HP-GPC) is a useful alternative to SDS–gel electrophoresis. An attractive feature of HP-GPC is that the proteins are separated under nondenaturing conditions, allowing the study of protein–protein and protein–ligand interactions as well as the detection of enzymatic proteolysis.

TERMINOLOGY

The main regions of prothrombin are illustrated in Fig. 1.

The nomenclature used for prothrombin and its activation fragments, intermediates, and products is that recommended by the

[1] This paper was presented at the International Symposium on HPLC of Proteins and Peptides, November 16–17, 1981, Washington, D. C.

[2] Contribution No. 542 from the American Red Cross Blood Services Laboratories. Supported in part by National Institutes of Health Grant HL 19282 and Biomedical Research Support Grant 5 SO7 RR05737.

[3] Abbreviations used: ECV-P, the prothrombin activator protein from *E. carinatus* venom; SDS, sodium dodecyl sulfate; HP-GPC, high-performance gel-permeation chromatography; DAPA, dansyl arginine-*N*-(3-ethyl-1,5-pentanediyl)amide.

KOSOW ET AL.

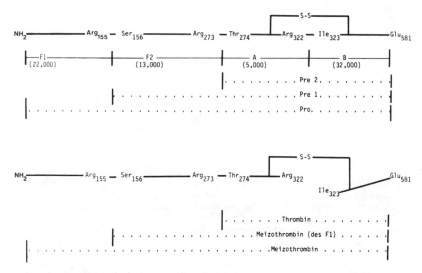

FIG. 1. Schematic representation and nomenclature of prothrombin, regions and cleavage products. F1, Fragment 1; F2, Fragment 2; A, thrombin A chain; B, thrombin B chain; Pre 2, prethrombin 2; Pre 1, prethrombin 1; and Pro, prothrombin.

International Committee on Thrombosis and Haemostasis (5). Fragment 1 is the peptide that arises from the amino terminal region of prothrombin by cleavage of the bond between Arg_{155}–Ser_{156}. The remainder of the molecule is called prethrombin 1. Further cleavage at Arg_{273}–Thr_{274} releases Fragment 2 and prethrombin 2. If this cleavage occurs in the absence of the cleavage at Arg_{155}, then Fragment 1 · 2 is released. Cleavage at Arg_{322}–Ile_{323} does not release a peptide but produces a two-chain molecule. If Fragments 1 and 2 are also released, thrombin is produced. Thrombin is composed of an A chain (M_r 5000) and a B chain (M_r 32,000). If cleavage at Arg_{322} occurs in the absence of any other cleavage, meizothrombin is the product; if Fragment 1 is cleaved but no cleavage of Fragment 2 occurs, the two-chain molecule is meizothrombin (des F1). The two-chain molecules have esterolytic activity, whereas the single-chain molecules are enzymatically inactive.

MATERIALS AND METHODS

Fresh frozen human cryosupernatant plasma was supplied by the Baltimore Regional Blood Service–American Red Cross. Crude lyophilized *E. carinatus* venom was obtained from the Miami Serpentarium. TSK Type SW 3000 gel-permeation columns were obtained from Toyo Soda Manufacturing Company. Cohn fraction III was kindly supplied by the Michigan Department of Health. Dr. Edward E. Somers, Reilly Tar and Chemical Corporation, provided us with a gift of 4-ethylpiperidine.

Human prothrombin was prepared from cryosupernatant plasma by adsorption and elution from DEAE–Sephadex followed by chromatography on sulfated dextran by the method of Miletich *et al.* (6). Human α-thrombin, prethrombin 2, and Fragment 2 were prepared from Cohn fraction III as described by Orthner and Kosow (7). Fragment 1 and prethrombin 1 were prepared by the method of Owen *et al.* (8). ECV-P was purified by the method of Rhee *et al.* (9). Dansyl arginine–*N*-(3-ethyl-1,5-pentanediyl)amide (DAPA) was synthesized by the method of Nesheim *et al.* (10).

HP-GPC analysis of prothrombin and its derivatives was performed on a 60-cm TSK SW 3000 gel-permeation column. The column was equilibrated and the proteins eluted at a flow rate of 1 ml/min with 0.1 M Tris–HCl, pH 7.2, containing 0.15 M NaCl using a Waters Associates Model 6000 delivery system. A Waters Associates Model 440 wavelength absorbance detector equipped with a 280-nm filter was used to detect the protein peaks.

Protein samples were electrophoresed in the reduced or unreduced state according to the procedure of Weber and Osborn (11). Bovine serum albumin, ovalbumin, chymotrypsin, and ribonuclease were used as molecular weight markers. Prothrombin, prethrombin 1, prothrombin Fragment 1,

and thrombin were also used as standards to identify the products of prothrombin activation. The protein bands were visualized by staining with Coomassie blue (11).

The samples were prepared by removing aliquots of the reaction mixtures at the specified time intervals and adding them to an equal volume of 8 M urea, 1% SDS, 0.02 M EDTA, and 0.01 M NaPO$_4$, pH 7.0. The mixture was immediately placed in a boiling-water bath for 5 min. For the reduced samples, 1% β-mercaptoethanol was then added. An aliquot containing about 30 μg of protein was applied to 10% acrylamide gels. Molecular weights of the activated products were determined by comparison with protein standards.

The formation of meizothrombin was measured by a modification of the DAPA titration method of Nesheim (10). The reaction mixture contained prothrombin, 10 μM; DAPA, 10 μM; NaCl, 0.15 M; Tris–HCl, pH 7.4, 0.02 M in a final volume of 0.25 ml.

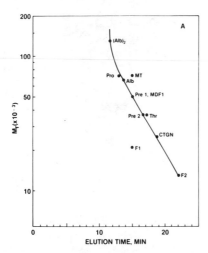
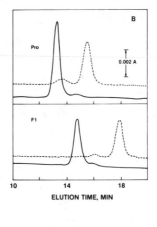

FIG. 2. (A) The relationship between elution time and log (molecular weight of protein). (B) Elution diagrams for prothrombin and Fragment 1 in the absence (——) and presence (– – –) of 2 mM Ca(II). Alb, albumin; (Alb)$_2$, albumin dimer; MT, meizothrombin; MDF1, meizothrombin (des F1); and CTGN, chymotrypsinogen. Other abbreviations are as in Fig. 1. Conditions are as described in the text. Between 3 and 10 μg of protein was injected.

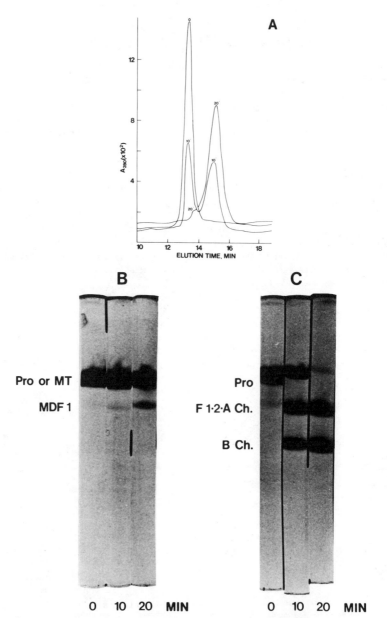

FIG. 3. Time course of prothrombin activation by ECV-P in the presence of DAPA. Prothrombin, 30 μM; DAPA, 450 μM; NaCl, 150 mM; Tris–HCl, pH 7.4, 20 mM; and ECV-P, 0.2 μM were incubated at room temperature. At the indicated times, 10 μl of the reaction mixture was (A) injected into the TSK 3000 column or (B,C) analyzed by SDS–polyacrylamide gel electrophoresis as described in the text. Electrophoresis was performed in the (B) absence or (C) presence of 1% β-mercaptoethanol.

The reaction was initiated by the addition of ECV-P to a final concentration of 0.07 μM. The change in fluorescence intensity due to the interaction of DAPA with the active site of meizothrombin was measured at 545 nm with excitation at 350 nm using a Perkin–Elmer Model 44B spectrofluorimeter.

RESULTS

The TSK Type SW 3000 gel-permeation column was evaluated by individually injecting 3–4 μg of chymotrypsinogen, albumin, and albumin dimer. As shown in Fig. 2A, there is a linear relationship between the logarithm of the molecular weight and the elution time for molecular weights between 10,000 and 70,000. The elution time of prothrombin and several of its intermediates is also shown in Fig. 2A. Prothrombin, prethrombin 1, prethrombin 2, and Fragment 2 elute at their expected elution times. The discrepancy between the actual molecular weight of Fragment 1 (22,000) and the molecular weight of 50,000 calculated from its elution time of 15 min may be due to formation of a dimer. Although the formation of a Fragment 1 dimer in the absence of Ca(II) has not been reported, both prothrombin and Fragment 1 are known to self-associate in the presence of Ca(II) (12,13). Therefore, we decided to investigate the effect of Ca(II) on the elution time of prothrombin and its intermediates. Prothrombin and Fragment 1 are retarded in the presence of Ca(II) (Fig. 2B), while the elution times of albumin, prethrombin 1, prethrombin 2, thrombin, chymotrypsinogen, and Fragment 2 were not altered by the presence of Ca(II). Fragment 1 has been reported to exist as a monomer in solution in the absence of Ca(II) (14) and a dimer in the presence of Ca(II) (12,13). Thus, the shorter elution time of Fragment 1 in the absence of Ca(II) than in its presence may be due to factors other than self-association.

When prothrombin is converted to an ac-

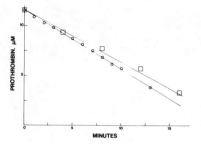

FIG. 4. Kinetics of prothrombin disappearance catalyzed by ECV-P in the presence of DAPA. Prothrombin, 10 μM; DAPA, 10 μM; NaCl, 150 mM; Tris–HCl, pH 7.4, 20 mM; and ECV-P, 0.07 μM were incubated at room temperature. At the indicated times the amount of prothrombin remaining was analyzed by (□) HP–GPC or by (○) fluorescence as described in the text.

tive protease by ECV-P in the presence of the thrombin inhibitor DAPA, the only bond cleaved is the Arg_{322}–Ile_{323} bond (9,15). This produces meizothrombin, which has the same molecular weight as prothrombin. When the reaction is analyzed by HP–GPC using a TSK 3000 column, one can detect the appearance of a new protein peak with an elution time corresponding to a molecular weight of about 50,000. This peak increases as the prothrombin peak disappears (Fig. 3A). When the reaction is analyzed by SDS–polyacrylamide gel electrophoresis performed in the presence or absence of β-mercaptoethanol, there is no indication of any product other than meizothrombin being formed after 20 min except for a trace amount of prethrombin 1 (Figs. 3B and C). By measuring the height of the prothrombin peak at various times after the addition of ECV-P, the rate of disappearance of prothrombin can be measured. When graphed as a zero-order plot, a zero-order rate of 5.1 min^{-1} was obtained (Fig. 4). When the rate of disappearance of prothrombin is measured by titrating thrombinlike active sites with DAPA as described under Materials and Methods, a similar rate of 5.8 min^{-1} is observed (Fig. 4). When ECV-P and prothrom-

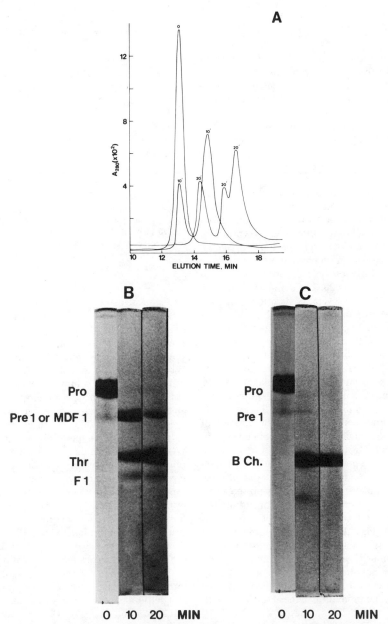

FIG. 5. Time course of prothrombin activation by ECV-P. Conditions were the same as described for Fig. 3 except no DAPA was present. The reaction was analyzed by (A) HP–GPC or (B,C) SDS–polyacrylamide gel electrophoresis. Electrophoresis was performed in the (B) absence or (C) presence of β-mercaptoethanol.

bin are incubated in the absence of DAPA, several reaction products are formed. Analysis of the reaction by HP-GPC shows that at 10 min only the peak corresponding to an apparent molecular weight of 50,000 appears (Fig. 5A), whereas SDS electrophoresis indicates that not only meizothrombin but also meizothrombin (des F1), Fragment 1, and prethrombin 1 are produced under these conditions (Figs. 5B and C). Because Fragment 1 elutes with meizothrombin and meizothrombin (des F1) from the TSK 3000 column (Fig. 2), the second peak is a mixture of these proteins.

DISCUSSION

The proteolysis of prothrombin by ECV-P is a complex reaction in which several products are formed. Analysis of the reaction by use of chromogenic substrates is complicated by the fact that not all of the products have esterolytic activity. The fluorogenic inhibitor DAPA provides a means of measuring the initial rate of the reaction but, by inhibiting the enzymatically active products, does not allow the investigation of the side reactions. Analysis by SDS–polyacrylamide gel electrophoresis allows the determination of the identity of the products of the reaction, but it is difficult to quantitate and is a slow procedure. The availability of HP-GPC offered a potential means of detecting and quantitating the products of the activation of prothrombin by ECV-P. Our initial investigations on the analysis of prothrombin and prothrombin activation products and intermediates indicated that, unlike the proteins used as molecular weight standards, the retention time of these proteins and peptides did not always correlate with their molecular weight. Fragment 1 elutes as a higher molecular weight protein in the absence of Ca(II), but in the presence of Ca(II) its elution time is that of a protein of its monomeric molecular weight of 22,000. Because Fragment 1 has been shown to be monomeric in the absence of Ca(II) at much higher protein concentrations than used here and to exist as a dimer in the presence of Ca(II) (12,13), these results indicate that the elution times are not solely due to molecular weight. Prothrombin and Fragment 1 have been shown to undergo Ca(II)-induced conformational changes by fluorescence (16,17), circular dichroism (18,19), and immunological methods (20,21). It is possible that these conformational changes are responsible for the longer elution times observed in the presence of Ca(II).

That factors other than molecular weight determine the elution time from the TSK 3000 column proved to be an advantage in analyzing the activation of prothrombin by ECV-P since one of the products, meizothrombin, has the same molecular weight as prothrombin but has a longer elution time. This enabled us to use the rate of disappearance of prothrombin as a measure of the reaction rate. In the presence of DAPA the reaction is zero order (Fig. 4C) because prothrombin was present at 30 μM, which is about 70-fold greater than its K_m of 0.45 μM (9). When the activation of prothrombin by ECV-P is carried out in the absence of DAPA, several products are formed that do not separate from each other on the TSK 3000 column. When we can define conditions for the separation of prethrombin 1, meizothrombin, and meizothrombin (des F1) from each other, it will be possible to determine the rate of formation of each product as well as the rate of disappearance of prothrombin.

REFERENCES

1. Franza, B. R., Aronson, D. L., and Finlayson, J. S. (1975) *J. Biol. Chem.* **250**, 7057–7068.
2. Morita, T., and Iwanaga, S. (1978) *J. Biochem.* **83**, 559–570.
3. Morita, T., Iwanaga, S., and Suzuki, T. (1976) *J. Biochem.* **79**, 1089–1108.
4. Kornalik, F., and Blomback, B. (1975) *Thromb. Res.* **6**, 53–63.
5. Jackson, C. M. (1977) *Thromb. Haem.* **38**, 567–577.

6. Miletich, J. P., Broze, G. J., and Majerus, P. W. (1980) *Anal. Biochem.* **105**, 304–310.

7. Orthner, C. L., and Koscow, D. P. (1980) *Arch. Biochem. Biophys.* **202**, 63–75.

8. Owen, W. G., Esmon, C. T., and Jackson, C. M. (1974) *J. Biol. Chem.* **249**, 594–605.

9. Rhee, M.-J., Morris, S., and Kosow, D. P., (1982) *Biochemistry* **21**, 3437–3443.

10. Nesheim, M. E., Prendergast, F. G., and Mann, K. G. (1979) *Biochemistry* **18**, 996–1003.

11. Weber, K., and Osborn, M. (1969) *J. Biol. Chem.* **244**, 4406–4412.

12. Pletcher, C. H., Resnick, R. M., Wei, G. J., Bloomfield, V. A., and Nelsesteun, G. L. (1980) *J. Biol. Chem.* **255**, 7433–7438.

13. Prendergast, F. G., and Mann, K. G. (1977) *J. Biol. Chem.* **252**, 840–850.

14. Heldebrandt, C. M., and Mann, K. G. (1973) *J. Biol. Chem.* **248**, 3642–3652.

15. Briet, E., Griffith, M. J., Soule, P. B., Jordan, S. H., Braunstein, K. M., and Roberts, H. R. (1981) *Thromb. Haem.* **46**, 122.

16. Nelsestuen, G. L. (1976) *J. Biol. Chem.* **251**, 5648–5656.

17. Marsh, H. W., Scott, M. E., Hiskey, R. G., and Koehler, K. A. (1979) *Biochem. J.* **183**, 513–517.

18. Bloom, J. W., and Mann, K. G. (1978) *Biochemistry* **17**, 4430–4438.

19. March, H. C., Robertson, P., Scott, M. E., Koehler, K. A., and Hiskey, R. G. (1979) *J. Biol. Chem.* **254**, 10,268–10,275.

20. Furie, B., and Furie, B. C. (1979) *J. Biol. Chem.* **254**, 9760–9771.

21. Madar, D. A., Hall, T. J., Reisner, H. M., Hiskey, R. G., and Koehler, K. A. (1981) *J. Biol. Chem.* **255**, 8599–8605.

High-Performance Liquid Chromatography of Proteins: Purification of the Acidic Isozyme of Adenylosuccinate Synthetase from Rat Liver[1]

FREDERICK B. RUDOLPH AND SANDRA W. CLARK

Department of Biochemistry, Rice University, Houston, Texas 77001

High-performance ion-exchange liquid chromatography was utilized for the purification of the acidic isozyme of adenylosuccinate synthetase from rat liver. Initial steps in the purification included ammonium sulfate fractionation and DEAE-cellulose and agarose–GTP affinity columns. The final steps were done on a SynChropak AX-300 anion-exchange support. The enzyme was purified 3000-fold with an overall yield of 10%. The enzyme preparation exhibited only one protein band on gel electrophoresis.

Adenylosuccinate synthetase[IMP:L-aspartate ligase(GDP-forming), EC 6.3.4.4] catalyzes the reaction

$$IMP + Aspartate + MgGTP^{2-} \rightarrow$$

$$Adenylosuccinate + MgGDP^{1-} + P_i .$$

It has been purified to homogeneity from rat skeletal muscle (1,2), rabbit heart and skeletal muscle (3), *Azotobacter vinelandii* (4), and Yoshida sarcoma ascites tumor cells (5). In addition, many studies have been performed with impure preparations from various sources. It has been suggested that the enzyme has two physiological roles: one in adenine nucleotide biosynthesis and one in the purine nucleotide cycle (1,2,5–8). Consistent with the dual role, two isozymes of the synthetase exist in normal rat tissues (2,7,8). The first isozyme, designated Type L by Matsuda *et al.* (7), is an acidic protein and is likely involved in adenine formation (2,7,8). The second isozyme, designated Type M (7), is a basic protein and is associated with the purine nucleotide cycle (2,7,8). The basic isozyme has been purified to homogeneity from rat and rabbit tissues (1–3). The

acidic form is very labile to dilute conditions and other factors and has been very difficult to purify. One report of a reasonably high-specific-activity preparation has appeared, but the purity of the enzyme was not discussed (9).

Recent developments in microparticulate column materials have allowed advances in chromatography of proteins (10,11). An anion-exchange support of polyethyleneimine coated to and crosslinked on the surface of silica has been developed (10) and is commercially available as SynChropak AX-300. This column support has been utilized in the final steps of a purification of the acidic isozyme of adenylosuccinate synthetase.

MATERIALS AND METHODS

Hepes,[2] L-aspartate, dithiothreitol, IMP, and GTP were purchased from Sigma. Bio-Gel P-2 was from Bio-Rad. DEAE-cellulose (DE-52) was obtained from Whatman. GTP–agarose (GTP linked to polyacrylic hydrazide agarose through the 2'- and 3'-ribose hydroxyl groups) was obtained from PL Laboratories. SynChropak AX-300, SynChrosorb AX, and

---

[1] This paper was presented at the International Symposium on HPLC, of Proteins and Peptides, November 16–17, 1981, Washington, D. C.

[2] Abbreviations used: Hepes, 4-(2-hydroxyethyl)-1-piperazineethanesulfonic acid; SDS, sodium dodecyl sulfate.

SynChropak ASC were purchased from SynChrom. Male rats (Sprague–Dawley derived) were from Timco Breeding Laboratories.

Adenylosuccinate synthetase activity was determined by observing the increase in absorbance at 280 nm accompanying the conversion of IMP to adenylosuccinate. The assays were carried out in a 1.0-ml total volume containing 20 mM Hepes, pH 7.0, 5 mM magnesium acetate, 0.06 mM GTP, 5 mM L-aspartate, and 0.15 mM IMP. The reactions were followed at 30°C in a Cary Model 118 spectrophotometer. One unit (U) of enzyme is defined as the amount that forms 1 nmol of adenylosuccinate/min ($\delta\epsilon = 11.7 \times 10^6$ cm^2/mol).

The protein concentration was determined either by use of the biuret method with bovine serum albumin as a standard or by uv absorption (3). Specific activity is expressed as units of enzyme activity per milligram of protein. The purity of the purified protein was determined by SDS–polyacrylamide gel electrophoresis (12).

High-performance liquid chromatography was performed with a Glenco high-pressure liquid chromatograph equipped with a 5-ml sample loop on the injector. The Syn-Chropak AX-300 anion-exchange column (0.46×25 cm) was packed with a Micromeretics Model 705 slurry column packer using isopropanol as solvent. A solvent conditioning column (0.46×25 cm) was dry-packed using SynChrosorb ASC and placed before the injector. A postinjector guard column (0.46×5 cm) was dry-packed with SynChropak AX. The entire column system was extensively washed with 10 mM potassium phosphate, pH 7.5, containing 1 mM EDTA and 1 mM dithiothreitol and then with the same buffer containing 0.4 M KCl. After this conditioning step it was reequilibrated in the first buffer. The absorbance of the column effluent was monitored continuously at 280 nm and 1-ml fractions were collected.

PURIFICATION METHOD

The acidic isozyme of adenylosuccinate synthetase was purified from rat liver by the following steps. All steps were done at 4°C except the AX-300 chromatography, which was done at room temperature.

1. Homogenization. Rats (typically 300–400 g) were decapitated and the livers removed and cooled on ice. The livers (300 g) were then minced in a blender (2–3 s) in 3.5 vol (1050 ml) of 0.04 M potassium phosphate, pH 7.5, containing 0.05 M potassium chloride, 0.1 M sucrose, 1 mM EDTA, and 1 mM dithiothreitol. The minced liver suspension was then homogenized twice in an 80-ml Potter–Elvehjem homogenizer. The homogenate was spun at 19,000g for 1 h, decanted, and the supernatant spun at 95,000g for 1 h.

2. Ammonium sulfate fractionation. The final supernatant from the homogenization was slowly brought with gentle stirring to 45% of saturation (258 g/liter) with solid ammonium sulfate. After stirring for 30 min, the solution was centrifuged for 30 min at 19,000g. The precipitate was discarded and the supernatant brought to 60% (90 g/liter) of saturation with ammonium sulfate. After stirring and centrifugation as above, the precipitate was collected and dissolved in a minimal volume of 10 mM potassium phosphate, pH 7.5, with 1 mM EDTA and 1 mM dithiothreitol (standard buffer). The enzyme solution was then desalted on a Bio-Gel P-2 column (500 ml bed volume) equilibrated in the standard buffer. Complete removal of the ammonium sulfate was assured by testing fractions with Nessler's reagent. The active fractions were pooled.

3. DEAE-cellulose fractionation. The basic isozyme of adenylosuccinate synthetase does not bind to DEAE-cellulose at neutral pH. The property allows quantitative separation of the two isozymes at this step. The pooled fractions from the P-2 column above were applied to a DEAE-cellulose column

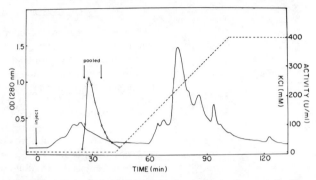

FIG. 1. Chromatograph of the acidic isozyme of adenylosuccinate synthetase on SynChropak AX-300. A 61-mg protein sample was injected onto the column, which was equilibrated with 10 mM potassium phosphate, pH 7.5, with 1 mM EDTA and 1 mM dithiothreitol. The flow rate was 0.7 ml/min and the pressure was 400 psi. A linear gradient between the equilibration buffer and 0.4 M KCl in the same buffer was run as indicated by the dashed line. The active fractions were combined as indicated by the arrows. Enzyme activity is indicated by (O) and the solid line indicates absorbance at 280 nm.

(6.3 × 9.5 cm, 300 ml bed volume) equilibrated in the standard buffer (flow rate of 2 ml/min, fraction size 5 ml). The column was eluted with the standard buffer until the 280-nm absorbance was below 0.05. The basic isozyme was in this fraction. The acidic isozyme was then eluted with 0.3 M KCl in the standard buffer. Active fractions were pooled and precipitated by dialysis against saturated ammonium sulfate containing the standard buffer. The precipitate was collected by centrifugation at 20,000g and then suspended in 0.03 M potassium phosphate, pH 7.5, with 0.25 M sucrose, 5 mM EDTA, and 1 mM dithiothreitol (sucrose buffer) and desalted on a Bio-Gel P-2 column (50 ml) equilibrated in the sucrose buffer.

4. GTP-affinity step. The agarose–GTP column (16 ml bed volume) was equilibrated in the sucrose buffer. The pooled fractions (8 ml) from the P-2 column above were applied to the column and it was eluted with the sucrose buffer until the 280-nm absorbance was below 0.05. The enzyme was then eluted stepwise with 1.0 M KCl in the sucrose buffer. It was found that this column had a capacity for about 500 mg of protein at this stage of purification. The pooled active fractions were precipitated with saturated ammonium sulfate as described in step 3. The centrifuged pellet was suspended in a minimal volume (0.5 ml) of the standard buffer and desalted on a Bio-Gel P-2 column (12 ml bed volume) equilibrated in the standard buffer.

5. SynChropak AX-300 chromatography. The SynChropak AX-300 column was equilibrated with the standard buffer and the sample (61 mg of protein) applied through the injection valve. The separations were performed at room temperature under the conditions described in Fig. 1. The enzyme elutes isocratically with this buffer as shown in Fig. 1. The gradient to 0.4 M KCl was run to illustrate the potential resolving power of the column even with this rather large sample. The pooled active fractions were precipitated and transferred to standard buffer as in step 4. This fraction was reapplied to the SynChropak AX-300 column under the same conditions listed in Fig. 1. The elution profile is shown in Fig. 2. The active fractions were pooled and concentrated as described above.

The results of the purification are summarized in Table 1. The final fractions from

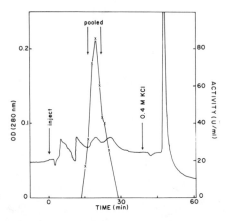

FIG. 2. A rechromatograph of the pooled material from the separation described in Fig. 1. The conditions were the same as for Fig. 1 except that a step gradient to 0.4 M KCl in buffer was used as indicated by the arrow.

the second AX-300 step exhibit one major protein band on SDS–gel electrophoresis compared to four bands prior to that step. A band representing less than 1–2% by densitometry was present. The kinetic constants and molecular weight of the protein are similar to those described previously (13). The enzyme is stable for up to 8 months when stored under nitrogen at −20°C as an ammonium sulfate precipitate.

DISCUSSION

The acidic isozyme of adenylosuccinate synthetase has not been previously purified to near homogeneity. An acidic enzyme has been purified from Yoshida sarcoma ascites tumor cells (5), but its properties are somewhat different from the enzyme from normal tissues. Fukutome has described purification of the acidic isozyme from rat liver to a specific activity of 1217, but purity of the sample was not described (13). The enzyme has proven to be unstable in dilute solutions and generally difficult to purify. The present purification scheme utilized step gradients on the DEAE-cellulose and GTP–agarose columns to avoid dilution. The AX-300 column allows a rapid separation under isocratic conditions. The column steps described in this report all give better than 50% recoveries of applied activity, and greater than 80% of the applied activity can be routinely recovered from the AX-300 column. Generally, the enzyme loses activity during concentration and equilibration for the next step in the purification. Use of Bio-Gel P-2 columns for buffer changes have afforded the greatest recoveries of a number of procedures that have been used.

It is interesting to note that a considerable amount of the protein applied does not elute

TABLE 1

PURIFICATION OF THE ACIDIC ISOZYME OF ADENYLOSUCCINATE SYNTHETASE FROM RAT LIVER

Fraction	Protein (mg)	Specific activity (U/mg protein)	Yield (%)
Ammonium sulfate fraction[a] (45–60% fraction)	4100	1.9[b]	—
DEAE-cellulose fraction	850	5.0	100
GTP–agarose affinity column	61	36.0	50
First SynChropak AX-300 column	1.2	890	25
Second SynChropak AX-300 column	0.15	3000	10[c]

[a] The activity of the homogenate is impossible to determine due to interfering activities.

[b] The ammonium sulfate fraction contains both isozymes of the synthetase in approximately equal amounts so the specific activity of the acidic form is approximately 1 U/mg protein. After the DEAE step only the acidic form is present so the yield is calculated from that step. Only a small loss of activity occurs prior to that step.

[c] About 50% of the activity was lost during concentration prior to the column.

from the second run on the AX-300 column until the 0.4 M KCl is used (Fig. 2). This indicates that a number of proteins are initially excluded from the AX-300 column in the first run (Fig. 1) due to competition for binding sites on the support. The capacity of the column for a given protein will depend to a large extent on the composition of the sample. The column will have greater capacity for the strongest binding protein in any given sample. Initial separations on DEAE-cellulose would be particularly useful to avoid saturation of the column or irreversible binding of components of the sample.

In studies not shown, we have chromatographed a number of other proteins on the SynChropak AX-300 support. These include hexokinase, adenylosuccinate lyase, several binding proteins, and peptide hormones. Hexokinase from rat brain was eluted as a homogeneous protein from the column and considerable purification has been achieved with the other proteins listed.

ACKNOWLEDGMENTS

This work was supported by Grant 14030 from the National Cancer Institute and Grant C-582 from the Robert A. Welch Foundation.

REFERENCES

1. Ogawa, H., Shiraki, H., Matsuda, Y., Kakiuchi, K., and Nakagawa, H. (1977) J. Biochem. 81, 859–869.
2. Baugher, B. W. (1980) PhD Thesis, Rice University.
3. Fischer, H. E., Muirhead, K. M., and Bishop, S. H. (1978) in Methods in Enzymology (Hoffee, P. A., and Jones, M. A., eds.), Vol. 51, pp. 207–213, Academic Press, New York.
4. Markham, G. D., and Reed, G. H. (1975) Arch. Biochem. Biophys. 184, 24–35.
5. Matsuda, Y., Shimura, K., Shiraki, H., and Nakagawa, H. (1980) Biochim. Biophys. Acta 616, 340–350.
6. Lowenstein, J. M. (1972) Physiol. Rev. 52, 382–419.
7. Matsuda, Y., Ogawa, H., Fukutome, S., Shiraki, H., and Nakagawa, H. (1977) Biochem. Biophys. Res. Comm. 78, 766–771.
8. Baugher, B. W., Montonaro, L., Welch, M. M., and Rudolph, F. B. (1980) Biochem. Biophys. Res. Commun. 94, 123–129.
9. Fukutome, S. (1977) Med. Biol. 94, 461–466.
10. Alpert, A. J., and Regnier, F. E. (1979) J. Chromatogr. 185, 375–392.
11. Regnier, F. E., and Gooding, K. M. (1980) Anal. Biochem. 103, 1–25.
12. Weber, K., Pringle, J. R., and Osborn, M. (1972) in Methods in Enzymology (Hirs, C. H. W., and Timasheff, S. N., eds.), Vol. 25, pp. 3–27, Academic Press, New York.
13. Clark, S. W., and Rudolph, F. B. (1976) Biochim. Biophys. Acta 437, 87–93.

Application of High-Performance Liquid Chromatographic Peptide Purification to Protein Microsequencing by Solid-Phase Edman Degradation[1]

JAMES J. L'ITALIEN[2] AND JAMES E. STRICKLER

Department of Internal Medicine, Yale University School of Medicine,
333 Cedar Street, New Haven, Connecticut 06510

Peptide purification via high-performance liquid chromatography (HPLC) and solid-phase sequencing were integrated to form a system allowing the determination of complete sequence information on a microscale without the use of radiolabels or modified phenylisothiocyanate. Mixtures of peptides (500 pmol to 10 nmol) resulting from proteolytic digestion or chemical cleavage were applied directly to reverse-phase columns. The columns, equilibrated in either 10 mM KP_i or 0.05% trifluoroacetic acid, were then developed using acetonitrile gradients. Eluates were monitored nondestructively by direct ultraviolet detection at both 214 and 254 nm. Each peak was collected as a discrete fraction, and purity was assessed by amino acid analysis prior to covalent attachment to a solid support for sequence analysis. Activation of the peptide carboxyl terminus via a water soluble carbonyldiimide was the solid-phase coupling method used 90% of the time. Coupling yields averaged 52% of starting material. Sequence analysis was performed in the range 100 pmol to 4 nmol of coupled peptide. Phenylthiohydantoin-amino acids were identified by reverse-phase HPLC using ultraviolet detection.

Elucidation of the primary structure of a protein beyond the amino terminus requires fragmentation of the protein and isolation of the resulting peptides. Many proteins of biochemical and biomedical interest remain uncharacterized due to difficulty in obtaining sufficient material to allow conventional techniques of fragment purification and subsequent sequence analysis. A generalized microsequencing method therefore must consider all aspects of the sequencing process from purification of protein fragments to the actual sequence analysis. This includes various aspects of sequenator function and identification of the resulting amino acid residues. The identification of amino acid-phen-ylthiohydantoins (PTHs)[3] on reverse-phase columns was one of the first applications of reverse-phase high-performance liquid chromatography (RP-HPLC) to the problem of protein sequencing (1,2).

It is now possible to detect PTHs by direct uv detection at 254 nm with quantitation of each cycle in the range 1–100 pmol (3,4). The more recent application of RP-HPLC to the problem of peptide purification (5–7) has had a profound effect on what has historically been the rate-determining step in protein sequence analysis by reducing the time required for the isolation of purified material while extending the lower working limits into

---

[1] This paper was presented at the International Symposium on HPLC of Proteins and Peptides, November 16–17, 1981, Washington, D. C.

[2] To whom all correspondence should be addressed. Present address: Molecular Genetics, Inc., 10320 Bren Road East, Minnetonka, Minn. 55343.

[3] Abbreviations used: PTH, phenylthiohydantoin; RP, reverse phase; VSG, variable surface glycoproteins; EF-Tu, elongation factor Tu; GnCl, guanidine hydrochloride; TFA, trifluoroacetic acid; SDS, sodium dodecyl sulfate; DMF, *N,N*-dimethylformamide; DITC-glass, *p*-phenylene diisothiocyanate-activated aminopropyl glass; Endo-H, *endo-β-N*-acetylglucosaminidase H.

the picomole range using totally nondestructive means of detection. Among the numerous advantages of RP-HPLC are its reproducibility and its ability to resolve complex mixtures of peptides as demonstrated by comparative peptide mapping (8,9) of mutant proteins containing only single amino acid substitutions.

In order to take full advantage of peptide purification by HPLC for sequencing purposes, a method was needed which would allow efficient sequencing of peptides in the size ranges which most commonly result from proteolytic digestion (generally less than 30 residues). Furthermore, the method should be compatible with trace amounts of HPLC buffer salts. Solid-phase immobilization of the peptides is a logical solution to both of these needs because it eliminates many of the technical problems associated with the automated sequencing of peptides while facilitating the removal of buffer salts which would interfere with the Edman degradation. This is accomplished simply by washing away the salts following immobilization of the peptide to the support prior to solid-phase Edman degradation (10). For the method to be generally applicable, however, the immobilization method should be specific for the carboxyl group of the C-terminal residue of the peptide and should allow identification of every residue in the polypeptide. In a recent study we showed that peptides isolated by RP-HPLC can be immobilized to aminopolystyrene following apparently selective activation of the carboxyl terminus of the peptide by a water-soluble carbodiimide, even in a presence of residual buffer salts (7). We are currently using this system to obtain complete sequence data on variable surface glycoproteins (VSG) of *Trypanosoma congolense*. In order to demonstrate the general applicability of the methodology we report here a more extensive, quantitative study on the immobilization of a large number of previously uncharacterized tryptic and *Staphylococcus aureus* protease peptides derived from one of these proteins. All peptide separations were performed by RP-HPLC

and immobilization was carried out without prior desalting steps. In addition, we demonstrate the sensitivity of the system with sequence information obtained on subnanomole quantities of a cyanogen bromide fragment of elongation factor Tu (EF-Tu) of *Escherichia coli*.

MATERIALS AND METHODS

Protein isolation and fragmentation. The variant specific glycoprotein from the Yale *Nannomonas* antigenic type 1.1 variant of *T. (Nannomonas) congolense* stock Lister 1/148 was isolated as previously described (11,12). The polypeptide elongation factor Tu was isolated from *E. coli* B cells as described by Miller and Weissbach (13). The lyophilized proteins were carboxyamidomethylated by a modification of the method of Konigsberg (14). The principal modifications were a reduction of the protein concentration from 1 to 2% (w/v) down to 0.05 to 0.1% (w/v), the use of only a 40-fold molar excess of dithiothreitol over cysteines, and reduction in the size of the reaction vessel to better accommodate the reduced volumes when the reaction was performed on 1 mg or less of protein. In some experiments the iodoacetamide was isotopically labeled with carbon-14 to a specific activity of 0.75 mCi/mmol by the addition of [1-^{14}C]iodoacetamide (sp act 10–25 mCi/mmol, New England Nuclear). Following carboxyamidomethylation the proteins were exhaustively dialyzed against deionized, distilled water and lyophilized.

Cyanogen bromide cleavage (15) was performed on both proteins in 70% formic acid at protein concentrations varying from 0.06 to 0.5% (w/v). A 100-fold molar excess of cyanogen bromide (Aldrich) over the methionine residues was then added and the reaction was allowed to proceed in the dark for 24 h at room temperature under a nitrogen atmosphere. Proteolytic digestions were carried out with trypsin-TPCK (EC 3.4.21.4, from Worthington) and *S. aureus* protease (EC 3.4.21.9, from Miles) as previously described (7). The *endo-β-N*-acetylglucosaminidase H digestion (protease-free from Miles) was performed on native VSG dissolved in 10 mM sodium phosphate buffer (pH 5.0) at an enzyme to substrate ratio of $1:10^4$ (w/w) for 16 h at 37°C. A second sample of VSG was treated to a sham digestion in 10 mM NaP_i buffer concurrently. Both samples were desalted by chromatography on Sephadex G-25 (1×40 cm) in deionized, distilled water. The protein peaks were collected and lyophilized. The lyophilized samples were carboxyamidomethylated and tryptically digested as described.

HPLC separation of peptides. Peptide separations were performed on an Altex Model 324 chromatography system (two Model 100A pumps, Model 421 microprocessor system controller, a dynamically stirred gradient mixing chamber, a Rheodyne Model 7125 sample injector, a Kipp and Zonen dual-pen chart recorder, a Model 153 Altex detector, and a Gilson Holochrome detector).

CNBr fragments were dissolved in 6 M GnCl (Pierce) and applied directly to a μBondapak C-18 column (Waters Assoc.) equilibrated with 0.05% trifluoroacetic acid (Sequemat) in water. The column was then developed using gradients of acetonitrile (Baker, HPLC grade) which was also 0.05% in trifluoroacetic acid. The purity of the fractions was assessed by 10 to 30% gradient SDS–polyacrylamide gel electrophoresis.

Proteolytic peptide mixtures were dissolved in 50% acetic acid (Baker reagent) and applied to a μBondapak C-18 column (preceded by a precolumn described in (7)) equilibrated in 10 mM potassium phosphate (pH 2.5). Peptides were eluted with acetonitrile gradients which were generally optimized for each mixture. The column eluates were monitored nondestructively by direct uv detection at both 206–214 and 254 nm which allowed assessment of the aromatic content of each peptide while providing a second fingerprint for comparative purposes. Each peak was collected as a discrete fraction, neutralized with triethylamine (redistilled), and dried prior to further characterization.

Immobilization of peptides. All peptides

less than 30 residues were immobilized to aminopolystyrene via activation of their C-terminal carboxyl group by a water soluble carbodiimide. The procedure used was modified from that of Wittmann-Liebold *et al.* (16) for coupling in the 1–5 nmol range. Five hundred microliters of 1.0 M pyridine chloride (pH 5.0) (pyridine redistilled from phthalic anhydride) was added to 10 mg of aminopolystyrene (Sequemat) and the resin was stirred at 30°C for 10 min or until a color transition of the resin (from brown to violet) was observed. The resin was then washed two times with 3 ml of water and two times with 500 μl of dimethylformamide (redistilled from P_2O_5). The lyophilized peptide was dissolved in 100 μl of 1.0 M pyridine chloride (pH 5.0) and transferred to the swollen resin. The original sample tube was rinsed with 100 μl of 1.0 M pyridine chloride followed by 100 μl of dimethylformamide (DMF) and both rinses were added to the peptide resin mixture. A fresh solution of 1-ethyl-3,3′-dimethylaminopropylcarbodiimide · HCl (Sigma) was prepared such that 3 mg of carbodiimide in 75 to 150 μl of water: DMF (1:4) was added to each tube. The reaction was allowed to proceed at 30°C for 1 h with stirring, after which the sample was centrifuged and the supernatant removed. The pH of the resin was raised to $\sim$9 by adding 200 μl of 1.0 M N-methylmorpholine (Pierce) and 200 μl of DMF, and allowing the resin to stir at 30°C for 10 min. The sample was then centrifuged, the supernatant was removed, and the excess amino groups were blocked by adding 150 μl DMF, 50 μl of N-methylmorpholine, and 50 μl of phenylisothiocyanate (Pierce). The blocking reaction was allowed to proceed at 30°C for 1 h. Excess reagents were removed by washing the support with 4 ml of methanol (Baker, reagent) three or four times. The last traces of methanol were removed by drying under vacuum (water aspirator) and the coupled peptide was stored under refrigeration until sequenced.

All polypeptides (proteins or their fragments) greater than 50 residues were immobilized to DITC glass which was either prepared as described by Machleidt and Wachter (17) or purchased from Sequemat. The polypeptide was solubilized in 100 μl of 6 to 8 M GnCl (Pierce) before being diluted to 3 to 4 M GnCl by the addition of 100 μl of water. The DITC-glass (50–100 mg) was then added to the sample followed by 100 μl of 4 M GnCl which was 5% in triethylamine (redistilled). The reaction was allowed to proceed at 50°C with occasional shaking. After 1 h, the sample was centrifuged, the supernatant removed, and the support washed once with 200 μl of 4 M GnCl · triethylamine. Excess DITC groups were then blocked by adding 200 μl of 4 M GnCl · triethylamine and 50 μl of 10 mM β-alanine (Sigma). The blocking reaction was carried out at 50°C for 45 min. The support was then washed two times with 4 M GnCl, three times with water, four times with methanol, and dried under vacuum as above. Other methods of immobilization used were as previously described (7).

Solid-phase Edman degradation. All polypeptides were sequenced by solid-phase Edman degradation (10) using a newly developed 65-min sequenator program (18) on an otherwise unmodified Sequemat Mini-15 solid-phase sequencer equipped with a P-6 Autoconvertor. The sequencer reagents used (described by Laursen and Machleidt (19)) were obtained from Baker (methanol and dichloroethane, both reagent grade, and acetonitrile, HPLC grade), Pierce (phenylisothiocyanate, sequenator grade in 5-ml ampoules), and Sequemat (aqueous coupling buffer and trifluoroacetic acid). The autoconvertor reagent was 1–2 M methanolic · HCl which was prepared from acetylchloride (Sequemat) and the wash solvent was dichloroethane–methanol (7:3, v/v) (both Baker reagents). Each cycle from the sequenator (after conversion) was dried twice under N_2 following the addition of 500 pmol of PTH-Norleucine. Each sample was then dissolved in 50 μl of methanol and applied to an Altex Ultrasphere RP-18 column (5-μm particle size) on a Waters liquid chromatography system consisting of two Model 6000A pumps,

a Model 710B WISP, a Model 720 system controller, a Model 730 data module, a Model 440 detector, and a Gilson Holochrome detector. The PTHs were eluted at room temperature with a series of linear gradients of acetonitrile (solvent B) into 100 mM sodium acetate (pH 4.1) (solvent A): 0–10 min (27–45% B), 10–15 min (45–50% B), 15–20 min (isocratic at 50% B), 20–25 min (50–60% B), 25–27 min (isocratic at 60% B), 27–33 min (60–27% B). Both PTH-Asp and PTH-Glu were detected as methyl esters resulting from conversion to PTHs in the methanolic HCl. The PTHs for all of the amino acids were detected at 254 nm except the dehydro derivatives of threonine and serine which were detected at 313 nm.

The coupling yields were determined by acid hydrolysis of an aliquot of the immobilized peptide in 5.7 N HCl for 6 h at 130°C *in vacuo.* The samples were then centrifuged, the supernatant extracted from the support, and dried under vacuum over NaOH pellets prior to analysis on a Beckman 121M amino acid analyzer. The resulting values were normalized to an internal standard which was added prior to hydrolysis and divided by the initial amount of peptide used for coupling to give the coupling yield.

RESULTS AND DISCUSSION

The methods outlined here, RP-HPLC peptide purification and solid-phase Edman degradation (10), are currently being used to determine the primary structure of one of the variable surface glycoproteins of *T. congolense.* To establish the linear order of the polypeptide chain both proteolytic and chemical means of fragmentation were employed on the intact, carboxyamidomethylated VSG to generate overlapping polypeptides which were purified and sequenced as described.

The separation profile of the cyanogen bromide fragments of CM-VSG is given in Fig. 1. Previous attempts to purify these fragments by either classical or HPLC gel-permeation chromatography were confounded

by the tendency of the peptides to self-associate and precipitate in the absence of a denaturant such as guanidine hydrochloride or
urea in the eluting buffer. We obtained purification of the fragments in good yield using
a reverse-phase column (μBondapak C18)
and a binary gradient of acetonitrile into
water with 0.05% trifluoroacetic acid in each
solvent (20). Reverse-phase supports and this
solvent system have a number of advantages
over other methods for the isolation of large
polypeptide fragments. These solvents are uv
transparent between 206 and 214 nm, allowing direct uv detection at high sensitivity
while maintaining peptide solubility without
the use of denaturing agents which are not
transparent below 238 nm. Thus, various
elution procedures can be explored using 0.1
to 1-nmol quantities of material until a satisfactory separation is obtained. In addition,
since the solvent system is completely volatile, we could carry out subsequent forms of
characterization, e.g., determination of fragment molecular weight by SDS–polyacrylamide gel electrophoresis, dansylation, or
proteolytic digestion, without desalting. This
is an important consideration when working
at material levels where every handling step
imparts significant loss. We prefer reverse-
phase chromatography for this type of separation because it is able to provide separation of fragments in a single chromatographic
step. Such a separation would otherwise require multiple steps using alternative methods, even other HPLC methods.

One nanomole of each of the purified
CNBr fragments was immobilized to DITC-
glass and sequenced. The coupling of each
fragment (which was ^{14}C-labeled) was quantitative as judged by counting the supernatant and washes after coupling. The limiting
factors in immobilization of peptides and
proteins to p-phenylene diisothiocyanate-activated aminopropyl glass are the presence
of lysine residues and the solubility of the
polypeptide during the coupling reaction.
Thus, although the presence and location of
lysine residues is beyond control, maximal
coupling efficiencies can be obtained by

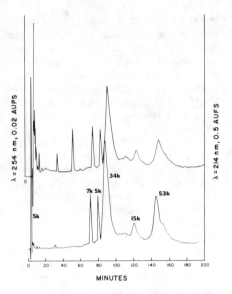

FIG. 1. Separation of CNBr fragments of VSG by
HPLC. Approximately 20 nmol (1 mg) of digest, dissolved in 50 μl 6 M GnCl, 0.5 M Tris–HCl, pH 8.1, 2
mM EDTA, were injected onto a Waters μBondapak
C18 column without a precolumn. The column was
equilibrated in 0.05% TFA, 30% acetonitrile and developed with an acetonitrile, 0.05% TFA gradient (30–45%
in 180 min, 45–60% in 30 min, and isocratic at 60%
acetonitrile for 10 min). The flow rate was 1 ml/min and
the initial pressure was 960 psi. Detection was at 254
nm through an Altex Model 153 detector (upper trace)
and at 214 nm through a Gilson Holochrome detector
(lower trace). The numbers above the peaks give the
apparent molecular weights of the peptides as judged by
SDS–polyacrylamide gel electrophoresis.

maintaining solubility of the polypeptide in
the coupling buffer or solvent. For example,
a hydrophobic tryptic peptide from bovine
white matter proteolipid was quantitatively
immobilized to DITC-glass in its purification
solvent chloroform–methanol (2:1, v/v) by
simply adding the support and triethylamine
(21). We have attempted to ensure the solubility of the HPLC-purified CNBr fragments by performing the immobilization reaction in the presence of guanidine hydrochloride. This presents an additional
advantage in that uncoupled polypeptide
material can be quickly and quantitatively

recovered from the supernatant following the coupling reaction because this coupling buffer is compatable with the HPLC system used for fragment purification. The effectiveness of this immobilization procedure is illustrated by Fig. 2 which gives the sequence of the 15,000 M_r fragment from Fig. 1 through 30 cycles starting with less than 1 nmol of polypeptide.

One of our principle interests in the structural analysis of this protein is the localization of posttranslational modifications, information which can only be obtained by analysis of the mature protein. In Fig. 3, we take advantage of the reproducibility of RP-HPLC for comparative mapping in order to localize N-linked carbohydrate in the protein. An aliquot of protein was divided into equal parts with one-half of the sample subjected to *endo-β-N*-acetylglucosaminidase H treatment. This enzyme selectively removes high mannose-type oligosaccharide sidechains coupled through GlcNAc–GlcNAc to Asn by splitting the GlcNac–GlcNAc bond (22,23). By gel electrophoresis and carbohydrate analyses, we have found that approximately 50% of the total carbohydrate of this VSG is sensitive to Endo-H (data not shown). After Endo-H digestion, both aliquots were car-

boxyamidomethylated and digested with trypsin. The peptide mixtures were applied to a Waters μBondapak C18 column. Comparison of the peptide profiles revealed two significant shifts in peak retention time (as highlighted by the dashed lines in Fig. 3). In both instances, the Endo-H-treated peptides exhibited an increase in their retention time relative to the non-Endo-H-treated sample. This was presumably due to the removal of N-linked oligosaccharide side chains. The precise location of this N-linked carbohydrate in these peptides is currently under investigation.

All of the major tryptic and *S. aureus* protease peptides greater than five residues have been sequenced by solid-phase Edman degradation following removal of an aliquot for amino acid analysis and/or dansylation. The method of choice for immobilization of these peptides was carbodiimide activation of the C-terminal carboxyl group for coupling to aminopolystyrene. The compatability of this method with peptides purified by HPLC is important in that this method may be employed for any peptide irrespective of the C-terminal residue. Historically, the first reports by Laursen of C-terminal activation by carbonyldiimidazole (24) had shown prob-

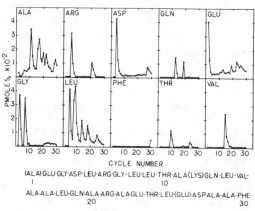

FIG. 2. Yields of PTH-amino acids from the N-terminal sequence analysis of the 15,000 M_r CNBr fragment of VSG. Each cycle was analyzed by HPLC as described. Integrated peak areas were converted to picomoles based on recovery of the 500-pmol norleucine standard added. Parentheses indicate uncertain residues confirmed by overlap analysis.

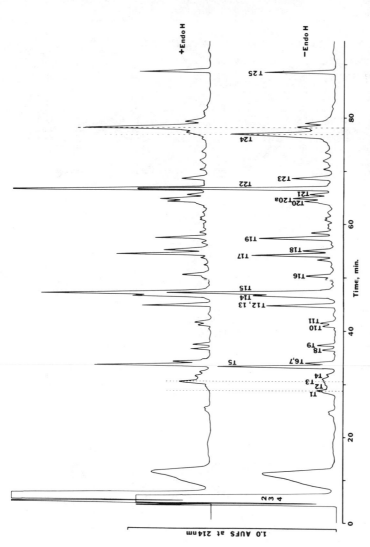

FIG. 3. Comparative tryptic HPLC maps of untreated and Endo-H-treated VSG. Approximately 6–8 nmol (300–400 μg) of each preparation was injected in 25% acetic acid onto a Waters μBondapak C18 column equipped with a precolumn described in Ref. (7). The column was equilibrated in 20 mM KF_2, pH 2.5, and developed through an acetonitrile gradient (0–30% in 60 min, 30–60% in 30 min, and isocratic at 60% for 10 min). The flow rate was 1 ml/min and the pressure at 100% aqueous buffer was 2300–2400 psi. Detection was at 254 nm through an Altex Model 153 cetector (trace not shown) and at 214 nm through a Gilson Holochrome detector. The numbers over the lower trace represent elution position and do not correspond to sequence position.

TABLE 1

CARBODIIMIDE IMMOBILIZATION OF PEPTIDES ISOLATED VIA REVERSE-PHASE HPLC
USING PHOSPHATE BUFFER SALTS

Peptide	Peptide No.[a]	No. of Residues[b]	Coupling yield[c] (%)
AIVGGWGNPTTPDESGLPTTFK	T22	22	50
SAKAFTKAIVGGWGNPTTPDE	S15	22	40
SLVFDIACLCTTSDSASGSTK	T23	21	32
SGDNGSGWLDNNGDNXGKPAK	T14	21	70
SGLPTTFKTNRADDCKLAGGD	S16	21	42
TNRADDCDLAGGDXKYXLXF	S12	21	ND[d]
VTPELISTXLVIFEGLIGTR	T25	20	23
SSXPPSTDAXTSBXGPXQXP	S5	20*	42
GDLRGLLTAKQLVAALQAR	S21	19+	29
DVFMLSFTEPSAVVTTLD	S20	19	37
DIFGTVATAQXCGXSTAR	T18	18	59
LSFTEPSAVVTTLDGXR	T20	17	70
AGEGLKEEDWLPCAGK	T19	16	44
YNALCRLYNIARAGE	S18	15	34
SASGTKYTCGPKSGD	S11	15	ND
DAPVNAAEYNALCR	T17	14	56
EDWLPCAGKAAE	S14ED	13	49
DWLPCAGKAAE	S14D	12	ND
TAASIEDVFMK	T21	11	70
TIKAQQLKYHE	S8	11	46
VELQNSASTR	T7	10	73
BLVAALQAR	T16	9	61
VLAAAETIK	T12	9	49
DAAFTIFRD	S17	9	30
IACLCTTSD	S10	9	48
GLIGTRAAA	S6	9+	ND
AWKNLRAD	S9	8	49,44
AGVRVTPE	S4	8	ND
DAPVNAAE	S3	8	58
KTAASIED	S1ED	8	50
GDIEWEK	T15	7	ND
SLLESAK	T11	7	32
MAEGDLR	T9	7	69
YAGGNGK	T1	7	87
SSKSLLE	S7	7	ND
KNLRMAE	S2	7	64
KTAASIE	S1E	7	47
LYNIAE	T13	6	49
GLLTAK	T10	6	58
GCIDYK	T8	6	53
YTCGPK	T4	6	76
TQIAWE	S13	6	62
ANFTK	T4	5	64
AQQLK	T3	5	74
EAWK	T5	4	51

Note. Average of 13 peptides with >15 residues was 41% with a range of 23 to 70. Average of 25 peptides between 4 and 14 residues was 57% with a range of 30 to 87.

[a] Peptide number corresponds to peak numbers shown in Figs. 3 and 4.

[b] Number of residues were determined both by amino acid composition and by sequence analysis.

[c] Coupling yield was determined as described under Materials and Methods.

[d] ND, not determined.

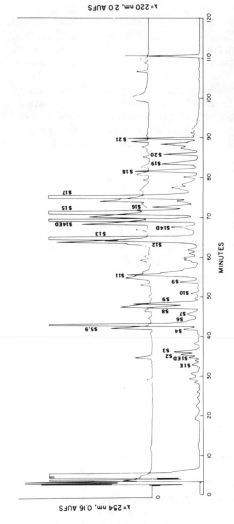

FIG. 4. Reverse-phase HPLC separation of *Staphylococcus aureus* protease peptides of VSG. Approximately 20 nmol (1 mg) of digest in 50% acetic acid, 4 M GnCl were injected onto a Waters μBondapak C-18 column equipped with a precolumn described in Ref. (7). The column was equilibrated in 20 mM KP_i, pH 2.5, and developed with an acetonitrile gradient (0–20% in 60 min, 20–30% in 20 min, 30–60% in 20 min, and isocratic at 60% for 10 min). The flow rate was 1 ml/min. Detection was at 254 nm (upper trace) and at 214 nm (lower trace). Numbers over the peaks in the lower trace represent elution position and do not reflect sequence position.

lems with side-chain coupling leading to termination of the sequenceable peptide. The problems were circumvented through the development of the DITC (25) and lactone (26,27) procedures which rely upon the specific chemistries either of internal residues or at the C-terminus for immobilization. In the case of DITC immobilization, however, gaps in the sequence at the attachment points result. Thus, there remained a need for a more universal method of attachment of peptides to solid supports. A number of carbodiimide coupling procedures have been proposed in recent years which were demonstrated using model peptides (10,28,29). There have been, however, no extensive quantitative data published on coupling yields of peptides from proteolytic digestion using these carbodiimide methods. In Table 1 we list the coupling yields, of the peptides isolated as per Figs. 3 and 4 and immobilized by the car-

bodiimide method described in this report. The average coupling yields for these 38 peptides was 52% with a range of 23 to 87% and a median value of 50%. These data clearly demonstrate the general usefulness of this coupling method on a wide variety of peptides. In every situation in which this method was applied the peptide was immobilized and the method appeared to be highly selective for the C-terminal carboxyl group of the peptide as both aspartic and glutamic acid residues were readily discernable and there was no detectable loss of peptide in the sequenator cycle following either of these residues. While this empirical observation has not yet been satisfactorily explained in terms of a chemical mechanism, it is given further credibility by the fact that carboxyl terminal Asp or Glu residues (resulting from *S. aureus* protease digestion) do not appear to be immobilized significantly better than peptides

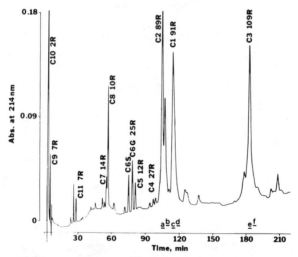

FIG. 5. HPLC separation of 3 nmol (125 µg) of EF-Tu CNBr fragments. The lyophilized fragments were solubilized in 6 M guanidine hydrochloride and applied to a µBondapak C18 column (Waters Assoc.) equilibrated in 0.05% trifluoroacetic acid in water (v/v). Peptides were eluted with the following successive linear gradients of acetonitrile (which was also 0.05% in TFA): 0–50 min (0–25% acetonitrile), 50–150 min (25–50% acetonitrile), 150–170 min (50–70% acetonitrile). The initial column back pressure was 900 psi at the 1.0 ml/min flow rate used for this chromatogram. The eluant was monitored at both 214 and 254 nm (trace not shown). Each peak is designated by a C-number which refers to CNBr fragment numbers previously reported (Ref. 30). The number preceding the R refers to the number of residues in that particular fragment. The bars with small letters over them refer to fractions taken for gel analysis (shown in Fig. 6).

of similar size and solubility containing C-terminal Lys or Arg. All of the peptides listed in Table 1 were sequenced using the 65-min sequenator program previously described (18) on an otherwise unmodified Sequemat Mini-15 sequenator.

To demonstrate the capability of the total sequencing approach which we have proposed here, a number of RP-HPLC peptide separations were performed at the 1–3 nmol level and then sequenced as described. The separation of the CNBr fragments of EF-Tu by RP-HPLC permitted the purification of all 11 of the peptides (which ranged in size from 2 to 109 residues) present in this complex mixture, in a single chromatographic step (Fig. 5). The one site of microheterogeneity present in the 25-residue peptide C-6 (Gly/Ser at position 393 in the complete sequence (30)) was separated into its two subspecies. Each of the large molecular weight peptides were recovered following HPLC purification as determined by comparison of the collected fractions with the total digest on SDS–polyacrylamide gel electrophoresis (Fig. 6). After HPLC purification of fragment C-3 1-nmol aliquots were digested with trypsin as described (6 samples were handled in parallel). Following digestion each mixture was dissolved in 50% acetic acid and applied directly to a μBondapak C-18 column. All of the resulting profiles were identical to the chromatogram shown in Fig. 7. Individual peaks were collected as discrete fractions, neutralized, and dried. Two samples of each peptide were hydrolyzed for amino acid analysis. The remaining four were immobilized to aminopolystyrene via carbodiimide activation. Two of the four coupled samples were hydrolyzed to give the coupling yields. All five of the peptides were sequenced as described. An internal standard was added to each cycle prior to drying and direct application of the sample to HPLC for identification. The actual tracing of one of these samples is shown in Fig. 8. The results of this study are given in Table 2. Recovery of the first three peptides following RP-HPLC was

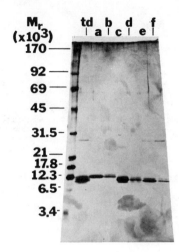

FIG. 6. SDS–polyacrylamide gel electrophoresis analysis of EF-Tu CNBr fragments separated by reverse-phase HPLC. The fractions shown in Fig. 5 were neutralized with triethylamine, acetonitrile was evaporated under a stream of N_2, and the samples were lyophilized. The samples were dissolved in 50 μl of 62.5 mM Tris–HCl, pH 6.8, 2.5% SDS, 5% 2-mercaptoethanol, 10% glycerol, and placed at 100°C for 3 min. The samples were then subjected to electrophoresis on a 10 to 30% linear polyacrylamide gradient slab gel (1:30, bis:acrylamide ratio) containing 0.1% SDS, for 16 h at 160 V constant voltage. The gel buffers and electrode buffers were those described by Maizel (32). The gel was fixed in 5% formaldehyde, 25.7% ethanol for 1 h at room temperature, and then in 50% methanol, 12% acetic acid for a second hour. The gel was stained with the silver stain of Merril et al. (33). The letters above the wells correspond to the fractions shown in Fig. 5; td = total digest.

quite good while the last two peptides showed a lower recovery which resulted in part because these peaks were shaved in order to minimize cross contamination. The coupling yields were between 50 and 60% while the sequenceable yields were found to be ~60% of the immobilized peptide. Thus on average, we were sequencing 30–36% of the initial amount of peptide. These levels proved sufficient to sequence through 16 cycles (with unambiguous identification of each cycle) starting with 500 pmol (before coupling) of the peptide ILELAGFLDSYIPEPER. These data compare favorably with the recently

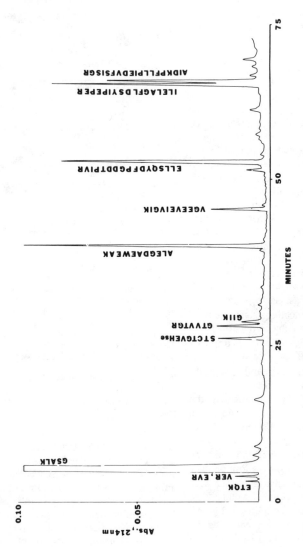

FIG. 7. HPLC separation of 1 nmol of EF-Tu CB-3 tryptic digest. The peptide mixture was dissolved in 50% acetic acid and applied to a μBondapak C18 column equilibrated in 20 mM potassium phosphate (pH 2.5). The peptides were eluted from the column using a series of linear gradients of acetonitrile (solvent B); 0–10 min (0–5% B), 10–55 min (5–35% B), 55–70 min (35–50% B). The initial column back pressure was 2000 psi at the 1 ml/min flow rate used to develop the chromatogram. The column eluate was monitored at 214 nm. The sequence of each peptide is given by the one-letter code adjacent to its corresponding peak. All 109 residues of this fragment are accounted for in this chromatogram.

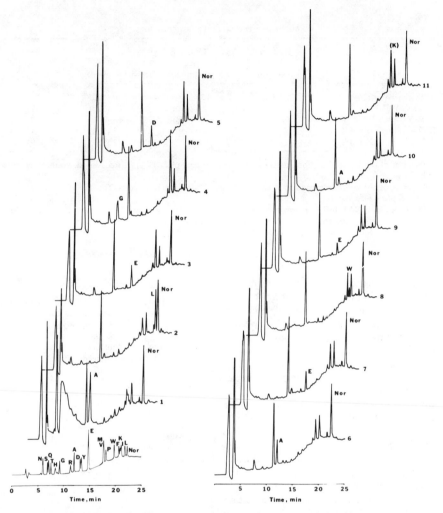

FIG. 8. Actual HPLC traces of PTH-amino acids from the sequencing of 500 pmol of the peptide ALEGDAEWEAK coupled to aminopolystyrene via carbodiimide as described under Materials and Methods. The PTH standard shown is 50 pmol except for PTH-Glu which is 100 pmol. The cycle number of each step is given immediately following each trace. Detection is by direct uv at 254 nm. In cycle 1 there is an unsymmetrical peak between 5 and 8 min on the trace. This peak is a pyridine peak which can be removed (as seen in the remaining cycles) by simply redrying the sample under nitrogen after adding a small volume of methanol. In each cycle there are three peaks resulting from the sequencing process which do not interfere with the identification of the PTHs.

described gas–liquid solid-phase sequenator (31) which required 1.4 nmol to sequence somatostatin (14 residues) and 500 pmol to sequence angiotensin (8 residues).

In summation, the procedures outlined here take advantage of the speed, sensitivity, resolution, and reproducibility of RP-HPLC for both peptide purification and PTH iden-

TABLE 2

MICROSEQUENCING OF TRYPTIC PEPTIDES

Peptide	Recovered from HPLC[a] (pmol)	Coupling yield[b] [% (pmol)]	Sequenceable[c] [% (pmol)]	Repetitive yield (%)
ALEGDAEWEAK	900	55 (500)	60 (300)	91 ± 1
VGEEVEIVGIK	900	50 (450)	60 (270)	91 ± 1
ELSVYDFPGDDTPIVR	900	61 (550)	64 (350)	91 ± 1
ILELAGFLDSYIPEPER	500	60 (350)	57 (200)	91 ± 1
AIDKPFLLPIEDVFSISGR	250	60[d]	63 (95)	91 ± 1

[a] Average of two sample injections of 1 nmol each onto reverse-phase column. Recovery determined by amino acid analysis.
[b] Average of two samples determined as described under Materials and Methods. The number in parentheses is the actual picomole amount.
[c] Sequenceable yield is based on area measurements of PTH-amino acid residues following automated Edman degradation, and is a measure of the amount of immobilized peptide capable of being sequenced.
[d] Determined from a duplicate experiment. In this experiment the amount of peptide coupled was below the limit of detection of the amino acid analyzer.

tification. The use of RP-HPLC as an analytical method for peptide separation allows for the development of a logical fragmentation strategy by permitting the assessment of fragmentation methods on small aliquots of material. Furthermore, these procedures eliminate handling steps (and thus loss of material) by direct application of peptide samples to HPLC and collection of peaks as discrete fractions following direct uv detection. The direct immobilization of HPLC-purified peptides, without prior desalting, was found to be efficient (even at the subnanomole level) while permitting the identification of every residue in the peptide by solid-phase Edman degradation. The methods described are capable of complete sequence determination of a protein on a microscale. They are also perhaps the most efficient means of providing an independent check on frame reading when used in conjunction with rapid DNA sequencing methods because of the potential of providing sequence information at multiple points in the polypeptide chain. Finally, these methods form the basis for the rapid screening of mutant proteins and post-translational modifications of mature protein through comparative peptide mapping (as presented here and in Ref. (9)).

REFERENCES

1. Zimmerman, C. L., Apella, E., and Pisano, J. J. (1977) Anal. Biochem. 77, 569–573.
2. Bhown, A. S., Mole, J. E., Weissinger, A., and Bennett, J. C. (1978) J. Chromatogr. 148, 532–535.
3. Johnson, N. D., Hunkapiller, M. W., and Hood, L. E. (1978) Anal. Biochem. 100, 335–338.
4. Lottspeich, F. (1980) Hoppe-Seyler's Z. Physiol. Chem. 369, 1829–1834.
5. Fullmer, C. S., and Wasserman, R. H. (1979) J. Biol. Chem. 254, 7208–7212.
6. Mahoney, W. C., and Hermondson, M. A. (1980) J. Biol. Chem. 255, 11,199–11,203.
7. L'Italien, J. J., and Laursen, R. A. (1981) J. Biol. Chem. 256, 8092–8101.
8. Wilson, J., Lam, H., Pravatmuang, P., and Huisman, T. (1979) J. Chromatogr. 179, 271.
9. Williams, K. R., L'Italien, J. J., Guggenheimer, R. A., Sillerud, L., Spicer, E., Chase, J., and Konigsberg, W. H. (1982) in Proceedings, IV International Conference on Methods in Protein Sequence Analysis (Elzinga, M., ed.), pp. 499–507, Humana Press, New Jersey.
10. Laursen, R. A. (1971) Eur. J. Biochem. 20, 89–102.
11. Strickler, J. E., Mancini, P., and Patton, C. L. (1978) Exp. Parasitol. 46, 262–276.
12. Onodera, M., Rosen, N. L., Lifter, J., Hotez, P. J., Bogucki, M. D., Cross, G. A. M., Konigsberg, W. H., and Richards, F. F. (1981). Exp. Parasitol. 52, 427–439.
13. Miller, D. L., and Weissbach, H. (1970) Arch. Biochem. Biophys. 141, 26–37.
14. Konigsberg, W. H. (1972) in Methods in Enzymology (Hirs, C. H. W., and Timasheff, S. N., eds.),

Vol. 25, pp. 185–188, Academic Press, New York.

15. Gross, E. (1967) *in* Methods in Enzymology (Hirs, C. H. W., ed.), Vol. 11, pp. 238–255, Academic Press, New York.

16. Wittmann-Liebold, B., Braver, D. and Dognin, J. M. (1977) *in* Solid Phase Methods in Protein Sequence Analysis (Previero, A., and Coletti-Previero, M.-A., eds.), p. 219–232, North-Holland, Amsterdam.

17. Machleidt, W., and Wachter, E. (1977) *in* Methods in Enzymology (Hirs, C. H. W., and Timasheff, S., eds.), Vol. 47, pp. 263–277, Academic Press, New York.

18. L'Italien, J. J., and Laursen, R. A. (1982) *in* Proceedings, IV International Conference on Methods in Protein Sequence Analysis (Elzinga, M., ed.), pp. 383–399, Humana Press, New Jersey.

19. Laursen, R. A., and Machleidt, W. (1980) *Methods Biochem. Anal.* **26,** 201–284.

20. Henderson, L. E., Sowder, R., and Oroszlan, S. (1981) *in* Chemical Synthesis and Sequencing of Peptides and Proteins (Liu, Schechter, Heinrikson, and Condliffe, eds.), pp. 251–260 Elsevier/North-Holland, Amsterdam.

21. Lees, M. B., Chao, B., Laursen, R. A., and L'Italien, J. J. (1982) *Biochim. Biophys. Acta* **702,** 117–124.

22. Tarentino, A. L., Plummer, T. H., Jr., and Maley, F. (1972) *J. Biol. Chem.* **247,** 2629.

23. Tarentino, A. L., and Maley, F. (1974) *J. Biol. Chem.* **249,** 811.

24. Laursen, R. A. (1966) *J. Amer. Chem. Soc.* **88,** 5344–5346.

25. Laursen, R. A., Horn, M. J., and Bonner, A. G. (1972) *FEBS Lett.* **21,** 67–70.

26. Horn, M. J., and Laursen, R. A. (1973) *FEBS Lett.* **36,** 285–289.

27. Wachter, E., and Worhahn, R. (1979) *Anal. Biochem.* **97,** 56–64.

28. Previero, A., Derancourt, J., Coletti-Previero, M.-A., and Laursen, R. A. (1973) *FEBS Lett.* **33,** 135–138.

29. Beyreuther, K. (1977) *in* Solid Phase Methods in Protein Sequence Analysis (Previero, A., and Coletti-Previero, M.-A., eds.), pp. 107–120. North-Holland, Amsterdam.

30. Laursen, R. A., L'Italien, J. J., Nagarkatti, S., and Miller, D. L. (1981) *J. Biol. Chem.* **256,** 8102–8109.

31. Hewick, R. M., Hunkapiller, M. W., Hood, L. E., and Dreyer, W. J. (1981) *J. Biol. Chem.* **256,** 7990–7997.

32. Maizel, J. J., Jr. (1971) *in* Methods in Virology (Maramorosch, K., and Koprowski, H., eds.), Vol. 5, pp. 179–247, Academic Press, New York.

33. Merril, C. R., Goldman, D., Sedman, S. A., and Ebert, M. H. (1981) *Science* **211,** 1437–1438.

Measurement of Endogenous Leucine Enkephalin in Canine Caudate Nuclei and Hypothalami with High-Performance Liquid Chromatography and Field-Desorption Mass Spectrometry[1]

Shigeto Yamada* and Dominic M. Desiderio†,[2]*

*Stout Neuroscience Mass Spectrometry Laboratory and †Department of Neurology, University of Tennessee Center for Health Sciences, 800 Madison Avenue, Memphis, Tennessee 38163

For the first time, endogenous amounts of Leu-enkephalin are measured in brain tissue with a technique preserving integrity of the entire molecular structure of the neuropeptide. Field-desorption mass spectrometry enables measurement of picomole amounts of endogenous, chemically underivatized Leu-enkephalin in canine caudate nuclei and hypothalami. The optimal sensitivity and resolution of high-performance liquid chromatography is coupled with maximal molecular specificity of field-desorption mass spectrometry to measure enkephalins in caudate nuclei and hypothalami from dog brains. This novel combination of two recent instrumental methodologies provides a firm molecular basis for calibrating the radioimmunoassay measurement of endogenous levels of biologically active brain neuropeptides.

Neuropeptide research activity has increased rapidly over the past several years following discovery of opioid peptides. Multiple opioid receptors have been found (1–4), biological activity of enkephalins has been studied (5–7), enkephalins have been quantified in cerebrospinal fluid (8,9) and a series of papers (10–16) has illustrated various methods by which biologically active peptides are extracted from biologic sources and endogenous levels are measured. Many measurement methods use radioimmunoassay (RIA)[3] techniques (17,18) and, in one example, a novel combination of gas chromatography–mass spectrometry, enzymatic, and RIA techniques was employed (13).

The recent development of field-desorption mass spectrometry (FD-MS) (19) has

significantly advanced application of mass spectrometric methods to peptide analyses (20–22). In very simple terms, microgram or lower amounts of a chemically underivatized peptide of 1000 to 2000 daltons are placed on a field-desorption emitter (a tungsten wire activated in an atmosphere of benzonitrile under high voltage), the emitter is placed in the mass spectrometer, and current is passed through the wire. The sample is subjected to a 10^6 V cm^{-1} electric field and one electron is removed from the peptide molecule. The positive ion desorbs from the positively charged wire surface and is accelerated by several thousand volts to the detector. Minimal energy is transferred to the molecule in this ionization mode and little or no peptide bond fragmentation occurs. Highly simplified mass spectra result (19–22) and consist mainly of protonated $(M + H)^+$ and cationized, mainly $(M + Na)^+$, molecular ions.

Field-desorption mass spectrometry has been utilized for quantification of only a few biologically active compounds in studies such as time-dependent thallium distribution in mouse organs (23), berberine chloride in

---

[1] This paper was presented at the International Symposium on HPLC of Proteins and Peptides, November 16–17, 1981, Washington, D. C.

[2] To whom all correspondence and reprint requests should be addressed.

[3] Abbreviations used: RIA, radioimmunoassay; FD-MS, field-desorption mass spectrometry; IS, internal standard; ehc, emitter heating current.

human urine (24), choline and acetylcholine in distinct rat brain regions (25), and cyclophosphamide in urine, serum, and cerebrospinal fluid of multiple sclerosis patients (26).

The purpose of this paper is to report data from extraction of biologically active peptides from a specific canine brain region, purification of these biologic extracts by simple column chromatographic techniques, high resolution of a mixture of endogenous neuropeptides and measurement of an enkephalin by reverse-phase high-performance liquid chromatography (RP-HPLC), and measurement of an underivatized neuropeptide with field-desorption mass spectrometric techniques using an internal standard (IS). This novel combination of methodologies is a unique development in the field of biologic neuropeptide analysis and provides for the first time a molecular basis for measurement of endogenous amounts of neuropeptides by combining off-line high resolution and sensitivity capabilities of high-performance liquid chromatography with molecular specificity of field-desorption mass spectrometry, where the latter monitors a mass unique to the neuropeptide being quantified.

MATERIALS AND METHODS

Tissue procurement. Dogs under pentobarbital anesthesia are exsanguinated via a femoral artery. The cranium is opened and the entire brain is removed. Caudate nuclei, hippocampus, pituitary, cerebellum, hypothalamus, cortex, spinal cord, olfactorum tubercule, thalamus, etc., are neuroanatomically identified and then neurosurgically excised. Tissue is placed (within 4 min) into liquid nitrogen. Tissue samples are stored at −70°C. Tissue weights given are wet weight. The weight of the pair of canine caudate nuclei is approx 1 g.

Internal standard and spiking. Two micrograms of Ala²-leucine enkephalin internal standard (Bachem, Torrance, Calif.) per gram of tissue in 20 μl Tris buffer (pH 7.4) is added to the homogenization flask. Leucine enkephalin (200 or 250 ng) is added; the appropriate amount is indicated in Table 1.

Sample preparation. Samples (2 g) are defrosted, mixed with 3 ml 1.0 N acetic acid, and homogenized 3 min at 0°C with a homogenizer (VirTis 23, Gardiner, N. Y.). Cells in this solution are disrupted for 3 min with an ultrasonic generator (Kontes Evanston, Ill., 300 W) with a 4.5-in. probe.

Tracer. Homogenate is diluted 10-fold with either acetic acid or acetone:HCl (80:20) and transferred to a polypropylene centrifuge tube. Tritiated leucine enkephalin (tyrosyl-3,5-³H-Gly₂-Phe-Leu, 0.1 μCi, New England Nuclear, Cambridge, Mass.) is added and samples equilibrated overnight at 4°C.

Protein precipitation. Proteins are precipitated with 10 vol of acetone:0.01 N HCl (80:20) and removed after centrifugation at 15,000g for 20 min at 0 to 5°C (Beckman J-21, Palo Alto, Calif.). Supernatant is removed, 5 vol of acetone:0.01 N HCl (60:40) is added, and the mixture recentrifuged (10). Supernatant is removed and evaporated in a stream of nitrogen and is ready for chromatography.

Porous polystrene–divinylbenzene columns. Residue is dissolved in 4 ml Tris–HCl buffer at pH 7.4 and placed on a polystyrene–divinylbenzene column (2 g Biobeads SM2, Bio-Rad, Richmond, Calif., 20–50 μm mesh). These columns are 50:50 copolymers of polystyrene:divinylbenzene. A peptide fraction is eluted with 8 ml methanol instead of 0.9 ml as mentioned by other workers (13).

Octadecylsiloxane minicolumn. This type of minicolumn is a commercially available (Sep-Paks, Waters, Milford, Mass.) prepackaged octadecylsiloxane-derivatized siliconaceous column (0.4 g) used for reverse-phase chromatography (27). The minicolumn is prepared by washing with 4 ml methanol, 4 ml water, 4 ml methanol, then 8 ml trifluoroacetic acid (0.5%) (28). The sample is dissolved in 4 ml trifluoroacetic acid (0.5%) and placed on the minicolumn. A peptide fraction is removed with 2 ml of acetonitrile:0.1% trifluoroacetic acid (80:20).

Radioactivity counting. Radiolabeled enkephalin is determined by counting radioactivity (Hewlett–Packard Tri-Carb 460C) during several purification steps: protein precipitation, before and after polystyrene–divinylbenzene and octadecylsiloxane columns, and at the end of high-performance liquid chromatography separation.

High-performance liquid chromatography. A Waters (Milford, Mass.) high-performance liquid chromatography apparatus is used and consists of a U6K injector, guard column packed with Corasil B (50-μm diameter), μBondapak C_{18} (10-μm diameter) reverse-phase analytic steel column (4 mm × 30 cm), two Model 6000A solvent delivery pumps, Model 600 solvent programmer, and a Model 450 variable wavelength uv detector (22). Analog signals are recorded on a Houston recorder or a Waters data module.

A novel, volatile buffer is used for peptide resolution. The organic modifier is acetonitrile. Dilute formic acid (0.04 M) is titrated to pH 3.15 with distilled triethylamine to form the triethylamine formate (TEAF) buffer (29,30). Aqueous solutions are filtered and degassed with filters (0.45-μm pore diameter, HAWP 04700, Millipore, Bedfore, Mass.). Organic solutions are treated similarly (FHIP 04700, 0.5-μm pore diameter).

Samples are dissolved in triethylamine formate buffer (100 μl). After high-performance liquid chromatographic resolution into individual peaks, fractions are collected by switching the "waste–recycle–collect" valve to "collect." The collected volume of solvent (several milliliters) is reduced by elution through an octadecylsiloxane minicolumn.

Field-desorption mass spectrometry. A Finnigan (Varian) MAT 731 (Bremen, West Germany) mass spectrometer of Mattauch-Herzog double-focusing geometry outfitted with a field-desorption/field-ionization/electron-ionization combination source is used. Nominal resolution is 1000, source temperature 90°C, emitter potential +8 kV, and counterelectrode −3kV. Field-desorption emitters are fabricated in our laboratory from 10-μm-diameter tungsten wire by activation at high temperature in a benzonitrile atmosphere under the influence of a high electric field. Carbon microneedle growth on the

emitter wire surface extends to a length of approx 30 μm.

The sample is dissolved in 100 μl methanol to wash down the sides of the reactivial, and a gentle stream of nitrogen reduces sample volume to less than 10 μl. The sample is carefully transferred to the emitter with assistance from a microsyringe–micromanipulator–stereomicroscope ensemble (31).

The peak-matching unit of the mass spectrometer is set to scan alternatively protonated molecular ions of leucine enkephalin at mass 556 and Ala^2-leucine enkephalin at 570. Emitter heating current (ehc) is increased manually to 16 mA, where peptides desorb optimally (22). An oscillographic recording of the entire desorption envelope is obtained manually as peak switching continues. Individual ion currents of the two (M + H)$^+$ ions are integrated manually.

RESULTS

Fresh canine brain tissue was obtained immediately after surgical exposure of the brain. Rapid neurosurgical excision, tissue freezing, and storage of brain tissue at −70°C minimize further peptide biosynthesis and catabolism, in contrast to a recent report which indicates stable methionine enkephalin and substance P-like immunoreactivities when a brain is cooled from 36 to 4°C over a 72-h period (32). Vogel and Altstein report a tissue protease cleaves the Tyr–Gly bond in enkephalins (33).

The combination of homogenization of canine caudate nuclei tissue and subsequent ultrasonic cell disruption is a procedure which optimizes extraction of cell cytosolic contents.

Ideal internal standards are peptides labeled with stable isotopes at several nonlabile positions. Appropriate stable isotopes are ^2H, ^{15}N, and ^{18}O. At least 3 isotopes per molecule are required to shift the protonated molecular ion away from endogenous peptide isotope peaks. The hydrophobicity properties during HPLC and the field-desorption properties during mass spectrometric measurement of the internal standard molecule must be compatible with endogenous leucine enkephalin.

Higher homologs such as Ala²-leucine enkephalin also serve well as internal standards. Even though leucine enkephalin and Ala²-leucine enkephalin are well resolved by the triethylamine formate buffer on high-performance liquid chromatography, the two fractions are readily collected in one vial. Overnight equilibration of the internal standard with disrupted cells is necessary before chromatographic separation.

Nonsilanized glassware is obligatory for any quantification method which utilizes field desorption mass spectrometry. When HPLC quantification is performed on peptide samples collected in silanized glassware, no interferences (shifted retention times) are observed. However, field-desorption mass spectra of peptides that are collected in silanized glassware show an increase in the molecular weight of the peptide and reflect apparent derivatization of a labile hydrogen.

The amount of total lipids in whole brain is 104 mg/g fresh brain tissue, whereas total protein is 110 mg/g (34). Acidified acetone was chosen for protein precipitation because of increased speed of solvent removal and sample preparation. Tracer recovery studies showed little difference of recovery of tritiated leucine enkephalin between precipitating proteins with acidified acetone or acetic acid (35).

It is emphasized that no study can determine exactly the efficiency of extracting endogenous compounds from a biologic matrix (36). Addition of a radiolabeled compound to a biologic sample followed by solvent extraction and monitoring of the radiolabel serves as one of the best estimates of the endogenous level of a compound, but may or may not actually reflect extraction of the compound of interest. Addition of a stable isotope-labeled internal standard provides a better estimate for mass spectrometric quantification methods. However, both addition methods may (and probably do) suffer from equilibration effects, nonquantitative trans-

port into (presumably totally) disrupted cells, protein–peptide interactions, etc. No methodology exists to resolve these potential unknown factors and appropriate assumptions must be made.

Polystyrene–divinylbenzene copolymer columns have a specific affinity for enkephalins, presumably because of the presence of the benzene rings in the copolymer matrix. Whereas milligram amounts of polystyrene–divinylbenzene suffice for retention of pure peptide standards, biologic matrix effects demand 2 g in the present scheme for sample extraction. Tracer studies show 65% recovery of leucine enkephalin. According to the manufacturer, the molecular weight operating range of polystyrene–divinylbenzene columns is 600–14,000.

Nearly quantitative (95%) amounts of leucine enkephalin are recovered from octadecylsiloxane minicolumns which are utilized to remove an interfering peak following elution from a polystyrene–divinylbenzene column.

Reverse-phase high-performance liquid chromatography methodology utilizing dilute triethylamine formate buffer provides high resolution of peptide separation, high sensitivity of detection, and reproducibility of retention time for neuropeptides (22).

Speed of isocratic HPLC separation is measured in minutes and is optimized by altering percentage of organic modifier. Femtomoles (10^{-15} mol) of somatostatin are readily detected at 194 nm (37). One reason for this high level of sensitivity is the high molar absorptivity of the peptide bond at this wavelength. The molar absorptivity of methionine enkephalin equals 10^5. Reproducibility of retention time for multiple injections of endogenous neuropeptides in a biologic matrix is a pivotal feature of using HPLC in this type of biological work. Retention time of multiple injections does not vary significantly (coefficient of variation using a paper chart recorder is 4%) in the present system (22). Even though an HPLC uv peak may not be observed, retention time can be relied upon and collections obtained for bioassay, radioimmunoassay, or mass spectrometric measurements.

Figure 1 is a photograph of the oscillographic paper recording during field desorption of peptide and internal standard to illustrate the type of primary data obtained in this study. Intervening peaks observed in Fig. 1 are due to the return of the oscilloscopic trace. Optimal accuracy of manual integration requires approx 30 scans across each peak per emitter loading.

Figure 2 is the high-performance liquid chromatogram of the extract from 3 g of hypothalamic tissue. Although the chromatogram appears complicated, it is less complex when compared to cortex tissue extracts. The uv detector is intentionally overloaded by injecting the majority ($0.80 \times 3 = 2.4$ g) of tissue extract. The internal standard peak is well resolved from other peaks and, even though peaks due to methionine enkephalin (5.4 min) and leucine enkephalin (7.7 min) are not resolved from their respective immediate neighbors, the latter peptide effluent is collected (6.5–8.5 min) and mixed with collected internal standard (13–14 min) to provide a sample for FD-MS.

Although this method at first seems to decrease the high resolution of peaks resolved with RP-HPLC, it must be remembered that the detector (a mass spectrometer) measures a unique molecular parameter (protonated molecular ion) and actually provides increased chromatographic resolution.

Individual ratios (R) of manually integrated ion currents due to leucine enkephalin and internal standard unspiked and spiked samples are given in Table 1. Two separate caudate nucleus and two hypothalamic tissue extracts were studied. Ion-current integrals are corrected for cross contributions from the two respective ion currents (contribution of 570 to 556 is 2% and 556 to 570 is 0.3%).

Use of both spiked and unspiked samples to calculate endogenous amounts of neuropeptides in biologic tissue bypasses the potential difficulty of differential extraction recoveries from different tissues.

DISCUSSION

The sensitivity of uv detectors for HPLC is high for standard solutions of underivatized neuropeptides such as somatostatin and ranges down to 615 fmol (1 ng, Ref. (31)). On the other hand, amounts quantified by HPLC in biologic samples are in the nanogram (picomole) level, a less sensitive level due to biologic matrix effects. High-performance liquid chromatographic resolution of endogenous neuropeptides from neighboring peaks is extraordinarily high, illustrated by high-performance liquid chromatograms of standards and biologic extracts (22,35,37). Nevertheless, a chromatographic system simply provides a measurable but nonstructural parameter (uv absorption, fluorescence, electrochemical potential, refractive index change, etc.) at a specific retention time. Retention time is very reproducible in HPLC (coefficient of variation 4%, Ref. (22)), and HPLC can be depended upon to collect, identify, and measure specific compounds extracted from biologic sources. Nevertheless, structures of compounds cannot be obtained by chromatographic procedures alone.

With RIA, the activity of usually only a portion of the molecule is what is actually quantified. Carboxy-directed antisera yielded higher results than amino-directed antisera for plasma neurotensin (38). Other neuropeptides are quantified by antisera directed to the middle portion. Furthermore, if a small molecule must be chemically conjugated to a larger carrier molecule for RIA, probability increases that molecular specificity of immunoreactivity toward the smaller molecule (the object of the quantification) will further decrease. Analytic methodology is required as an ultimate standard to calibrate RIA data (39). Data presented in this paper illustrate that the novel technique of field-desorption mass spectrometry is suited for this purpose. For example, down to 30 ng of underivatized synthetic leucine enkephalin is readily quantified (31). This quantification level is currently achieved with manual techniques (ehc control, oscillo-

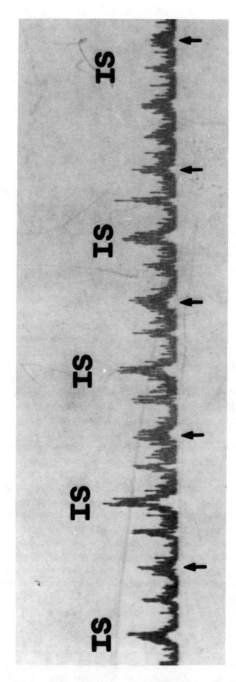

FIG. 1. Photograph of a section of oscillographic paper recording of ion currents of protonated molecular ions of leucine enkephalin (arrow) and Ala²-leucine enkephalin internal standard (IS) obtained during a field-desorption mass spectrometric measurement.

graphic scans, integrate-ion currents). A 10-fold (or greater) increase in overall sensitivity is predicted when microprocessor-controlled ehc increases and data-acquisition procedures (currently under development in our laboratory) are employed. When the molecular weight of the compound is less than 1000 to 2000 amu and field-desorption mass spectrometric techniques can be employed, there is total confidence that the molecular ion produced by FD-MS reflects the total, intact molecular structure of the neuropeptide under investigation. It must be remembered that, in those cases where even greater structural specificity is required, high-resolution mass spectrometry and the new technique of collision-activation mass spectrometry (40) are available to provide structural information. Both methods concomitantly decrease sensitivity while providing increased confidence that the compound being quantified is indeed the compound of interest and not a larger precursor (41–44) or metabolite which displays, in some cases, equivalent radioimmunoreactivity.

Field-desorption mass spectrometric measurements presented here for the first time are most compatible with the data of Lindberg and Dahl (45), who measured 520 ng leucine enkephalin/g of rat caudate nucleus. (These workers and others assume 10% protein in wet tissue.) Bovine caudate nucleus tissue was measured by Dupont et al. (14) and leucine enkephalin was 210 ng/g tissue. Leucine enkephalin in human caudate nucleus was measured by Kubeck (46) as 60 ng/

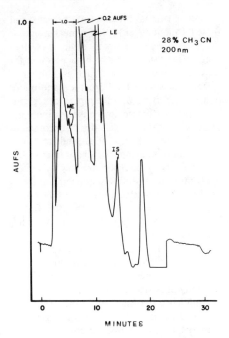

FIG. 2. Reverse-phase high-performance liquid chromatogram of 3 g hypothalamic tissue extract. Parameters are amount injected, 80 μl (80% sample); pH 3.2; 1 μg internal standard added/g tissue; 28% acetonitrile:72% triethylamine formate isocratic elution; detection at 200 nm; flow rate, 1.5 ml min^{-1}; collections made between 6.5 and 8.5 and 13 and 14 min.

g and by Emson et al. (32) as 14 ng/g tissue. Peralta et al. (13) measured methione enkephalin in rat tissue and found 130 ng/g tissue. Assuming one-seventh as much leucine enkephalin as methionine enkephalin,

TABLE 1

MEASUREMENTS OF LEUCINE ENKEPHALIN IN CAUDATE NUCLEUS AND HYPOTHALAMUS TISSUE EXTRACTS[a]

Tissue	Unspiked ratios[b]	Spiked ratios[b]	Amount (ng)
Caudate nucleus	0.340, 0.365 (0.353)	0.422, 0.618 (0.520)	423
	0.402, 0.314, 0.395 (0.370)	0.465, 0.657, 0.620 (0.581)	351
Hypothalamus	0.229, 0.089, 0.250 (0.189)	0.271, 0.390 (0.330)	268
	0.168, 0.125, (0.147)	0.282, 0.319, (0.300)	240

[a] In nanograms/gram tissue.
[b] Numbers in parentheses are the mean values.

19 ng/g of leucine enkephalin is present, and these RIA leucine enkephalin measurements in caudate nucleus tissue cover a 37-fold range. This difference can be accounted for by several considerations, such as differences in antibodies used for RIA, the handling of tissue (time to acquire), temperature, peptidase (47), synthetase, and enkephalinase activity, species differences, use of chromatography before RIA, age of animal, and possibly "emotional state" of the animal (48).

In the past, mass spectrometry was effectively utilized in peptide sequencing and was used for the first structural elucidation of a hypothalamic releasing factor (49). Today, high-performance liquid chromatography, field-desorption mass spectrometry, and appropriate stable isotope-labeled internal standard (50) enable measurement of endogenous levels of neuropeptides and more importantly provide a significant advancement in confidence that the compound of interest is what is being measured. It is realized that mass-spectrometric instrumentation is sophisticated to operate, costly to purchase, and is not available to the majority of laboratories performing neuropeptide research. Nevertheless, whenever possible, other types of quantification data should be substantiated with the structural certainty that mass spectrometry offers of the compound being quantified. Several brain peptides are now being shown to arise from larger precursor molecules and a general scheme of metabolic regulation is evolving (41–44,51). Chromatographic and FS-MS techniques outlined in this paper provide means to specifically relate structure to the quantifiable amount of a neuropeptide. In short, for the first time, a method is available to calibrate RIA data. It is clear that high-performance liquid chromatography and field-desorption mass spectrometry will not replace radioimmunoassay due to the advantages of the latter in low cost, high speed, and high sensitivity.

In summary, this is the first report of the combination of chromatographic, HPLC, and FD-MS techniques for measurement of endogenous amounts of biologically active underivatized neuropeptides extracted from canine brain tissue extracts. Data presented in this paper compare favorably with the data of others who quantified enkephalins in regions of the brain with RIA. Work continues along several fronts. Sensitivity of detection will be improved by using microprocessors for reasons stated above. Quantification of methionine and leucine enkephalins in several brain regions will be performed to supplement data obtained for caudate nuclei. Other neuropeptides amenable to HPLC and FD-MS techniques described here will be quantified. For peptides having a molecular weight greater than 1000 to 2000 amu, appropriate enzymology will be employed to reduce the peptide length to a size compatible with FD-MS methods.

ACKNOWLEDGMENTS

The authors gratefully acknowledge assistance from J. C., financial assistance from NIH (GM NS 26666), helpful discussions with Dr. F. S. Tanzer, technical assistance of J. Trimble, typing assistance of K. Smith and D. Cubbins, and cooperation of Dr. R. A. White (Departments of Pharmacology and Neurosurgery) in obtaining dog brains.

REFERENCES

1. Chang, K.-J, and Cuatrecasas, P. (1979) *J. Biol. Chem.* **254**, 2610–2618.
2. Yang, H.-Y. T., Di Giulio, A. M., Fratta, W., Hong, J. S., Majane, E. A., and Costa, E. (1979) *Neuropharmacology* **19**, 209–215.
3. Terenius, L. (1977) *Psychoneuroendocrinology* **2**, 53–58.
4. Chang, K.-J., Cooper, B. R., Hazum, E., and Cuatrecasas, P. (1979) *Mol. Pharmacol.* **16**, 91–104.
5. Smith, D. G., Massey, D. E., Zakarian, S., and Finnie, M. D. A. (1979) *Nature (London)* **279**, 253–254.
6. Clements-Jones, V., Lowry, P. J., Rees, L. H., and Besser, G. M. (1980) *Nature (London)* **283**, 295–297.
7. Rossier, J., Vargo, T. M., Minick, S., Ling, N., Bloom, F. E., and Guillemin, R. (1977) *Proc. Nat. Acad. Sci. USA* **74**, 5162–5165.
8. Sarne, Y., Azov, R., and Weissman, B. A., (1978) *Brain Res.* **151**, 399–403.
9. Clements-Jones, V., Lowry, P. J., Rees, L. H., and Besser. G. M. (1980) *J. Endocrinol.* **86**, 231–243.

10. Ueda, H., Shiomi, H., and Takagi, H. (1980) *Brain Res.* **198**, 460–464.

11. Stern, A. S. Lewis, R. V., Kimura, S., Rossier, J., Gerber, L. D., Brink, L., Stein, S., and Udenfriend, S. (1979) *Proc. Nat. Acad. Sci. USA* **76**, 6680–6683.

12. Wahlstrom, A., Johansson, L., and Terenius L. (1976) Opiates and Endogenous Opioid Peptides, pp. 49–55, Elsevier, Amsterdam/New York.

13. Peralta, E., Yang, H.-Y. T., Hong, J., and Costa, E. (1980) *J. Chromatogr.* **190**, 43–51.

14. Dupont, A., Lepine, J., Langelier, P., Merand, Y., Rouleau, D. Vaudry, H., Gros, C., and Barden, N. (1980) *Regulatory Pept.* **1**, 43–52.

15. Gros, C., Pradelles, P., Rouget, C., Bepoldin, O., Dray, F., Fournie-Zaluski, M. C., Roques, B. P., Pollard, H. Llorens-Cortes, C., and Schwartz, J. C. (1978) *J. Neurochem.* **31**, 29–39.

16. Akil, H., Watson, S. J., Sullivan, S., and Barchas, J. D. (1978) *Life Sci.* **23**, 121–126.

17. Yalow, R. S., (1980) *Ann. Rev. Biophys. Bioeng.* **9**, 327–345.

18. Skelley, D. S., Brown, L. P., and Besch, P. K. (1973) *Clin. Chem.* **19**, 146–186.

19. Beckey, H. D., and Schulten, H.-R. (1975) *Angew. Chem. Int. Ed.* **14**, 403–415.

20. Frick, W., Barofsky, E., Daves, G. D., Barofsky, D. F., Chang, D., and Folkers, K. (1978) *J. Amer. Chem. Soc.* **100**, 6221–6225.

21. Przybylski, M., Luderwald, I., Kraas, E., and Voelter, W., (1979) *Z. Naturforsch.* **34b**, 736–743.

22. Desiderio, D. M., Stein, J. L., Cunningham, M. D., and Sabbatini, J. Z. (1980) *J. Chromatogr.* **195**, 369–377.

23. Achenbach, C., Hauswirth, O., Heindrichs, C., Ziskoven, R., Kohler, F., Bahr, U., Heindricks, A., and Schulten, H.-R. (1980) *J. Toxicol. Environ. Health* **6**, 519–528.

24. Miyazaki, H., Shirai, E., Ishibashi, M., Hosoi, K., Shibata, S., and Iwanaga, M. (1978) *Biomed. Mass Spectrom.* **5**, 559–565.

25. Lehmann, W. D., Schulten, H.-R., and Schroder, N., (1978) *Biomed. Mass Spectrom.* **5**, 591–595.

26. Bahr, U., Schulten, H.-R., Hommes, O. R., and Aerts, F. (1980) *Clin. Chem. Acta* **103**, 183–192.

27. Bennett, H. P. J., Browne, C. A., and Solomon. S. (1980) *Biochemistry* **20**, 4530–4538.

28. Fasco, M. J., Cashin, M. J., and Kaminsky, L. S. (1979) *J. Liq. Chromatogr.* **2**, 565–575.

29. Desiderio, D. M., Cunningham, M. D., and Trimble, J., (1981) *J. Liq. Chromatogr.* **4**, 1261–1268.

30. Desiderio, D. M., and Cunningham, M. D., (1981) *J. Liq. Chromatogr.* **4**, 721–733.

31. Desiderio, D. M., Yamada, S., Sabbatini, J. Z., and

Tanzer, F. (1981) *Biomed. Mass Spectrom.* **8**, 10–12.

32. Emson, P. C., Arregui, A., Clement-Jones, V., Sandberg, B. E. B., and Rossor, M. (1980) *Brain Res.* **199**, 147–159.

33. Vogel, Z., and Altstein, M. (1977) *FEBS Lett.* **80**, 332–336.

34. Diem, K., and Lentner, C. (1970) *Sci. Tables*, p. 520, Ciba-Geigy, Basel.

35. Desiderio, D. M., Yamada, S., Tanzer, F. S., Horton, J., and Trimble, J. (1981) *J. Chromatogr.* **217**, 437–452.

36. Lawson, A. M., Lim, D. K., Richmond, W., Samson, D. M., Setchell, K. D. R., and Thomas, A. C. S. (1980) *in* Current Developments in the Clinical Applications of HPLC, GC, and MS (Lawson, A. M., Lim, C. K., and Richmond, W., eds.), pp. 144–145, Academic Press, New York.

37. Desiderio, D. M., and Cunningham, M. D., (1981) *J. Liq. Chromatogr.* **4**, 721–733.

38. Carraway, R., Hammer, R. A., and Leeman, S. E. (1980) *Endocrinology* **107**, 400–406.

39. Granstrom, E., and Kindahl, H. (1978) *Advan. Prostagl. Thrombox. Res.* **5**, 119–210.

40. Desiderio, D. M., and Sabbatini, J. Z. (1981) *Biomed. Mass Spectrom.* **8**, 565–568.

41. Lewis, R. V., Stern, A. S., Kimura, S., Rossier, J., Stein, S., and Udenfriend, S. (1980) *Science* **208**, 1459–1461.

42. Jones, B. N., Stern, A. S., Lewis, R. V., Kimura, S., Stein, S., Udenfriend, S., and Shively, J. E. (1980) *Arch. Biochem. Biophys.* **204**, 392–395.

43. Blalock, J. E., and Smith, E. M. (1980) *Proc. Nat. Acad. Sci. USA* **77**, 5972–5974.

44. Brazeau, P., Ling, N., Esch, F., Bohlen, P., Benoit, R., and Guillemin, R. (1981) *Regulatory Pept.* **1**, 255–264.

45. Lindberg, I., and Dahl, J. L. (1981) *J. Neurochem.* **36**, 506–512.

46. Kubeck, M. J., and Wilber, J. F., (1980) *Neurosci. Lett.* **18**, 155–161.

47. Roques, B. P., Fournie-Zaluski, M. C., Soroca, E., Lecomte, J. M., Malfroy, B., Llorens, C., and Schwartz, J. C. (1980) *Nature (London)* **288**, 286–288.

48. Carr, D. B. (1981) *Lancet*, February 14, p. 390.

49. Burgus, R., Dunn, T. F., Desiderio, D. M., Ward, D. N., Vale, W., Guillemin, R. (1970) *Nature (London)* **226**, 321–325.

50. D. M. Desiderio, and M. Kai, submitted.

51. Patey, G., De La Baume, S., Schwartz, J. C., Gros, C., Roques, B., Fournie-Zaluski, M. C., and Soroca-Lucas, E. (1981) *Science* **212**, 1153–1155.

Fluorescent Techniques for the Selective Detection of Chromatographically Separated Peptides[1]

Timothy D. Schlabach and C. Timothy Wehr

Varian Instrument Group, Walnut Creek Instrument Division, 2700 Mitchell Drive, Walnut Creek, California 94598

Fluorescence detection has been used for the detection of peptides that contain tyrosine or tryptophan. The strong native fluorescence of these groups makes this method several times more sensitive than detection at 254 nm. Furthermore, when combined in series with a lower wavelength absorbance detector—220 nm, for example—the fluorescence-to-absorbance ratios can be used to distinguish peptides having a single tyrosine residue from those having a tryptophan residue. Reaction detection can also be used to specifically identify peptides with lysine residues. The ϵ-amino group in lysine reacts far better with *o*-phthalaldehyde than any other peptide group including the primary amino terminus. The reagent is continuously added to the column effluent, generating a chromatogram of the peptides with lysine. Not only is this postcolumn reaction specific, it also detects as little as 10 pmol of some peptides.

Rapid peptide separations have become routine with efficient columns (1,2) and high-pressure liquid chromatographs. Eluting peptides are typically detected by absorbance around 220 nm or below (2). However, some solvent systems absorb at these wavelengths and require other means for detection. Unblocked peptides can be detected at the nanogram level by postcolumn derivatization with fluorescamine in the presence of strongly absorbing solvents such as pyridine (3,4). Either method measures peptide concentration but does not provide clues to peptide identity.

For the analysis of an unknown peptide mixture, peptide identity must be established by sequence analysis. However, if there is prior knowledge about the peptide mixture, amino acid analysis or N-terminal identification might suffice. Unfortunately, all these techniques are off-line and greatly increase the time and expense of the analysis.

On-line methods do exist for detecting key residues directly as peptides elute from the column. Absorbance ratioing is a simple technique to identify peptides with aromatic amino acid residues. The peptide chromatogram is monitored at a wavelength where all peptides absorb, such as 220 nm, and also at 254 nm, where only peptides with aromatic amino acids are detected (5). Typically, two detectors are used in series.

Detection at 254 nm does have a couple of drawbacks. First, sensitivity is much poorer than detection at lower wavelengths. Second, it is difficult to establish the identity of the aromatic amino acid. Thus, peak identification could be impaired if two or more closely related peptides all contained a residue that absorbed at 254 nm.

These deficiencies may be overcome by substituting a fluorescence detector. Both tyrosine and tryptophan fluoresce strongly, so detection limits should be lower than with 254-nm detection. Because tryptophan fluoresces with greater intensity and somewhat downfield from tyrosine (6), it should be possible to choose conditions where the former will yield peaks of greater intensity. Thus, it

[1] This paper was presented at the International Symposium on HPLC of Proteins and Peptides, November 16–17, 1981, Washington, D.C.

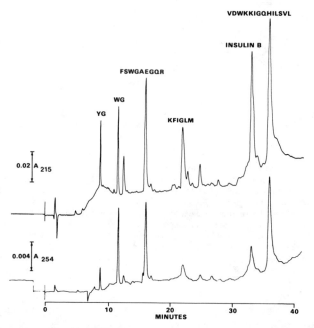

FIG. 1. Comparison of peptide profiles observed at 215 and 254 nm. Detection at 215 nm is shown in the upper trace and that at 254 nm in the lower. The sample contained about 2 μg each of tyrosylglycine, trytophylglycine, allergic peptide, eledoisin-related peptide, insulin B chain, and polistes mastoparan in order of their elution. The eledoisin peptide is an amide.

may be possible to distinguish between tryptophan and tyrosine peptides.

Lysine is another key residue of great interest in peptide mapping. Unfortunately, spectroscopic methods can't distinguish lysine from a host of other amino acids. However, chemical methods have been used to determine the lysine content of proteins (7). This method is based on the reaction of o-phthalaldehyde (OPA)[2] with primary amine groups, forming a highly fluorescent isoindole ring (8). When OPA was studied for peptide detection, it was found to be surprisingly unreactive with the primary amino terminus but was reported to react with peptidyl lysine groups (9). If suitable conditions can be found, it should be possible to perform

the OPA assay as a continuous postcolumn reaction. This would generate a chromatogram of lysine-containing peptides.

MATERIALS AND METHODS

Samples and reagent. Peptides were obtained from Peninsula Laboratories, San Carlos, California. Tryptic peptides from human growth hormone were a gift from Genentech, San Bruno, California.

The o-phthalaldehyde reagent was obtained from Varian, Palo Alto, California. The freshly prepared reagent contained 0.25 mg/ml o-phthalaldehyde, 0.05% w/w mercaptoethanol in 0.1 M potassium borate, pH 10.4.

Apparatus. All experiments were performed with a Varian Model 5060 ternary liquid chromatograph equipped with a UV-

---

[2] Abbreviations used: OPA, o-phthalaldehyde; MSH-INH, melanotropin, releasing inhibitor factor.

5 selectable wavelength detector. Postcolumn reactions were performed with the Varian System I postcolumn reaction system including the Fluorichrom fluorescence detector that comes with the appropriate excitation and emission filters for the OPA reaction, 340 and 450 nm, respectively. For native peptide fluorescence, a deuterium source was used along with 220- and 330-nm interference filters for excitation and emission, respectively.

Chromatographic methods. Peptides were separated on MicroPak MCH-5 or SP-3 columns, 15 × 0.4 cm. Both are C-18 reverse-phase columns. Peptide elution began with 0.05% v/v trifluoroacetic acid in distilled water. The strong solvent was 90% v/v acetonitrile and distilled water containing the same amount of trifluoroacetic acid. The flow rate was 0.8 ml/min. The gradient ramped linearly to 20% strong solvent, 5 min

after injection. During the next 25 min, the gradient was linearly incremented to 40% strong solvent. Over the last 10 min, 60% strong solvent was reached.

For OPA reaction detection, the OPA reagent was added at 0.5 ml/min and reacted for about 25 s in the reaction coil. The total flow rate was 1.3 ml/min.

RESULTS AND DISCUSSION

Tyrosine/Tryptophan Detection

Aromatic amino acids have characteristic uv-absorption bands between 250 and 290 nm (10). Peptides containing such aromatic groups can be readily identified by absorbance in the characteristic region. To distinguish those peptides with aromatic amino acids, the peptide separation is monitored at two wavelengths, usually with two detectors

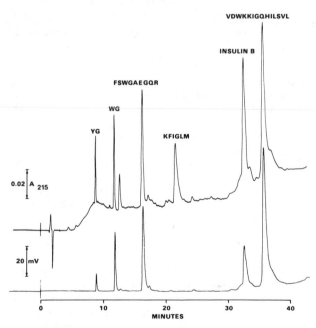

FIG. 2. Comparison of peptide detection with native fluorescence and absorbance at 215 nm. Fluorescence excitation was at 220 nm and emission at 330 nm. Fluorescence detection is shown in the lower chromatogram. The peptide sample is described in Fig. 1.

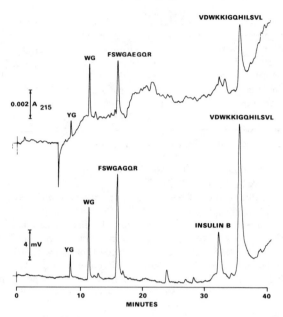

FIG. 3. Relative sensitivity of native fluorescence and absorbance at 254 nm for peptide detection. The upper chromatogram shows absorbance and the lower fluorescence. The previously described sample was diluted sixfold.

in series. Typically, 215 nm is chosen for universal peptide detection and 254 nm for selective detection of the aromatic amino acid residues. As seen in the peptide chromatogram shown in Fig. 1, peptides containing at least one phenyl, tyrosyl, or tryptophyl residue can be detected at 254 nm. Comparing the peptide chromatogram observed at 254 nm with that observed at 215 nm reveals differences in both the absolute and relative magnitudes of the peaks. The response at 215 nm is at least 6 times greater than at 254 nm. More importantly, the most prominent peaks observed in the 254-nm profile all have a tryptophan residue. The next largest peak, insulin B chain, has two tyrosine residues, and the smaller peaks have a single tyrosine or phenylalanine residue.

Fluorescence has been reported to be a very sensitive technique for detecting tryptophan and tyrosine with sensitivities as low as 3 ng/ml for the former and 5 ng/ml for the latter (11). Excitation was around 280 nm for both, and maximal tryptophan emission was at 365 and 310 nm for tyrosine in alkaline solution. In slightly acidic solutions emission maxima are slightly blue-shifted to 354 and 303 nm, respectively (12). Excitation at shorter wavelengths, such as 220 nm, was reported to improve detection more than threefold for a decapeptide containing one tryptophan residue (13). This is not unexpected because tryptophan absorbance is higher at 220 nm and the excitation spectrum usually closely follows the absorbance spectrum. Because tyrosine also absorbs more strongly in the low uv, excitation at 220 nm was chosen to provide the best sensitivity for tyrosine and tryptophan residues.

By collecting emission at 330 nm where both tryptophan and tyrosine emission overlap, fluorescence detection should also be selective for peptides with these two groups. Separation of the same peptide mixture in Fig. 2 reveals that the fluorescence chromatogram is similar to that observed at 254 nm with some notable exceptions. The peptide containing only phenylalanine is not observed in the fluorescence profile. The baseline is certainly flatter and less noisy. As before, the tryptophan peptides are the dominant peaks.

Direct comparison of 254-nm and fluorescence detection is observed in Fig. 3 for a sixfold dilution of the previous sample.

Fluorescence detection is more than 6 times more sensitive than 254 nm for the two peptides containing tryptophan. The fluorescence detection limit for these peptides is about 10 ng (6–10 pmol). Insulin B chain peptide is not even discernable in the 254-nm chromatogram, although this peptide has two tyrosine residues. The fluorescence detection limit for this peptide is about 40 ng (12 pmol).

It is evident from Fig. 2 that the ratio of absorbance peaks at 215 nm to fluorescence is different for peptides with tyrosine than

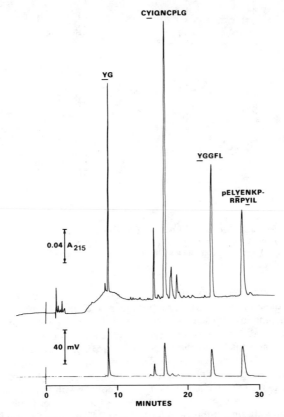

FIG. 4. Fluorescence and absorbance detection of peptides containing tyrosine. The upper chromatogram was recorded at 215 nm and fluorescence in the lower. The sample contained about 5 μg each of tyrosylglycine, oxytocin, Leu-enkephalin, and neurotensin in order of their elution. Oxytocin has a C-terminal amide group. The peptides were chromatographed on the MCH-3 column.

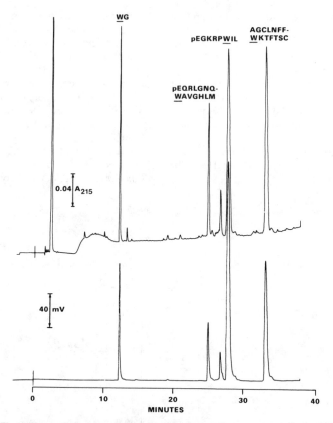

FIG. 5. Fluorescence and absorbance detection of peptides containing tryptophan. The chromatographic and detector conditions are as described in Fig. 4. The sample contained about 5 µg of tryptophylglycine, bombesin (an amide), xenopsin, and somatostatin in order of their elution.

those with tryptophan. There is also substantial variation in these ratios among a group of peptides containing just tyrosine, as seen in Fig. 4. The lowest ratio was found for oxytocin and may be attributable to the additional absorbance of the cystine group. Neurotensin has a ratio which is 3 times higher. This is likely due to the presence of a second tyrosine residue.

There is similar scattering in this ratio for peptides with a single tryptophan residue. In the peptide separation shown in Fig. 5, xenopsin has a ratio which is more than twice that of bombesin. Perhaps the adjacent proline residue in the former peptide better situates the tryptophan group for fluorescence.

When the two previous separations are compared, it is apparent that fluorescence-to-absorbance ratios determined from the respective peak heights are uniformly higher for peptides with tryptophan residues. Ratio values for a number of peptides are listed in Table 1. The group average for peptides containing a single tyrosine group is 195, whereas the average for peptides with a lone tryptophan residue is 707. There is less than a 1%

TABLE 1

NATIVE PEPTIDE FLUORESCENCE[a] TO ABSORBANCE[b] RATIOS

Peptide	Sequence	Fluorescence/ absorbance (mV/AU)	Number of residues	
			Tyr	Trp
Tyrosylglycine	YG	240	1	
Leu-enkephalin	YGGFL	214	1	
Met-enkephalin	YGGFM	247	1	
Proctolin	RYLPT	215	1	
Pressinoic acid	CYFQNC	102	1	
Angiotensin II	DRVYIHPF	235	1	
Oxytocin[c]	CYIQNCPLG	115	1	
Neurotensin	pELYENKPRRPYIL	347	2	
Insulin B chain	FVNQHLCGSH			
	LVGALYLVCG			
	ERGFFYTPKT	302	2	
Tryptylglycine	WG	537		1
Allergic peptide	FSWGAEGQR	716		1
Xenopsin	pEGKRPWIL	1170		1
Ranatensin[c]	pEVPQWAVGHFM	557		1
Bombesin[c]	pEQRLGNQWAVGHLM	435		1
Somatostatin	AGCKNFFWKTFTSC	636		1
Polistes mastoparan	VDWKKIGQHILSVL	898		1

[a] Excitation, 220 nm; emission, 330 nm.
[b] At 215 nm.
[c] Amide at the carboxy terminus.

probability that random error could account for the differences between these groups. Thus, this ratio should prove to be useful for distinguishing between peptides with a single tyrosine residue and those with a solitary tryptophan.

TABLE 2

FLUORESCENCE RESPONSE OF PEPTIDES IN THE OPA POSTCOLUMN REACTION

Peptide	Sequence	Fluorescence (MV/nmol)
Lysylvaline	K.V	180.0
Tyrosylglycine	Y.G	3.2
Tryptylglycine	W.G	3.0
Glycylglycylglycine	G.G.G	12.0
MSH-INH[a]	P.L.G	1.0
Tuftsin	T.K.P.R	240.0
Kentsin	T.P.R.K	210.0
Met-enkephalin	Y.G.G.F.M	3.3
ERP[a]	K.F.I.G.L.M	42.0
Xenopsin	pE.G.K.R.P.W.I.L	190.0
LHRH[a]	pE.H.W.S.Y.G.L.R.P.G	2.5

[a] Amide at the carboxy terminus.

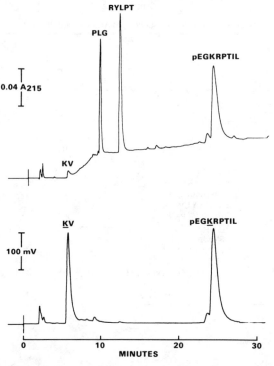

FIG. 6. Direct absorbance detection coupled with OPA reaction detection. Absorbance was monitored at the column exit prior to the postcolumn reaction. Fluorescence excitation was at 340 nm and emission at 450 nm. It was monitored at the reaction coil exit and is shown in the lower chromatogram. The sample contained about 5 μg each of lysylvaline, MSH-inhibitor (also an amide), proctolin, and xenopsin in order of their elution.

Lysine Specific Detection with OPA

Peptides lacking aromatic amino acid residues can only be detected at low wavelengths where amide groups begin to absorb. Alternatively, such peptides can be detected with a postcolumn reaction. Fluorescamine reacts with free amino groups including the amino terminus in peptides to produce a fluorescent product (14). Primary amines also form a strongly fluorescent product, an isoindole (6), when reacted with OPA (8). Unlike fluorescamine, OPA does not react appreciably with the amino terminus in peptides (9). It does, however, react with the ε-amino group of lysine residues in peptides (9).

The reactivity of 10 peptides with OPA is seen in Table 2. Peptide solutions were directly injected into the postcolumn reaction system. Only those peptides with lysyl groups produced strong fluorescence in the OPA reaction. Typically, peptides with lysine produced more than 50 times the fluorescence of those lacking lysine. The lone exception was glycylglycylglycine. This tripeptide may have reacted more strongly with OPA because of less steric hindrance to the formation of the ternary structure. Certainly, the amino terminus did not prove very reactive for peptides having more than just the peptide backbone. The reactivity of ε-amino group with OPA may be due to the flexible tetramethylene arm that allows the ternary reaction product to form.

The relative response of the OPA reagent to lysine- and non-lysine-containing peptides can be seen in Fig. 6. Lysylvaline and the octapeptide xenopsin both were clearly evident with OPA detection, whereas the other peptides were not. The detection limit for xenopsin with OPA is about 15 ng, or 15 pmol. Both MSH-inhibitor peptide and protolin have a free amino terminus, but failed to react to any great extent with OPA.

The peptide separation shown in Fig. 7 reveals the same effect. Only those peptides containing lysine were clearly represented in the OPA chromatogram. The three peptides with lysine ranged from 9 to 12 residues in length and had lysine in the third or fourth position. The peptide with the highest OPA response relative to the peak absorbance had a blocked amino terminus, indicating that the free amino group is probably of little significance to the OPA produced fluorescence. The detection limits for serum thy-

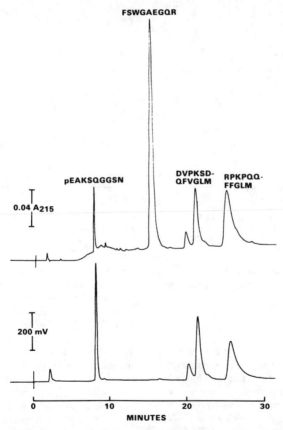

FIG. 7. Direct absorbance detection coupled with OPA reaction detection for larger peptides. The chromatographic and detector conditions are as in Fig. 6. The sample contained about 4 µg each of serum thymic factor, allergic peptide, kassinin (also an amide), and substance P in order of their elution.

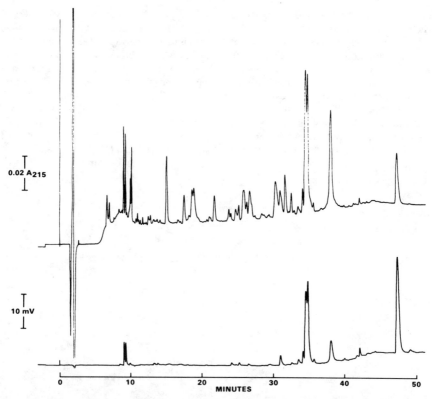

FIG. 8. Separation of tryptic peptides from human growth hormone with absorbance and native fluorescence detection. The upper chromatogram is absorbance and the lower fluorescence. About 20 μg of tryptic digest was separated on a MCH-3 column.

mic factor (pEAKSQGGSN), kassinin (DVPKSDQFVGLM), and substance P range from 5 to 10 pmol.

Application to Tryptic Peptides

Combined absorbance and fluorescence detection is seen in Fig. 8 for the separation of tryptic peptides from human growth hormone. The 3-μm particle column was used to enhance resolution of this complex mixture. This 191-residue peptide was reported to yield 17 identifiable tryptic peptides by HPLC (15). These tryptic peptides ranged from 3 to 22 residues in length. Five of these

peptides have at least one tyrosine and one has a lone tryptophan. The latter peptide can be readily identified at 48 min in Fig. 8 by its high fluorescence-to-absorbance ratio of about 945. The other ratios fall between 110 and 290. Four of the peptides with tyrosine were reported to elute toward the end of the chromatogram (15) and may be represented by peaks at 32, 35, 36, and 38 min. Apparently, a small tyrosyl peptide may have eluted as a doublet around 9 min.

Lysine at the C-terminal position could prove less reactive because of the ionic attraction between the carboxyl group and the amino group. Free lysine reacts poorly with

OPA unless a surfactant is added to the reagent (16). Although OPA did react with a tetrapeptide having lysine at the carboxy terminus, this reactivity might not hold for larger peptides. Several tryptic peptides of human growth hormone have lysine at the C-terminus, so the tryptic digest was chromatographed as before except that the OPA reaction was substituted for native fluorescence. The OPA reaction resulted in several more peaks in Fig. 9 than for native fluorescence in the previous figure. Eight tryptic peptides were reported to contain lysine, and four of those have a solitary lysine at the car-

boxy end (15). Eight major peaks are observed in Fig. 9 along with three smaller peaks. The peaks at 35 and 36 min most likely correspond with tryptic peptides 14 and 15 because they both contain tyrosine and lysine (15). The latter peptide has 21 residues and a lone lysine at the C-terminus. Although not all the peaks can be matched up with those reported because of selectivity differences between our columns, there is a strong likelihood that most of the C-terminal lysyl peptides are represented in the OPA chromatogram.

Native fluorescence and OPA detection

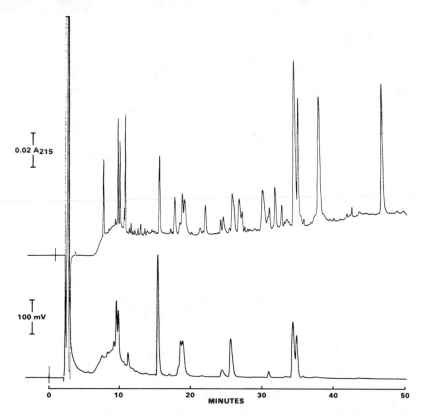

FIG. 9. Separation of tryptic peptides from human growth hormone with absorbance and OPA reaction detection. The upper chromatogram is absorbance and the lower fluorescence resulting from OPA reaction. The sample and chromatographic conditions are as described in Fig. 8.

can be used separately or in series to aid peptide identification in tryptic digests. These techniques should also prove valuable for a variety of peptide analyses. The former technique can be used in conjunction with either peptide sequencing or amino acid analyses. The latter technique requires a stream splitter for collection of unreacted peptide. With these techniques, three key residues can be identified directly as the peptide elutes from the column. As little as 10 pmol of peptide can be analyzed for these residues. Thus, the presence of tyrosine, tryptophan, and lysine can be readily determined in peptide peaks at very low levels with rapid on-line techniques.

ACKNOWLEDGMENT

We gratefully acknowledge Bill Kohr, Rodney Keck, and Richard Harkins for supplying the tryptic digest of biosynthetic human growth hormone and for their helpful comments in preparing this manuscript.

REFERENCES

1. Hearn, M. T., and Grego, B. (1981) J. Chromatogr. 203, 349–363.
2. Rivier, J. E. (1978) J. Liq. Chromatogr. 1, 343–366.
3. Rubinstein, M., Chen-Kiang, S., Stein, S., and Udenfriend, S. (1979) Anal. Biochem. 95, 117–121.
4. Bohlen, P., and Kleeman, G. (1981) J. Chromatogr. 205, 65–75.
5. Bishop, C. A., Hancock, W. S., Brennon, S. O., Carell, R. W., and Hearn, M. T. (1981) J. Liq. Chromatogr. 4, 599–612.
6. Guilbault, G. G. (1973) in Practical Fluorescence, p. 499, Dekker, New York.
7. Goodno, C. C., Swaisgood, H. E., and Catignani, G. L. (1981) Anal. Biochem. 115, 203–211.
8. Roth, M. (1971) Anal. Chem. 43, 880–882.
9. Joys, T. M., and Kim, H. (1979) Anal. Biochem. 94, 371–377.
10. Wetlaufer, D. B. (1962) Advan. Protein Chem. 17, 303–390.
11. Duggan, D., Bowman, R., Brodie, B., and Udenfriend, S. (1957) Arch. Biochem. Biophys. 68, 1–11.
12. Meyers, M. L., and Seybold, P. G. (1979) Anal. Chem. 51, 1609–1612.
13. Krol, G. J., Mannan, C. A., Pickering, R. E., Amato, D. V., Kho, B. T., and Sonnenschein, A. (1977) Anal. Chem. 49, 1836–1839.
14. Udenfriend, S., Stein, S., Bohlen, P., Dairman, W., Leimgruber, W., and Weigele, M. (1972) Science 178, 871–872.
15. Kohr, W. J., Keck, R., and Harkins, R. N. (1982) Anal. Biochem. 122, 348–359.
16. Benson, J. R., and Hare, P. E. (1975) Proc. Nat. Acad. Sci. USA 72, 619–622.

High-Performance Liquid Chromatographic Purification of Peptide Hormones: Ovine Hypothalamic Amunine (Corticotropin Releasing Factor)[1]

J. Rivier, C. Rivier, J. Spiess, and W. Vale

The Salk Institute, P.O. Box 85800, San Diego, California 92038

Amunine, or corticotropin releasing factor (CRF), a 41-peptide amide, has been isolated from ovine hypothalami. Tissues were extracted, defatted, and partitioned prior to being sized by gel permeation. The active zone followed by a sensitive *in vitro* assay (CRF-mediated corticotropin release from pituitary cells in monolayer cultures quantitated by a specific corticotropin radioimmunoassay) could be purified by reverse-phase high-performance liquid chromatography. Column characteristics (i.e., properties of the derivatized silicas including pore size), selection of mobile phases, and temperature effects are discussed.

Guillemin and Rosenberg (1) and Saffran and Schally (2) independently postulated and demonstrated in 1955 that corticotropin (ACTH)[2] release from the pituitary was stimulated by a hypothalamic substance, which they named corticotropin releasing factor, CRF. Since that time CRF has gained the label "putative" or "elusive" because it had not been characterized for so long.

We recently purified a 41-amino acid peptide from ovine hypothalami (3,4) sequenced it (3,5,6) using a spinning cap microsequencer, and synthesized it (3,7) by solid phase. The synthetic replicate exhibited high potency and intrinsic activity to release immunoliclike ACTH and β-endorphin from pituitary cells maintained in monolayer cultures (3,8) or *in vivo* in several rat prepara-

tions (9). This peptide, which we named amunine (3,5), has the following sequence:

H-Ser-Gln-Glu-Pro-Pro-Ile-Ser-Leu-Asp-Leu-Thr-Phe-His-Leu-Leu-Arg-Glu-Val-Leu-Glu-Met-Thr-Lys-Ala-Asp-Gln-Leu-Ala-Gln-Gln-Ala-His-Ser-Asn-Arg-Lys-Leu-Leu-Asp-Ile-Ala-Ala-NH_2. So far, amunine has exhibited all the properties previously attributed to CRF. The main difficulties encountered by us and others ((3) and references therein) in carrying out this project were associated with (i) the bioassay, because a number of substances, including vasopressin and ACTH (10,11), can lead to false positive or negative; (ii) the complexity of the peptide/protein mixes generated during extraction; (iii) the occurrence of losses due to degradation (chemical as well as enzymatic) or adsorption to column supports or test tubes in which fractions were collected; (iv) the small amount of purified peptide that was available for sequence analysis; and finally, (v) the total synthesis for comparative chemical and biological studies. We wish to report here in detail the purification of ovine CRF using HPLC techniques developed in our laboratory (12,13).

---

[1] This paper was presented at the International Symposium on HPLC of Proteins and Peptides, November 16–17, 1981, Washington, D.C.

[2] Abbreviations used: ACTH, human adrenocorticotropin; CRF, corticotropin releasing factor; TEAP, triethylammonium phosphate; TEAF, triethylammonium formate; RP, reverse phase; TFA, trifluoroacetic acid; Gn·HCl, guanidine hydrochloride; LRF, luteinizing hormone releasing factor; F, sheep hyphothalami fragments.

EXPERIMENTAL

Both HPLC systems and buffer compositions (triethylammonium phosphate, TEAP; or formate, TEAF) have been described earlier (13). Columns, solvent, flow rate, temperature, and load are described in the figure legends illustrating each RP-HPLC purification step. Buffer A was the aqueous buffer; buffer B contained 60% CH_3CN in buffer A. When 0.1% TFA was used as buffer A, buffer B also contained 0.1% TFA, thus allowing for flat baselines at 210 nm. Chart speed was 1 cm/min. The wavelength at which the eluant was monitored is shown on the ordinate. Column cuts were shown above the abscissa and the active zone is hatched. The gradient shape is shown by a dotted line. Other parameters are reported in the figure legend. An in vitro assay based on the ability of a putative CRF to stimulate the secretion of corticotropin by primary cultures of rat anterior pituitary cells (3,11) was used to follow CRF-like activity. Direct ACTH radioimmunoassay of the fractions prior to incubation with the pituitary cells indicated which zones contained ACTH and helped us discriminate between ACTH-like molecules and CRF-like activities. Similarly, the secretory rate at maximal concentration of added substance allowed us to discriminate between CRF-like activity due to vasopressin versus that due to CRF (11).

RESULTS AND DISCUSSION

As part of a project directed toward the isolation of luteinizing hormone releasing factor (LRF) in the Laboratories for Neuroendocrinology at The Salk Institute, 490,000 sheep hypothalami (from now on referred to as *fragments,* abbreviated F) were extracted in a mixture of ethanol:acetic acid:chloroform, defatted with a mixture of ether:petroleum ether, and partitioned in an n-butanol:pyridine:0.1% acetic acid (5:3:11) system as reported by Burgus et al. (14). The lower phase (2 kg) of this last step contained a fraction of the total CRF-like activity exhibited by the defatted starting material. A succession of classical purification steps including ultrafiltration or dialysis against 2 N acetic acid and batch gel filtration on Sephadex G-50 eluted at 4°C with 2N acetic acid yielded two zones exhibiting CRF-like activity. The low-molecular-weight fraction eluting at 2 V_e/V_0 was later found to have the composition of [Arg^8]vasopressin. The large-molecular-weight fraction eluting at 1.3 V_e/V_0 was suspected to be CRF because it elicited a much higher secretory rate at maximum concentration of added substance than did [Arg^8]vasopressin and showed encouraging properties in a series of in vitro studies (11). Treatment of the larger molecular weight active fraction (guanidine hydrochloride (Gn · HCl)/AcOH, pH 2.5, at 90°C for 5 min, and chromatography on Bio-Gel P-10 using 4 M Gn · HCl/AcOH, pH 2.5, as eluant) gave one single peak of activity which was only partially retarded, thus excluding the possibility that the activity had been associated noncovalently with larger molecular weight carriers. The pool of this zone from several columns was the starting material for the RP-HPLC purification of ovine CRF. The protein content of this highly viscous pool was estimated to be 1.0 to 1.5 g/300,000 hypothalami on the basis of absorbance at 210 nm (0.75 ml of this stock solution corresponded to 1550 F).

Early attempts to purify CRF by RP-HPLC using μBondapak C_{18} or even μBondapak CN and 0.01 N NH_4OAc, pH 4.0/ CH_3CN buffer as described in (12) resulted in complete loss of CRF biological activity. This disappointing observation was the main incentive to search for a buffer system that would have minimal absorbance at 200 to 210 nm and be nontoxic or volatile so that bioassays could be carried out without any further manipulation of the samples except for lyophilization. The trialkylammonium buffers and TEAP (13), in particular, were found to indeed increase recovery as well as overall column performance. The role of the added alkylamine to the mobile phase was to competitively inhibit the participation of the solute in the ion-exchange or adsorption reactions with the nonbonded silanols on the stationary phase. This early observation was subsequently documented by Sokolowsky and Wahlund (15), who studied the effect of the addition of several alkylamines to the mobile phase on peak tailing and retention behavior of tricyclic antidepressant amines.

The first step in the HPLC purification scheme of ovine CRF was a rough separation with emphasis on recovery rather than resolution. Because more than 100 runs were necessary to process the active guanidine fraction from the P-10 gel-permeation column, emphasis was put on repeatability of the OD pattern so that each column would not have to be tested for biological activity. This goal was met, and a typical profile where biological activity was found in the hatched fraction (i.e., fraction 5 in the assay reported in Table 1) is shown in Fig. 1. The pool of the active zone of a hundred such runs generated a fraction which we labeled 28-125-00 (150,000 fragments in 8.75 ml). In this step Gn · HCl was removed and approximately 100-fold purification was achieved.

The fact that CRF eluted at a higher concentration of acetonitrile than did other releasing factors (see Table 2) at acidic pH was an indication of its overall hydrophobicity under those conditions. This may explain its propensity to adsorb to glassware, polypropylene tubes, and other surfaces including RP-HPLC packing materials. Addition of an organic modifier or, more adequately, keeping the CRF-containing HPLC fractions in polypropylene tubes at $-20°C$ in the original mobile phase that eluted them, allowed us to retain biological activity throughout our purification schemes.

Three independent schemes of purification using three different aqueous buffers/acetonitrile gradients and four different columns (some of which became available as the project progressed) could lead to a CRF preparation that would be pure enough for

TABLE 1

REPRESENTATIVE ASSAY OF COLUMN FRACTIONS GENERATED AS IN FIG. 1

Substance tested	Dose (F/ml)	ACTH secreted (pg/ml)	SE
Control	—	155	39
Starting material	1.0	2175	212
Column fractions[a]			
4	0.3	130	8
4	3.0	337	64
5	0.3	883	173
5	3.0	1850	54
6	0.3	217	31
6	3.0	527	11

[a] Fraction 3 tested in another assay was found to be inactive.

microsequence determination (i.e., >85% purity) (5). These are summarized in Figs. 2 through 4.

At a time when column characteristics were poorly understood and desired features for high resolution and recovery of peptides were neither recognized nor available, our attempt to isolate CRF from the 28-125-00 pool was mostly empirical. As mentioned earlier, however, we developed the TEAP buffer to partially solve the problem of free silanols. Many columns other than those reported in the figure legends, including Supelco Sil C_{18} (5 μm), DuPont ODS and CN (5 μm) columns, and several solvent systems made of different organic modifiers (CH_3CN, 1-propanol, 1-butanol, 2-propanol, or mixes thereof) and TEAP or NH_4OAc at different concentrations (0.1 to 0.5 N) or pH (2.25 to 7.2), were tested without success. In most cases either biological activity was lost or no separation was achieved. It is only when we used a μBondapak C_{18} (but not Supelco Sil or DuPont C_{18}) immersed in ice water, thereby following initial gel-filtration conditions, that the TEAP/CH_3CN buffer system would yield acceptable resolution and recovery of biological activity. Optimized gradient conditions on the C_{18} column are shown in Fig. 2A. Similarly, it took lowering of the

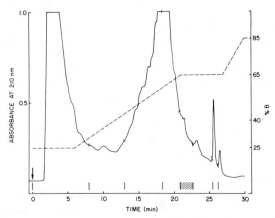

FIG. 1. First HPLC step to generate the pool 28-125-00. Load: 1550 F in 750 μl of the pool after Gn · HCl P-10 column. Conditions: μBondapak CN column (0.46 × 30 cm), buffer A: TEAP, pH 3.0; buffer B: 60% CH₃CN in A; flow rate: 1.5 ml/min at room temperature.

temperature to find isocratic conditions on the μBondapak CN columns (Fig. 2B) that would achieve some further purification. At that stage we became aware of the fact that

TABLE 2

HPLC RETENTION TIMES OF SYNTHETIC RELEASING FACTORS IN THREE SYSTEMS

Peptide[a]	System		
	I[b]	II[c]	III[d]
TRF	169	389	104
LRF	719	787	586
SS-14	937	1009	810
[Met(O)²¹]CRF	1147	1290	950
CRF	1189	1320	982

Note. Retention times in seconds.

[a] TRF, thyrotropin releasing factor; LRF, luteinizing hormone releasing factor; SS-14, somatostatin-14; CRF, corticotropin releasing factor.

[b] Vydac C₁₈ (5 μm) 0.46 × 25-cm column; buffer A: TEAP, pH 2.25; buffer B: CH₃CN (60%) in buffer A; flow rate: 2 ml/min; gradient 0–80% B (in 20 min); V_e = 3.3 ml.

[c] Same as sytem I; buffer A: 0.1% TFA; buffer B: 60% CH₃CN, 40% of 0.24% TFA.

[d] Same as system I; μBondapak CN (10 μm) 0.46 × 30-cm columns; V_e = 3.3 ml.

the nonvolatile TEAP buffer was interfering in the Edman degradation used in microsequencing. Desalting using the "volatile" TEAF buffer (13,16) at room temperature and a sharp gradient of CH₃CN is illustrated in Fig. 2C and yielded approximately 10 nmol of highly purified ovine CRF.

The main drawbacks of this original isolation scheme were a relatively low overall yield due to adsorption on the columns and the need for making sharp cuts which emphasized purity versus recovery. Because of our inability to monitor column effluent at 210 nm at the desalting step (CRF did not seem to adsorb at 255 or 280 nm and TEAF would not be transparent in the uv at the concentration used), we had to follow CRF by biological testing or amino acid analysis while arbitrarily fractionating the column effluent.

Most of these difficulties were later resolved by using large-pore-size silica (300–300 Å) that had been properly derivatized (C₁₈) and end capped, i.e., Perkin–Elmer HC-ODS SIL-X-1 (10 μm) (Fig. 3A and B) or Vydac (5 μm) (Fig. 4).

One should note that a fairly sharp cut of the active zone shown in Fig. 3A could be

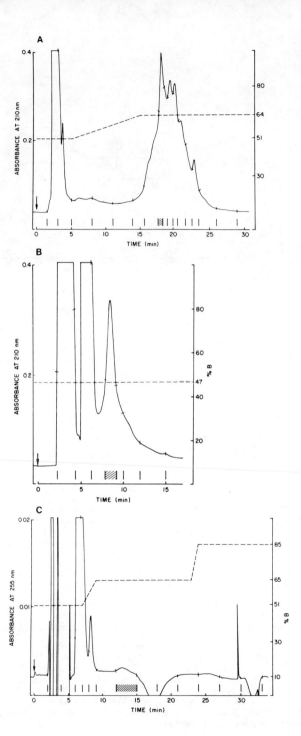

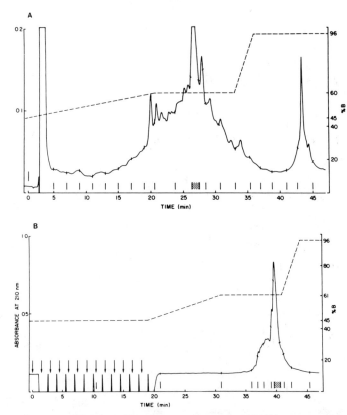

FIG. 3. The second scheme and two-step purification of CRF from pool 28-125-00. (A) Load: 1400 F in 80 µl of pool 28-125-00. Conditions: Perkin–Elmer ODS-HC SIL-X-1 column (0.26 × 25 cm); buffer A: TEAP, pH 2.25; buffer B: 60% CH₃CN in A; flow rate: 0.7 ml/min at room temperature. (B) Load: 15,000 F in 13 ml of a pool of the active fraction shown in (A). It was injected 1 ml at a time every 90 s in a Rheodyne injector having a 2-ml loop. Conditions: Perkin–Elmer ODS-HC SIL-X-1 column (0.26 × 25 cm); buffer A: 0.1% TFA; buffer B: 60% CH₃CN, 40% of 0.24% TFA; flow rate: 0.7 ml/min at room temperature.

further resolved by using a 0.1% TFA (Fig. 3B). Because the same column was being used in both cases, we do not think that this selectivity was due to differences in column load or gradient shape, but was rather attributable to the different eluotropic properties of TEAP versus 0.1% TFA buffer. This seems to hold true in many instances and has been

FIG. 2. The first scheme and three-step purification of CRF from pool 28-125-00. (A) Load: 2150 F in 125 µl of pool 28-125-00. Conditions: µBondapak C₁₈ (0.46 × 25 cm); buffer A: TEAP, pH 3.0; buffer B: 60% CH₃CN in A; flow rate: 1.2 ml/min at 0°C. (B) Load: 7500 F in 1 ml of pool of the active fraction shown in (A). Conditions: µBondapak CN (0.46 × 30 cm); buffer A: TEAP, pH 2.25; buffer B: 60% CH₃CN in A; flow rate: 1.2 ml/min at 0°C. (C) Load: 12,000 F in 1 ml of pool of the active fraction shown in (B). Conditions: µBondapak CN (0.46 × 30 cm); buffer A: TEAF, pH 3.0; buffer B: 60% CH₃CN in A; flow rate: 1.2 ml/min at room temperature.

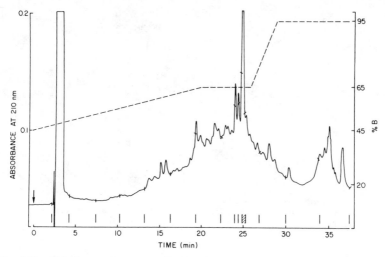

FIG. 4. The third scheme and one-step purification of CRF from pool 28-125-00. Load: 1400 F in 80 µl of pool 28-125-00. Conditions: Vydac C_{18} (5 µm) No. 1422S-3 column (0.45 × 25 cm); buffer A: 0.1% TFA; buffer B: 60% CH_3CN, 40% of 0.24% TFA; flow rate: 1.2 ml/min at room temperature.

taken advantage of in many purifications of synthetic peptides[3] as well as naturally occurring substances (4,16).

Whereas acceptable recovery of CRF-like activities could be obtained in the 100-Å (µBondapak) support by cooling the column (thus driving the equilibrium of mass transfer from the solid support toward the eluant), one could do the reverse and increase resolution (as illustrated in Figs. 3 and 4) by using end-capped C_{18}, 330-Å silica at room temperature. Higher resolution was obtained on 5-µm particles (Vydac) (Fig. 4) than on 10-µm particles (Perkin–Elmer) (Fig. 3) as one would theoretically expect. The use of solvent systems having a relatively high concentration of alkyl ammonium salt also be-

came less critical due to the absence of residual silanols. This allowed for the purification and desalting of 28-125-00 in one step using 0.1% TFA/CH_3CN as eluant (Fig. 4). Rechromatography of purified ovine CRF is shown in Fig. 5. Amino acid and sequence analyses of this material were consistent with a structure having 41 residues (3,5). A synthetic replicate $[Met(O)^{21}]CRF$ could be made (3,7) and exhibited the physicochemical and biological properties of the natural product as it was isolated.

CONCLUSION

The ability to follow and purify CRF activity from ovine hypothalami was the direct result of a reliable CRF assay and of a better understanding and constant improvement of HPLC technology. It was shown, indeed, that what may have seemed to be the most untractable mixture could be resolved provided that an optimum use of column and eluant characteristics be taken advantage of. At the present time, with the availability of hundreds

[3] Since our original report (13) we have found that TEAF only partially lyophilizes and leaves some colored residue (probably coming from the formic acid used since the TEA was extensively purified before use) after drying. This may detract from its usefulness in some cases, whereas in others, its selectivity versus that of TEAP allows for separations not otherwise possible.

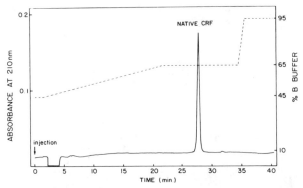

FIG. 5. Rechromatography of purified native ovine CRF from Fig. 4. Load: 1500 F of native ovine CRF in 1.2 ml: estimated amount when comparing integrated surface areas under peak with that of a synthetic standard at different loads was estimated to be 7 μg. Conditions: see Fig. 4.

of type of column from different manufacturers and the availability of several buffer systems, it became obvious that the conditions described here were not the only ones that could be used to obtain high recovery of a highly purified CRF preparation from ovine extracts. General characteristics of the column packings (such as C_{18} properly endcapped, large pore size, and small particle size) and compatible solvent systems that will exhibit different selectivities were, however, critical in the optimization of the separation which we had to achieve.

This CRF isolation project focused on purifying enough material for chemical characterization (amino acid analysis and sequencing). Because the starting material available to us only represented a small portion of the total activity in all fractions, we cannot address the issue of how much of the total CRF-like activity is represented by this 41-amino acid peptide. It is, however, an important question that we hope to answer by quantitative CRF radioimmunoassay of fresh sheep hypothalamic extracts (W. Vale, J. Vaughan, C. Douglas, G. Yamamoto, and J. Rivier, in preparation).

To minimize the amount of material to be worked up for biochemical analysis, we are still challenged to obtain a yield as high as possible during any purification scheme. This is complicated by the nature of the original extract, the presence of degradative enzymes, and chemical reactivity of certain amino acids toward oxidation, reduction, hydrolysis, etc. It is, however, tempting in view of recent developments in HPLC technology and availability of enzyme inhibitors and chemical scavengers to believe that the time will come when hypothalamic releasing factors or other brain peptides present in very low concentration could be isolated in high yield, thus precluding the necessity for the tedious work up of large quantities of starting material.

ACKNOWLEDGMENTS

Research supported in part by NIH Grants AM 26741, AM 20917, AM 18811, and HD 13527. Research conducted in part by The Clayton Foundation for Research, California Division. Dr. W. Vale, Dr. J. Spiess, Dr. C. Rivier, and Dr. J. Rivier are Clayton Foundation Investigators. We acknowledge the expert technical assistance of J. Vaughan, G. Yamamoto, C. Douglas, R. Wolbers, and R. McClintock, and thank L. Wheatley for manuscript preparation.

REFERENCES

1. Guillemin, R., and Rosenberg, B. (1955) *Endocrinology* 57, 599.

2. Saffran, M., and Schally, A. V. (1955) *Canad. J. Biochem. Physiol.* **33**, 408.

3. Vale, W., Spiess, J., Rivier, C., and Rivier, J. (1981) *Science* **213**, 1394–1397.

4. Rivier, J., Rivier, C., Branton, D., Millar, R., Spiess, J., and Vale, W. (1982) *in* Peptides: Synthesis, Structure, Function, Proceedings of the Seventh American Peptide Symposium (Rich, D. H., and Gross, E., eds.), pp. 771–776, Pierce Chemical Company, Rockford, Ill.

5. Spiess, J., Rivier, J., Rivier, C., and Vale, W. (1981) *Proc. Nat. Acad. Sci. USA* **78**, 6517–6521.

6. Spiess, J., Rivier, J., Rivier, C., and Vale, W. (1982) *in* Methods in Protein Sequence Analysis (Elzinga, M., ed.), pp. 131–138, Humana Press, New York.

7. Rivier, J., Spiess, J., Rivier, C., and Vale, W., in preparation.

8. Vale, W., Grant, G., Amoss, M., Blackwell, R., Guillemin, R. (1972) *Endocrinology* **91**, 562.

9. Rivier, C., Brownstein, M., Spiess, J., Rivier, J., and Vale, W. (1982) *Endocrinology,* **110**, 272–278.

10. McCall, S. M. (1957) *Endocrinology* **60**, 664; Por-

tanova, R., and Sayers, G. (1973) *Proc. Soc. Exp. Biol. Med.* **143**, 661; Schally, A. V., *et al.* (1978) *Biochem. Biophys. Res. Commun.* **82**, 582 (1978); Knudsen, J., *et al.* (1978) *ibid.* **80**, 735.

11. Vale, W., and Rivier, C. (1977) *Fed. Proc. Fed. Amer. Soc. Exp. Biol.* **36**, 8.

12. Burgus, R., and Rivier, J. (1976) *Peptides,* pp. 85–94.

13. Rivier, J. (1978) *J. Liq. Chromatogr.* **1**, 343–367.

14. Burgus, R., Amoss, M., Brazeau, P., Brown, M., Ling, N., Rivier, C., Rivier, J., Vale, W., and Villarreal, J. (1976) *in* Hypothalamus and Endocrine Functions pp. 355–372.

15. Sokolowski, A., and Wahlund, K.-G. *J. Chromatogr.* **189**, 299–316.

16. Marki, W., Spiess, J., Tache, Y., Brown, M., and Rivier, J. (1981) *J. Amer. Chem. Soc.* **103**, 3178–3185.

17. Rivier, J., Spiess, J., Perrin, M., and Vale, W. (1979) *in* Biological/Biomedical Applications of Liquid Chromatography II. (Hawk, G., ed.), pp. 223–241 Dekker, New York.

Isolation and Purification of *Escherichia coli* Heat-Stable Enterotoxin of Porcine Origin[1]

R. Lallier,* F. Bernard,* M. Gendreau,† C. Lazure,‡§ N. G. Seidah,‡ M. Chrétien,‡ and S. A. St-Pierre†§[2]

Département de Pathologie et Microbiologie, Faculté de Médecine Vétérinaire, Université de Montréal, St-Hyacinthe; †Département de Physiologie et Pharmacologie, Faculté de Médecine, Université de Sherbrooke; Sherbrooke; ‡Institut de Recherche Clinique, 110 ouest Avenue des Pins, Montréal; and §Chercheur Boursier, Conseil de la Recherche en Santé du Québec, Québec, Canada

This paper describes the isolation and the purification of *Escherichia coli* heat-stable enterotoxin. The toxin was produced by the *E. coli* strain F11 (P155) of porcine origin. It was purified to homogeneity by successive operations of desalting on XAD-2 resin, gel filtration on Bio-Gel P4, and high-pressure liquid chromatography on cyanopropyl and octadecylsilane columns. After chemical characterization by quantitative amino acid analysis and determination of its sequence, the heat-stable enterotoxin was shown to be a monomeric octadecapeptide with the following primary structure: H-Asn-Thr-Phe-Tyr-Cys-Cys-Gly-Leu-Cys-Cys-Asn-Pro-Ala-Cys-Ala-Gly-Cys-Tyr-OH.

Various strains of *Escherichia coli* with enterotoxigenic activity have been shown to cause watery diarrhea in neonatal animals (1–6), human infants (7), and travelers (8) following colonization of the small bowel and release of enterotoxins. Two main groups of enterotoxins have been described so far. One of the toxins is a heat-labile (LT),[3] immunogenic, high-molecular-weight protein showing a mechanism of action and antigenic properties that are similar to the *Vibrio cholerae* enterotoxin (9). Another toxin (ST) appeared to be a nonimmunogenic, lower molecular weight, heat-stable molecule (10). It appears in reports from various investigators that ST is produced by all enterotoxigenic strains, whereas only certain strains of porcine or human origin can secrete LT (11).

Conditions that are required for the production and purification of the heat-labile toxin have been determined (12–14) by several authors. However, the purification of ST was made difficult by the complexity of the various growth media used for toxin production and by the laborious and expensive pig gut loop assay used for its detection. Since then, the suckling mouse (15) and the rabbit gut loops tests (16) have been reported and rapid progress has followed.

Several reports have been published during the last 10 years concerning the preparation, isolation, and characterization of substances displaying ST activity (11,17,18). Most of these reports tended to show that ST has a proteinic nature and contained several cysteine residues. However, the material used for the characterization of ST was heterogeneous and the results so obtained were misleading. However, Gianella and collaborators have recently described the isolation and purification to apparent homogeneity (19) of a substance displaying ST activity, using classical column chromatography tech-

[1] This paper was presented at the International Symposium on HPLC of Proteins and Peptides, November 16–17, 1981, Washington, D. C.

[2] To whom correspondence should be addressed.

[3] Abbreviations used: LT, heat-labile enterotoxins; ST, heat-stable enterotoxins; CA, Casamino Acids; YE, yeast extract.

niques. Later on, the toxin, originating from a *E. coli* strain of human origin, was characterized and shown to contain 18 amino acids, including six cysteine residues (20).

This paper describes the isolation, purification, and chemical characterization of heat-stable enterotoxin produced by a strain of *E. coli* of porcine origin. The final purification steps were performed by high-pressure liquid chromatography.

MATERIALS AND METHODS

Bacterial Strain and Growth

E. coli strain F11 (P155) was derived from nonenterotoxigenic avian strain F11 (serotype 018ab:k?:H14) by conjugation with the porcine strain P155 from H. W. Smith (10). This strain has the ability to produce both heat-labile and heat-stable enterotoxins.

The production of enterotoxins was tested in either a complex or a semisynthetic medium. Brain heart infusion (Difco Laboratories) was used as the complex medium. The semisynthetic medium (CA) was prepared as described by Finkelstein (21) and contained the following (in grams per liter): Na_2HPO_4, 5.0; K_2HPO_4, 5.0; NH_4Cl, 1.18; Na_2SO_4, 0.089; $MgCl_2 \cdot 6H_2O$, 0.042; $MnCl_2 \cdot 4H_2O$, 0.004; $FeCl_3 \cdot 4H_2O$, 0.005; Casamino Acids (Difco), 30.0. The pH of this medium was 7.3. CA–YE medium was CA medium containing 0.6% yeast extract (Difco). These media were prepared both with and without glucose.

Growth was performed in brain heart infusion with agitation for 4 h at 37°C. Then, 300 ml of the same medium used for fermentation was inoculated with 0.3 ml of the 4-h culture. After overnight incubation without agitation at 37°C, 250 ml were transferred into 5 liters of fresh medium in a Model 19 fermentor (New Brunswick Scientific Co.). Except when otherwise noted, the cells were allowed to grow for 7 h at 37°C with vigorous agitation (500 rpm) and forced aeration (5 liters/min), and the pH was automatically controlled. Samples were withdrawn at hourly intervals for measurement of growth and enterotoxin production. Growth was evaluated by optical density at 540 nm and by plate counts. Cells were removed by centrifugation at 10,000g for 20 min.

Concentration and Desalting

The cell-free solution (5 liters) containing the toxin was cooled down to 4°C. The cold solution was stirred vigorously while 3.25 kg of $(NH_4)_2SO_4$ was added in small portions until the salt was completely dissolved. The toxin was allowed to precipitate overnight in the cold room at 4°C. Most of the supernatant was decanted and the concentrated suspension centrifuged at 12,000g for 45 min. An aliquot of the supernatant was then tested in the suckling mouse assay (15) and shown to contain no significant enterotoxigenic activity. The brown solid obtained after centrifugation was collected and dissolved in water (500 ml). An aliquot of this solution was tested in the suckling mouse assay to evaluate the amount of isolated enterotoxin.

The toxin solution was passed over a column (2.5 × 40 cm) containing 200 g of XAD-2 resin (Bio-Rad). The resin was first washed with water (3 × 1 liter) and then with mixtures of methanol:acetic acid 1% containing increasing proportions of methanol, as previously described by Gianella *et al.* (19). The toxin was eluted out of the resin with the mixture 99% MeOH–1% acetic acid (500 ml). A sample of this solution was tested for ST activity in the suckling mouse assay. The methanol was evaporated *in vacuo* to a light brown solid.

Gel Filtration

The crude toxin solution (5 ml) was applied to a Bio-Gel P4 (Bio-Rad) column (2.5 × 90 cm) equilibrated with 5% acetic acid and operated at a flow rate of 12 ml/h. The eluate was collected in 10-ml fractions. The elution pattern was monitored at 270 nm. The tubes containing enterotoxigenic activity were detected by the suckling mouse test. These tubes were pooled and the solvent lyophilized.

High-pressure Liquid Chromatography

After gel filtration, the toxin was further purified by two successive operations of semi-preparative high-pressure liquid chromatography (HPLC), using a Waters system. The apparatus consisted of two Model 6000A pumps, a Model UK6 injector, and a Model 660 programmer for gradient elution. Detection of components was performed using a Schoeffel Model 770 variable wavelength flow spectrophotometer, operated at 210 nm.

(i) HPLC on cyanopropylsilane columns. The crude toxin was completely separated from colored contaminants, using a 0.4 × 30 cyanopropylsilane column, equilibrated with a 0.2 M triethylamine phosphate buffer at pH 3.0. The flow rate was at 1.5 ml/min and the components were detected at 210 nm. The toxin was eluted with a 25-min linear gradient from 0 to 80% acetonitrile, as shown in Fig. 1. The tubes corresponding to the active fraction were detected with the suckling mouse test and pooled. Acetonitrile was eliminated by evaporation *in vacuo* on a flash evaporator and the aqueous solution containing triethylamine phosphate lyophilized.

(ii) HPLC on octadecylsilane columns. The phosphate-containing toxin was then applied on a Waters μBondapak C_{18} (0.7 × 30 cm) column equilibrated with 0.01 M ammonium acetate buffer at pH 4.15. The toxin was eluted with a 1-h linear gradient from 0 to 100% methanol at a flow rate of 2 ml/min, as shown in Fig. 2. The active fraction was detected by

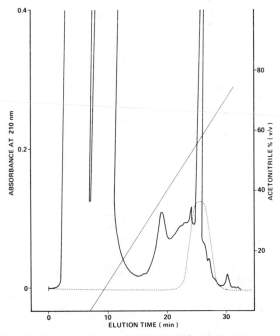

FIG. 1. HPLC purification of crude enterotoxin (following gel filtration) on cyanopropylsilane (Waters), 0.4 × 30 cm, CH_3CN/0.2 M triethylamine phosphate buffer, pH 3.0, 1.5 ml/min. —, Absorbance at 210 nm; . . . , biological activity (suckling mouse test).

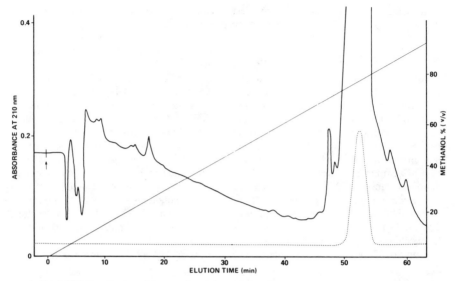

FIG. 2. HPLC purification of enterotoxin activity peak from Fig. 1 on octadecylsilane (μBondapak C$_{18}$, Waters), 0.7 × 30 cm, MeOH/0.01 M ammonium acetate buffer, pH 4.15, 2 ml/min. —, Absorbance at 210 nm; . . . , biological activity (suckling mouse test).

the suckling mouse test and the corresponding tubes were pooled together. Methanol was evaporated *in vacuo* and the aqueous solution lyophilized.

(iii) HPLC analysis of the purified toxin. A 20-μg aliquot of the toxin was dissolved in water, the sample was applied to the 0.7 × 30-cm μBondapak C$_{18}$ column, and the chromatography was run under the same conditions as described above.

Test Systems for Enterotoxin Detection

(i) Suckling mouse test. Swiss albino mice (1–3 days old) were separated from their mothers just before use and randomly divided into groups of three. Samples containing ST were assayed by making twofold dilutions in physiological saline (containing 1 drop of 2% Evan's blue dye), and 0.1 ml of each sample was administered via the oroesophagal route. After the mice were maintained at room temperature for 4 h, they were

killed and their responses were determined by the method of Dean *et al.* (15). The last dilution that gave a response greater than 0.09 was considered as the endpoint and was expressed as the number of units per 0.1 ml.

(ii) Rabbit gut loop test. Jejunal loops, as described by Larivière *et al.* (22), were used. Young adult rabbits weighing 1.5 kg were starved for 24 h before surgery. Samples (2 ml) were injected into 4- to 6-cm loops, and the rabbits were sacrificed after 6 h. Only results from rabbits with appropriate positive (response greater than 0.5 ml of fluid/cm of loop) and negative control loops were accepted.

Characterization of the Toxin

(i) Amino acid analysis. Amino acid analysis was performed on the native peptide following 22 h hydrolysis at 105°C in 5.7 N HCl in the presence of 0.1% mercaptoethanol. The separation of the amino acids was done on a modified 120C Beckman

amino acid analyzer using a Beckman W3 resin (23).

(ii) Microsequencing. Starting from 45 μg of the native peptide, the cysteine residues were reduced with dithiothreitol and alkylated using [^{3}H]iodoacetic acid (New England Nuclear), according to Crestfield *et al.* (24). The ^{3}H-labeled peptide was sequenced on an updated Beckman Model 890B instrument using a 0.3 M Quadrol program in the presence of 3.0 mg Polybrene and 2.5 mg sperm whale apomyoglobin, as described in another publication (25).

RESULTS

Preliminary experiments were done with brain heart infusion broth cultures. Forced aeration and agitation considerably increased ST production, and maximal yield of ST was obtained after 7 h of incubation. Approximately the same amount of ST was produced in CA medium without glucose as in brain heart infusion medium. However, a high variation in toxin production was observed with bacterial cultures grown in different batches of Casamino Acids, but the bacterial cell growth was unaffected by the batches. CA medium was chosen as a starting medium, but each new lot of Casamino Acids was carefully controlled for ST production, and only batches with which the production of toxin was high were retained. Other conditions necessary for optimizing toxin production have been described in a previous article (26). The highest yield of ST was obtained by using CA–YE medium containing 0.2% glucose and maintained between pH 7.2 and 7.8 in a fermentor under forced aeration and agitation for 7 h.

Concentration of ST from the cell-free supernatant by acetone precipitation was unsuccessful, giving at best less than 20% recovery of enterotoxigenic activity. On the other hand, precipitation of ST out of the supernatant with 90% ammonium sulfate overnight, in the cold (4°C), allowed for a nearly quantitative (90–95%) recovery of the activity (1.1 $\times$ 10^6 mouse units).

At this point, various attempts to desalt the toxin solution were made. Ultrafiltration steps were lengthly and shown to be mostly useless. The most successful technique for that purpose was similar to the one recently described by Staples *et al.* (19) and used adsorption on XAD-2 hydrophobic resin. After this step, the salt and most of the dark-brown pigmented material were eliminated, leaving the toxin contaminated mostly with proteinic substances. The latter inactive material was separated from the active toxin fraction by gel-filtration step on Bio-Gel P4. Two major peaks were then obtained: the first one, displaying no significant activity, near V_0; and the second, active one, around 0.75 V_c (26). After lyophilization, a light-brown solid was obtained that contained 9 $\times$ 10^5 mouse units of enterotoxigenic activity. From this point, various attempts to further purify the toxin by using classical chromatographic techniques mostly led to loss of material without significant improvement of specific activity. The semipreparative purification of the crude fraction of ST obtained after gel filtration on Bio-Gel P4 is shown in Figs. 1 and 2. Two consecutive operations of reverse-phase HPLC had to be used to maximize separation and recovery of active material and to remove the last traces of colored impurities from the toxin fraction. The first step of HPLC purification using a cyanopropylsilane analytical (0.4 $\times$ 30-cm) column and a triethylamine phosphate–acetonitrile solvent system is illustrated in Fig. 1. It is seen that the biological activity is contained in the area of a sharp peak coming out at 26-min elution time, at a 60% concentration of acetonitrile. This rather high concentration of organic solvent indicates the hydrophobic character of the biologically active molecule. Large amounts of polar impurities (elution time up to 15 min) are separated from ST during this chromatography step, including the remnants of colored contaminants. The nature of these contaminants has not been investigated. Furthermore, the recovery of biologically active material has not been cal-

248

LALLIER ET AL.

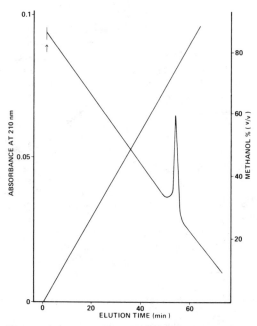

FIG. 3. Analytical HPLC of enterotoxin obtained as shown in Fig. 2 on μBondapak C₁₈, 0.4 × 30 cm. Same conditions as in Fig. 2.

culated after this step because of the large amount of salt that could bring misleading results.

A final step of reverse-phase HPLC using a C₁₈ semipreparative column (0.7 × 30 cm) afforded analytically pure material, as shown in Figs. 2 and 3. This time, methanol was used as the organic solvent, together with a volatile buffer containing ammonium acetate, to avoid further desalting operations. Once again, the fraction containing ST is eluted at a high percentage of solvent and comes out as a sharp peak, with minor impurities on both sides. Quantity was sacrificed for quality, and only the fraction contained in the center of that peak was conserved for characterization, to yield 710 μg (corresponding to approx 6.5 × 10⁵ units of ST activity) of homogeneous fluffy white material (see Fig. 3). Although it contained slight amounts of contaminants (as demon-

strated by HPLC), the rest of the material corresponding to the active peak shown in Fig. 2 displayed full ST specific activity. This fraction was not used for chemical characterizaton, but was pooled, from one batch to another, for recycling or for immunization purposes. Moreover, it was not accounted for in the evaluation of the overall yield. Our calculation of the yield at each purification step is summarized in Table 1. Only the recovery in terms of mouse units has been considered in this calculation, the weight having been judged as a misleading factor because of the presence at every intermediate step of various amounts of salts originating from the buffers. The overall yield for the complete purification of ST, starting from the cell-free supernatant, thus corresponds to approximately 50%.

The amino acid composition of the native peptide is shown in Table 2. It can be seen

TABLE 1

SUMMARY OF THE STEPS USED FOR THE PRODUCTION, ISOLATION, AND PURIFICATION
OF *E. coli* HEAT-STABLE ENTEROTOXIN B

Conditions	Activity[a] (mouse units)
Five liters CA–YE medium containing 0.2% glucose, pH 7.2–7.8, 37°C; forced aeration and agitation; 7 h.	1.2×10^6
Centrifugation of the cells at 10,000g for 20 min.	1.2×10^6
Precipitation of the toxin (and other proteins) with $(NH4)_2 SO_4$ at 4°C; centrifugation of the solid at 12,000g, 45 min.	1.1×10^6
Desalting of the toxin solution (500 ml H_2O) on XAD-2.	1.1×10^6
Gel filtration on Bio-Gel P4 in 5% acetic acid.	9×10^5
Reverse-phase HPLC on cyanopropylsilane (0.2 M triethylamine phosphate buffer, pH 3.0–acetonitrile)	ND[b]
Reverse-phase HPLC on octadecylsilane (0.01 M ammonium acetate buffer, pH 4.15–methanol)	6.5×10^5 [c]

[a] As determined by the method of Dean *et al.* (15).
[b] Not determined because of the presence of interfering nonvolatile salts.
[c] Activity also determined by the rabbit gut loop system (22).

that the molecule is presumed to be 18 amino acids long, to contain Cys residues, and not to contain any Val, Met, Ile, Lys, His, nor Arg residues. The proposed sequence of ST is shown in Fig. 4. This sequence agrees well with the amino acid composition given in Table 2 and provides evidence for Tyr[18] as being the C-terminal residue. Moreover, C-terminal Tyr[18] was confirmed after hydrazinolysis of a toxin sample, according to the procedures of Bradbury (27).

DISCUSSION

This paper describes the preparation and the purification of a heat-stable enterotoxin secreted by the porcine strain F11 (P155) of *E. coli*. Standard procedures were established to obtain maximum yields of the enterotoxin by optimizing conditions for growth media, isolation, and purification of the toxin. These conditions are summarized in Table 1, together with the yield in mouse units at each step, for a 5-liter batch of medium.

After adsorption on XAD-2 resin and gel

filtration, the light-brown-colored powder containing the toxin was first purified by semipreparative reverse-phase HPLC on a

TABLE 2

AMINO ACID COMPOSITION OF THE REDUCED AND
ALKYLATED TOXIN

Amino acid	Analysis[a]	Nearest integer	Found sequence
Asx	2.25	2	2
Thr	0.9	1	1
Glx	1.1	1	1
Pro	0.9	1	1
Gly	1.2	1	1
Ala	2.0	2	2
Leu	1.1	1	1
Tyr	1.5	2	2
Phe	1.0	1	1
Cm-Cys	6.1	6	6
Presumed total		18	18

[a] After 22 h hydrolysis at 105°C using HCl 5.7 N and 0.1% mercaptoethanol. Values expressed according to underlined amino acid.

H - Asn - Thr - Phe - Tyr - Cys - Cys - Glu - Leu - Cys -
    1   2   3   4   5   6   7   8   9
Cys - Asn - Pro - Ala - Cys - Ala - Gly - Cys - Tyr - OH
 10   11   12   13   14   15   16   17   18

FIG. 4. Proposed complete amino acid sequence of heat-stable enterotoxin of porcine origin.

cyanopropylsilane column, out of which colored impurities eluted in less than 15 min at 0% acetonitrile, and the fraction containing ST, 15 min later, at 60% acetonitrile, indicating the hydrophobic character of the peptide. Up to this point of the purification of ST, the yield in terms of activity was nearly quantitative. As can be seen in Fig. 1, the peak containing ST activity in the chromatogram is poorly resolved and the corresponding active region is widespread on each side of that peak. The fraction that was selected for further purification work corresponded to the sharp peak (25 min elution time) only, as shown in Fig. 1. The shoulders on both sides of the peak were temporily discarded for later use, even though they contained a significant amount of activity. However, we are still not certain at this point whether this activity corresponds to ST_a only or to a mixture of ST_a and ST_b, E. coli strain F11 (P155) being known to produce both these forms.

The resulting colorless solid was applied to a C_{18} reverse-phase column in an ammonium acetate buffer–methanol system and purified to apparent homogeneity, as demonstrated by analytical HPLC in the same system (Fig. 3). Once again, the last purification step afforded important losses of ST, since only a thin fraction of the activity peak (see Fig. 2) was selected for toxin characterization. Starting with 5 liters of medium, the weight of pure ST_a, obtained after this last purification step was 710 μg, which corresponds to an overall yield of ST of approximately 50%, as based upon the original activity content of the supernatant solution. The activity of porcine ST_a in terms of mouse units thus corresponds to more than 900 units/μg.

From its physical and biological characteristics, the molecule corresponds to the heat-stable toxin ST_a (methanol-soluble, mouse positive) previously described by Burgess et al. (28). Quantitative amino acid analysis and microsequencing work have demonstrated the homogeneity of the material. On the other hand, attempts to assign the three disulfide bridges of ST have so far been unsuccessful. Although a limited number of possibilities exists for the pairing of the cysteine residues, total synthesis of ST has to wait for these positions to be known. In the meanwhile, the large quantities of ST that are necessary for biological studies and, eventually, for immunization purposes have to be supplied by isolating the natural product. Methods are presently being developed in our laboratory for scaling up the production of pure porcine ST_a and heat-stable enterotoxins from other strains of human, bovine, and porcine origin.

ACKNOWLEDGMENTS

This work has been supported by grants from the Natural Sciences and Engineering Research Council of Canada (NSERC) and from the Medical Research Council of Canada (MRC). We thank Mrs. Danielle Laurendeau for typing the manuscript.

REFERENCES

1. Gyles, C. L. (1971) Ann. N. Y. Acad. Sci. 176, 314–322.
2. Kohler, E. (1968) Amer. J. Vet. Res. 29, 2263–2274.
3. Kohler, E. (1971) Ann. N. Y. Acad. Sci. 176, 212–219.
4. Smith, H. W., and Gyles, C. (1971) J. Med. Microbiol. 3, 403–409.
5. Smith, H. W., and Linggood (1972) J. Med. Microbiol. 5, 243–250.
6. Whipp, S., Moon, H., and Lyon, N. (1975) Infect. Immun. 12, 240–244.
7. Ryder, K., Wachsmith, I., Buxton, A., DuPont, H., Mason, E., and Barrett, F. (1976) N. Engl. J. Med. 295, 849–853.
8. Morris, G., Merson, M., Sack, D., Wells, J., Martin, W., De Witt, W., Feeley, J., Sack, R., and Bessudo, D. (1976) J. Clin. Microbiol. 3, 486–495.
9. Gyles, C. L. (1974) J. Infect. Dis. 129, 277–283.
10. Smith, H. W., and Gyles, C. L. (1970) J. Med. Microbiol. 3, 387–401.

11. Robertson, D. C., and Alderete, J. F. (1980) *in* Cholera and Related Diarrheas, 43rd Nobel Symposium (Ouchterlony, O., and Holmgren, J. H., eds), pp. 115–126, Karger, Basel.

12. Domer, F. (1975) *J. Biol. Chem.* **250,** 8712–8719.

13. Finkelstein, R. A., La Rue, M. K., Johnston, D. W., Vasil, M. L., Cho, G. J., and Jones, J. R. (1976) *J. Infect. Dis.* **133,** S120–S137.

14. Shenkein, I., Green, R. F., Santos, D. S., and Maas, W. K. (1976) *Infect. Immun.* **13,** 1710–1720.

15. Dean, A. G., Ching, Y. C., Williams, R. G., and Harder, L. B. (1972) *J. Infect. Dis.* **125,** 407–411.

16. Evans, D. G., Evans, D. J., Jr., and Pierce, N. F. (1973) *Infect. Immun.* **7,** 873–880.

17. Alderete, J. F., and Robertson, D. C. (1978) *Infect. Immun.* **19,** 1021–1030.

18. Kapitany, R. A., Forsyth, G. W., Scott, A., Mckenzie, S. F., and Worthington, R. W. (1979) *Infect. Immun.* **26,** 173–179.

19. Staples, S. J., Asher, S. E., and Gianella, R. A. (1980) *J. Biol. Chem.* **255,** 4716–4721.

20. Chan, S. K., and Gianella, R. A. (1981) *J. Biol. Chem.* **256,** 7744–7746.

21. Finkelstein, R. A., Atthasampunna, P., Chulasamaya, M., and Charunmethee, P. (1966) *J. Immunol.* **96,** 440–449.

22. Larivière, S., Gyles, C. L., and Barnum, D. A. (1972) *Canad. J. Comp. Med.* **36,** 319–328.

23. Fauconnet, M., and Rochemont, J. (1978) *Anal. Biochem.* **91,** 403–409.

24. Crestfield, A. M., Moore, S., and Stein, W. H. (1963) *J. Biol. Chem.* **238,** 622–627.

25. Lazure, C., Seidah, N. G., Chretien, M., Lallier, R., and St-Pierre, S., submitted.

26. Lallier, R., Larivière, S., and St-Pierre, S. (1980) *Infect. Immun.* **28,** 469–474.

27. Bradbury, J. H. (1958) *Biochem. J.* **68,** 482–486.

28. Burgess, M. N., Bywater, R. J., Cowley, C. M., Mullan, N. A., and Newsome, P. M. (1978) *Infect. Immun.* **21,** 526–539.

α-N-Acetyl-β-Endorphin$_{1-26}$ from the Neurointermediary Lobe of the Rat Pituitary: Isolation, Purification, and Characterization by High-Performance Liquid Chromatography[1]

H. P. J. BENNETT,[2] C. A. BROWNE, AND S. SOLOMON

Endocrine Laboratory, Royal Victoria Hospital, and Departments of Biochemistry and Experimental Medicine, McGill University, Montreal, Quebec, Canada

The neurointermediary lobes from 190 rat pituitaries were homogenized in an acidic medium which inhibits peptidase activity and maximizes the solubilization of undamaged peptides. Octadecylsilyl-silica (ODS-silica) was used to extract the supernatant of the tissue homogenate. The ODS-silica eluate, now largely protein and salt free, was subjected to reversed-phase high-performance liquid chromatography (HPLC) employing 0.1% trifluoroacetic as counter ion. The column eluates were monitored for β-endorphin immunoreactivity. Five immunoreactive components were observed. The most hydrophobic of these was repurified on the same HPLC column using 0.13% heptafluorobutyric acid as counter ion. Characterization of the purified peptide by gel permeation HPLC, amino acid analysis, and tryptic fragmentation indicated that it corresponded in structure to α-N-acetyl-β-endorphin$_{1-26}$. Amino acid analysis of the native peptide and its trypsin and carboxypeptidase fragments indicated that an alanyl residue occupies position 26. This finding is in contrast to the sequence predicted for the β-lipotropin/ corticotropin precursor by recombinant DNA techniques which suggests that the 26th residue of the β-endorphin molecule should be valine.

In recent years many important advances have been made in the use of reversed-phase high-performance liquid chromatography (HPLC)[3] for the isolation and purification of both natural and synthetic peptides and proteins (1). This progress was made possible by the availability of efficient reversed-phase column packing materials and the introduction of novel ion-pairing reagents and tissue extraction techniques. We have recently used an efficient tissue extraction procedure using

octadecylsilyl-silica (ODS-silica) in combination with reversed-phase HPLC to purify peptides from the rat pituitary (2,3). The high efficiency of the extraction procedure and the resolving power of HPLC facilitated discovery of phosphorylated forms of corticotropin (ACTH) (3) and corticotropin-like intermediary lobe peptide (4), a diacetylated form of α-melanotropin (3), and a γ_3-melanotropin with an unexpected lysyl residue at position one (5). One characteristic of reversed-phase HPLC of peptides is that the order of elution is frequently unrelated to molecular weight. This paper illustrates how gel permeation HPLC can be used to provide this information. We also show how reversed-phase HPLC can be used to identify the products of carboxypeptidase digestion. These various HPLC techniques have been brought together in an overall strategy in the isolation and characterization of a variant of β-endorphin from

---

[1] This paper was presented at the International Symposium on HPLC of Proteins and Peptides, November 16–17, 1981, Washington, D. C.

[2] To whom reprint requests should be addressed: Endocrine Laboratory, Room L2.05, Royal Victoria Hospital, 687 Pine Avenue West, Montreal, Quebec H3A 1A1, Canada

[3] Abbreviations used: HPLC, high-performance liquid chromatography; ODS-silica, octadecylsilyl-silica; TFA, trifluoroacetic acid, HFBA, heptafluorobutyric acid; BSA, bovine serum albumin.

253

the intermediary lobe of the rat pituitary. The existence of this peptide had previously been predicted by a biosynthesis study (6) and has recently been isolated from rat neurointermediary lobe tissue (7). Structural analysis suggests that the amino acid sequence of this peptide differs from that predicted by recombinant DNA techniques (8).

MATERIALS AND METHODS

HPLC solvents were prepared as described previously (3,9). C_{18} Sep Pak cartridges, C_{18} μBondapak, and I-125 HPLC columns and a Waters Associates HPLC system were all obtained from Waters Scientific, Mississauga, Ontario, Canada. Column eluates were monitored continuously for uv absorbance at 210 nm and 278 nm using LC 55 and LC 75 variable wavelength spectrophotometers connected in series (Perkin–Elmer, Montreal, Quebec, Canada).

Peptide Isolation and Purification

Male Sprague–Dawley rats (250–400 g, Canadian Hybrid Farms, Stanstead, Quebec, Canada) were decapitated and the pituitaries dissected *in situ* into anterior and neurointermediary lobes. Pituitary tissue was collected and extracted in batches of 10 to 60 lobes over a period of 3 months. Reversed-phase extraction of tissue homogenates with ODS-silica cartridges and purification of peptides by reversed-phase HPLC using trifluoroacetic acid (TFA) and heptafluorobutyric (HFBA) as hydrophobic counter ions were performed exactly as described previously (2,3,5).

Characterization of Peptides

Molecular weight determination. A Waters Associates I-125 gel permeation HPLC column was used to determine the molecular weight of the purified peptide. The column was eluted at a flow rate of 1 ml/min with 40% aqueous acetonitrile containing 0.1% TFA. Calibration was achieved by injection of a mixture of 0.5 μg of each of the following

in 25 μl of 40% acetonitrile containing 0.1% TFA: bovine serum albumin (BSA), bovine cytochrome c (both from Sigma), ovine β-endorphin (from Drs. Ling and Guillemin), and somatostatin and met-enkephalin (both from Ayerst Laboratories). In order to determine the retention characteristics of the isolated endorphin-related peptide, a 25-μl portion of the final product was injected (i.e., 0.75 μg).

Trypsin and carboxypeptidase Y digestions. One-eighth (i.e., 18 μg) of the isolated endorphin was dried for trypsin (Sigma) digestion. In parallel digestions this material and synthetic ovine β-endorphin (70 μg) were incubated at 37°C for 18 h in 100 μl of 50 mM ammonium bicarbonate buffer (pH 8.4) with a peptide to enzyme ratio of 50:1 (w/w). One quarter (i.e., 36 μg) of the isolated endorphin was dried for carboxypeptidase Y (Sigma) digestion. This material was incubated in 100 μl of 50 mM trisodium citrate buffer (pH 5) for 2.5 h at 25°C with a peptide to enzyme ratio of 50:1 (w/w). Digestion products were separated by reversed-phase HPLC.

Total enzymatic digestion. One-eighth (i.e., 18 μg) of the isolated endorphin was subjected to total enzymatic digestion by a mixture of carboxypeptidase Y, trypsin, and protease, exactly as previously described (4). A parallel digestion of 25 μg of synthetic ovine β-endorphin was also undertaken. Digestion products were analyzed by reversed-phase HPLC using a μBondapak C_{18} column eluted isocratically with 0.1% TFA at a flow of 1.5 ml/min. The column eluate was monitored at 280 nm. Under these conditions tyrosine eluted between 11 and 12 min and N-acetyltyrosine eluted between 34 and 36 min. Acid hydrolysis of peptide samples were prepared and their amino acid compositions determined as previously described (3).

RESULTS

The neurointermediary lobes from 190 rat pituitaries were homogenized and extracted using ODS-silica as described under Materials and Methods. This initial extract was subjected to reversed-phase HPLC using a solvent system containing 0.1% TFA (Fig. 1). Four major peaks of β-endorphin immunoreactivity were observed. Peak 4 was purified further using the same HPLC column but this time using a solvent system containing 0.13% HFBA (Fig. 2). The main peak of β-endorphin immunoreactivity was found to correspond to a uv-absorbing peak (*) which was almost completely resolved from several other uv-absorbing materials. Complete purification of the endorphin-related peptide was achieved by rechromatographing this peak using the same column eluted with solvents containing 0.1% TFA (Fig. 3). The amino acid composition of the purified endorphin-related peptide shown in Table 1 indicated that the primary sequence is very similar to the 1–26 sequence of bovine β-endorphin (i.e., β-endorphin lacking the carboxyl-terminal five amino acids). Molecular weight determination using the Waters I-125 column confirmed that the isolated peptide was considerably smaller than β-endorphin$_{1-31}$ (Fig. 4). The calculated value of 2400 was even smaller than the value expected for β-endorphin$_{1-26}$ of 2800. Further character-

ization of this endorphin-like peptide was achieved through trypsin and carboxypeptidase digestion. Figure 5 shows the reversed-phase HPLC of the fragments resulting from trypsin digest of standard synthetic ovine β-endorphin$_{1-31}$ (upper panel) and the purified endorphin-related peptide (lower panel). Each fragment was subjected to amino acid analysis including those components which were unretained by the column (i.e., fragments 1 and 6). The amino acid compositions (Table 1) and elution positions (Fig. 5) of fragments 3 and 5 are identical to those of 7 and 9, respectively. This indicates that the sequence β-endorphin$_{10-24}$ is identical in both molecules. The results indicate that the isolated endorphin differs from β-endorphin$_{1-31}$ in modifications to both termini. The amino acid compositions of fragments 4 and 8 are identical and correspond to β-endorphin$_{1-9}$. However, the elution position of fragment 8 is considerably later than fragment 4. This strongly suggests that the isolated endorphin is N-acetylated at the amino-terminal tyrosine residue as has been reported for several endorphins isolated from rat neurointermediary lobe tissue (6,7,10). The nature of the tyrosine released from the

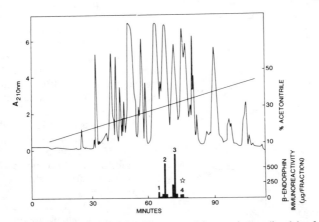

FIG. 1. Reversed-phase HPLC of an ODS-silica extract of the neurointermediary lobes from 190 rat pituitaries. The column was eluted with a solvent system containing 0.1% TFA throughout. The continuous line shows the uv absorbance at 210 nm and the bars show β-endorphin immunoreactivity in micrograms per fraction.

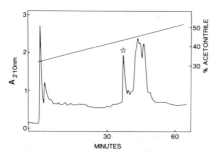

FIG. 2. Reversed-phase HPLC of fractions 76–79 (i.e., peak 4) from the initial chromatography of an extract of 190 rat neurointermediary lobes (Fig. 1). The column was eluted with a solvent system containing 0.13% HFBA throughout. The main peak of β-endorphin immunoreactivity was observed in fractions 41 and 42 which correspond to the peak marked with an asterisk.

isolated endorphin upon total enzymatic digestion was compared with that released from synthetic ovine β-endorphin. The synthetic ovine endorphin released uv-absorbing material which coeluted by reversed-phase HPLC with standard free tyrosine. The isolated endorphin gave rise to material which co-eluted with N-acetyl-tyrosine. A peptide corresponding to fragment 2 (β-endorphin$_{25-28}$) is absent from the digest of the isolated rat endorphin. Fragments 1 and 6, which were unretained by the HPLC column, correspond to β-endorphin$_{29-31}$ and β-endor-

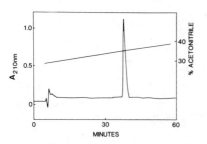

FIG. 3. Final purification of Peak 4 which had been partially purified from an extract of 190 rat neurointermediary lobes (Fig. 2). The reversed-phase HPLC column was eluted with a solvent system containing 0.1% TFA throughout.

phin$_{25-26}$, respectively. These results indicate that the carboxyl-terminal five amino acids are missing from the isolated rat endorphin.

Carboxypeptidase Y digestion of the isolated endorphin provided more positive information about the carboxyl-terminal sequence. (Fig. 6) The difference between fragments 14 and 15 (the undigested peptide) is a single alanyl residue and this clearly places this amino acid at the carboxyl terminus. Fragments 11, 12, and 13 resulted from endopeptidase digestion of the isolated endorphin and correspond to the 18–26, 1–13, and 1–14 sequences, respectively. The amino acid composition of the fraction unretained by the column (i.e., fragment 10) indicated that it contained a mixture of the 15 to 17 sequence of β-endorphin and free alanine and leucine. The tripeptide again appears to be a product of endopeptidase activity while the free alanine and leucine result from carboxypeptidase digestion of residues 14 and 26, respectively. The peptide mapping information is summarized in Fig. 7. These results indicate that the structure of the isolated rat endorphin is α-N-acetyl-β-endorphin$_{1-26}$.

DISCUSSION

Reversed-phase chromatographic techniques have been used to isolate a variant of β-endorphin from rat pituitary tissue (Figs. 1, 2, and 3). The usefulness of these techniques for isolating pituitary peptides has been discussed fully elsewhere (2,3). Two further applications of HPLC are illustrated in this study. These developments are the use of gel-permeation HPLC for molecular sizing and the use of reversed-phase HPLC for analyzing the products of carboxypeptidase Y digestion. They form part of a general approach for the rapid isolation and characterization of pituitary peptides.

While reversed-phase HPLC is the method of choice for peptide purification, retention times are not a reliable index of molecular weight. Recently, Rivier has demonstrated the usefulness of gel-permeation HPLC col-

TABLE I

AMINO ACID COMPOSITIONS OF THE ISOLATED ENDORPHIN AND THE TRYPSIN AND CARBOXYPEPTIDASE FRAGMENTS IN FIGS. 5 AND 6

		Asx	Thr	Ser	Glx	Prc	Gly	Ala	Val	Met	Ile[a]	Leu	Tyr	Phe	His	Lys
Isolated endorphin-related peptide (peak 4)	(α-N-acetyl-β-endorphin$_{1-26}$)[b]	2.1	2.9	1.7	2.2	1.1	1.7	2.0	1.2	0.8	1.4	2.0	1.0	2.0	—	3.0
Tryptic fragments of synthetic β-endorphin$_{1-31}$ (Fig. 5)	1. (β-endorphin$_{29-31}$)	1.0	—	—	1.1	—	0.9	1.0	—	—	—	—	—	—	—	1.0
	2. (β-endorphin$_{25-28}$)	0.9	—	—	—	—	—	1.0	—	—	—	—	—	—	0.9	1.0
	3. (β-endorphin$_{20-24}$)	—	—	0.8	—	—	—	1.0	—	—	1.4	—	—	—	—	1.0
	4. (β-endorphin$_{1-9}$)	—	1.0	0.8	1.1	1.0	1.9	—	—	0.8	—	—	1.0	1.0	—	1.0
	5. (β-endorphin$_{10-19}$)	—	1.6	0.9	1.1	1.0	—	—	1.1	0.8	—	1.8	—	1.0	—	1.0
Tryptic fragments of the isolated endorphin (Fig. 5)	6. (β-endorphin$_{25-26}$)	0.9	—	—	—	—	—	1.1	—	—	—	—	—	—	—	—
	7. (β-endorphin$_{20-24}$)	1.0	—	—	—	—	—	1.0	—	—	1.4	—	—	—	—	1.0
	8. (α-N-acetyl-β-endorphin$_{1-9}$)	—	1.0	0.8	1.2	—	1.9	—	—	0.5	—	—	1.0	1.1	—	1.0
	9. (β-endorphin$_{10-19}$)	—	1.8	0.9	1.4	1.1	—	—	1.1	—	—	2.0	—	1.0	—	1.0
Carboxypeptidase Y fragments of the isolated endorphin (Fig. 6)	10. (β-endorphin$_{15-17}$) + Ala, Leu	—	0.9	—	—	—	—	0.9	1.1	—	—	1.3	—	—	—	—
	11. (β-endorphin$_{18-26}$)	2.2	—	—	—	—	1.6	1.9	—	—	1.5	—	—	1.1	—	1.9
	12. (α-N-acetyl-β-endorphin$_{1-13}$)	—	2.1	1.6	2.1	1.4	1.8	—	—	0.7	—	—	1.0	1.1	—	0.9
	13. (α-N-acetyl-β-endorphin$_{1-14}$)	—	2.0	2.3	2.0	1.0	1.8	—	—	0.7	—	1.5	0.9	1.1	—	0.8
	14. (α-N-acetyl-β-endorphin$_{1-25}$)	2.1	3.1	1.8	2.1	1.3	1.6	0.9	1.1	0.8	1.5	2.1	0.9	2.0	—	3.0
	15. (α-N-acetyl-β-endorphin$_{1-26}$)	2.2	3.2	2.0	2.1	1.4	1.5	2.1	1.2	0.8	1.5	2.2	0.9	2.1	—	3.0

[a] Isoleucine values are low due to poor hydrolysis of the hydrophobic Ile$_{22}$–Ile$_{-3}$ sequence.
[b] Probable identity of each peptide.

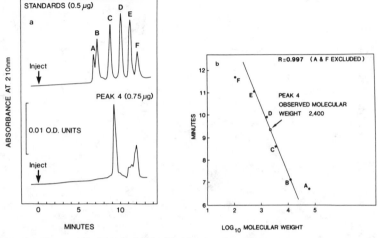

FIG. 4. Gel permeation of HPLC of the purified β-endorphin related peptide (peak 4) (a, lower panel). The column (Waters-I-125) was eluted with 40% acetonitrile containing 0.1% TFA as described under Materials and Methods. The column was calibrated (a, upper panel) with (A) bovine serum albumin, (B) bovine cytochrome c, (C) ovine β-endorphin$_{1-31}$, (D) somatostatin, and (E) methionine enkephalin. (F) TFA or the salt peak. (b) The molecular weight of the purified endorphin was estimated using the retention times of standards B, C, D, and E.

umns eluted with triethylammonium phosphate and formate buffers containing acetonitrile (11). In contrast to his experience, we have found that the Waters I-125 gel-permeation HPLC column eluted with 40% aqueous acetonitrile containing 0.1% TFA provides an excellent system for estimating the molecular weights of peptides and small proteins in the range of 500 to 15,000. Like Rivier (11), we find that this system underestimates the molecular weights of peptides with pronounced secondary structure (e.g., insulin and the neurophysins). The molecular constraints brought about by the large number of sulfur bridges presumably reduces the overall molecular diameters of these peptides compared with so-called linear peptides such as β-endorphin. This effect is reflected in the operating range suggested by the manufacturer for the Waters I-125 column i.e., 2000 to 80,000 M_r for globular proteins and 1000 to 30,000 M_r for random coil proteins. These molecular weight ranges were esti-

mated for aqueous solvent systems. The triethylamine buffer systems (11) and the TFA solvent system described here contain acetonitrile and both alter the properties of the I-125 column such that proteins of molecular weight above 60,000 (e.g., bovine serum albumin) are excluded and the operating range is extended downwards (to approximately 500 M_r for the TFA system). Rivier (11) has suggested that this shift in operating range is due to solvation effects which decrease the effective pore size of the column and/or increase the apparent molecular diameter of the peptides. Despite this alteration in column performance the linear relationship between retention time and log molecular weight still holds (Fig. 4b). We have found that the TFA solvent system generates highly symmetrical peaks (Fig. 4a). It is probably that the combination of the acetonitrile and the ion-pairing properties of TFA inhibit nonpolar and polar interactions of peptides with the column support. Gel-permeation

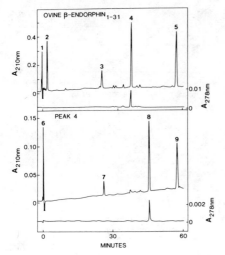

FIG. 5. Tryptic fragments of synthetic ovine β-endorphin₁₋₃₁ (upper panel) and the purified rat endorphin-related peptide (lower panel) separated by reversed-phase HPLC. In each instance the column was eluted over 1 h with a linear gradient of from 0.8 to 24% acetonitrile containing 0.1% TFA throughout at a flow rate of 1.5 ml per minute. The upper trace in each panel shows uv absorbance at 210 nm and the lower trace shows absorbance at 278 nm. Amino acid compositions of fragments 1 to 9 are given in Table 1.

chromatography predominates and is not complicated by other types of interaction which would result in peak tailing.

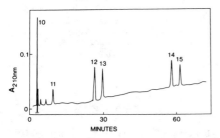

FIG. 6. Products of carboxypeptidase Y digestion of the purified rat endorphin-related peptide were separated by reversed-phase HPLC. The column was eluted over 1 h with a linear gradient of from 20 to 40% acetonitrile containing 0.1% TFA throughout at a flow rate of 1.5 ml per minute. The continuous line shows uv absorbance at 210 nm. Amino acid compositions of fragments 10 to 15 are given in Table 1.

The molecular weight estimate for the isolated endorphin using this technique (Fig. 4b) indicated that the peptide was considerably smaller than β-endorphin₁₋₃₁. The amino acid composition of the peptide and of its tryptic peptides (Fig. 5 and Table 1) suggested that it resembled bovine β-endorphin₁₋₂₆ with an alanine residue at position 26. According to the amino acid sequence predicted by recombinant DNA techniques (8) the carboxyl-terminal amino acid of rat β-endorphin₁₋₂₆ should be valine. We have used the resolving power of reversed-phase HPLC to separate the products of carboxypeptidase Y digestion of the isolated endorphin. The results clearly show that the carboxyl-terminal residue is alanine and not valine (Fig. 6 and Table 1). The carboxyl-terminus of this peptide appears not to be very labile to carboxypeptidase Y digestion. In a similar experiment using porcine ACTH as substrate it was found that the last 9 or 10 residues were released from this peptide under the same digestion conditions (Bennett, Browne, and Solomon, unpublished observations). Partly because the carboxypeptidase digestion is slow, an alarming degree of endopeptidase activity was observed in the present study (Fig. 6, and Table 1). Because the endorphin structure is very familiar we were still able to interpret the results. Our experience suggests that nearly all amino acids have either a positive or negative effect upon retention time (12). Reversed-phase HPLC should therefore be generally useful in interpreting amino- and carboxypeptidase digestions. However, care should be taken to check for endopeptidase contamination in exopeptidase preparations.

The tryptic fragmentation pattern of the isolated peptide suggested that the amino-terminal tyrosine residue was acetylated and this was confirmed by further analysis of the free amino acid by reversed-phase HPLC. While assuming that the core of the isolated endorphin is identical to that of other mammalian endorphins we concluded that the isolated peptide was α-*N*-acetyl-β-endor-

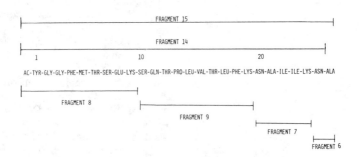

FIG. 7. Tryptic and carboxypeptidase Y peptide map of isolated rat α-N-acetyl-β-endorphin$_{1-26}$. The primary sequence 1 to 25 has been assumed to be identical to the corresponding bovine, ovine, and camel peptide.

phin$_{1-26}$ (Fig. 7 and Table 1). It constitutes yet another acetylated form of β-endorphin. Acetylated forms of β-endorphin$_{1-31}$ and β-endorphin$_{1-27}$ have previously been purified from porcine pituitaries (10). The question of the existence of some of these β-endorphin derivatives has recently become quite controversial. In a biosynthesis study, Mains and Eipper found that β-endorphin$_{1-31}$, acetyl-β-endorphin$_{1-31}$, acetyl-β-endorphin$_{1-27}$, and acetylβ-endorphin$_{1-26}$ are the major opiate peptides synthesized by the rat intermediary pituitary (6). In another study Zakarian and Smyth have suggested that nonacetylated forms of β-endorphin$_{1-27}$ and $_{1-26}$ are also major products of this tissue (13). Using the techniques described in this paper we are presently characterizing the other β-endorphin immunoreactive peptides shown in Fig. 1. Preliminary results indicate that peak 1 is β-endorphin$_{1-31}$, peak 2 is acetyl-β-endorphin$_{1-31}$, and peak 3 is acetyl-β-endorphin$_{1-27}$. Nonacetylated forms of β-endorphin$_{1-27}$ and $_{1-26}$ appear to be minor products of the rat intermediary pituitary. These findings are in general agreement with the biosynthesis study (6).

This study has illustrated how pituitary peptides can be isolated, purified, and char-acterized using only HPLC. These techniques are ideally suited for characterizing closely related peptides differing only in post-translational modifications.

ACKNOWLEDGMENTS

The authors would like to thank Susan James and Isabel Lehmann for their expert technical assistance. They are indebted to Dr. M. van der Rest and Mr. E. Wan of the Shriners Hospital, Montreal for performing the amino acid analyses and to Dr. N. Ling and Dr. R. Guillemin of the Salk Institute, La Jolla, California, and Dr. R. Deghenghi of Ayerst Laboratories, Montreal, for supplying synthetic peptide standards. This work was supported by Medical Research Council of Canada Grants MT-1658 and MA-6733, Fond de la Recherche en Santé du Québec Grant 800205, U. S. Public Health Service Grant HDO4365. H. P. J. Bennett is a recipient of a scholarship from the Fond de la Recherche en Santé du Québec.

REFERENCES

1. Hearn, M. T. W., and Hancock, W. S. (1979) *Trends Biochem. Sci.* **4**, N58–N62.
2. Bennett, H. P. J., Hudson, A. M., Kelly, L., McMartin, C., and Purdon, G. E. (1978) *Biochem. J.* **175**, 1139–1141.
3. Bennett, H. P. J., Browne, C. A., and Solomon, S. (1981) *Biochemistry* **20**, 4530–4538.
4. Browne, C. A., Bennett, H. P. J., and Solomon, S. (1981) *Biochemistry* **20**, 4538–4546.

5. Browne, C. A., Bennett, H. P. J., and Solomon, S. (1981) *Biochem. Biophys. Res. Commun.* **100,** 336–343.

6. Eipper, B. A., and Mains, R. E. (1981) *J. Biol. Chem.* **256,** 5689–5695.

7. Evans, C. J., Weber, E., and Barchas, J. D. (1981) *Biochem. Biophys. Res. Commun.* **102,** 897–904.

8. Drouin, J., and Goodman, H. M. (1980) *Nature (London)* **288,** 610–613.

9. Bennett, H. P. J., Browne, C. A., and Solomon, S. (1980) *J. Liquid Chromatog.* **3,** 1353–1365.

10. Zakarian, S., and Smyth, D. G. (1979) *Proc. Nat. Acad. Sci. USA* **76,** 5972–5976.

11. Rivier, J. E. (1980) *J. Chromatog.* **202,** 211–222.

12. Browne, C. A., Bennett, H. P. J., and Solomon, S. (1982) *Anal. Biochem.* **124,** 201–208.

13. Zakarian, S., and Smyth, D. G. (1982) *Nature (London)* **296,** 250–252.

Index

A

α-*N*-Acetyl-β-endorphin$_{1-26}$, purification and
   isolation by HPLC, 253–261
ACTH peptides, HPLC of, 67–71
Adenovirus type 2, proteins encoded by,
   HPLC studies on, 73–80
Adenylsuccinate synthetase
   HPIEC of, 5
   HPLC of acidic isozyme of, 189–193
Albumin, HPIEC of, 5
Alcohol dehydrogenase isoenzymes, HPLC of,
   39–46
Alkaline phosphatase(s)
   HPIEC of, 5
   as protein probe, physical parameters of,
     158
Allergic peptide, fluorescent detection of, 227
Amino acids, retention coefficients for, 202
Aminopropyltrimethoxy silane (APS), as
   chromatographic exchanger, 1
Amyloid P-component, reverse-phase HLPC
   of, 29–39
Amunine, purification by HPLC, 233–241

Analytical liquid chromatograph, serum
   apolipoprotein studies using, 119–127
Angiotensin II, fluorescent detection of, 227
Apolipoproteins
   HPIEC of, 5
   serum, preparative size-exclusion
     chromatography of, 119–127
Aryl sulfatase isoenzymes, HPIEC of, 5

B

Bart's hemoglobin
   detection of, 17
     by HPLC, 20–22
Bombesin, fluorescent detection of, 227
Bovine serum albumin (BSA)
   HPLC of, reverse-phase, on large-pore
     silicas, 162–172
   in studies of silicas for HPLC, 86–93

C

Carbonic anhydrase
   HPIEC of, 5
   HPLC of, 97–100

Cation-exchange resins, protein recovery from, 96

Caudate nuclei, leucine enkephalin in, HPLC studies on, 211–219

Ceruloplasmin
  glycopeptides
    amino acid sequence of, 51
    HPLC of, 47–54

Chymotrypsin purification, by reverse-phase HPLC, 23–27

Chymotrypsinogen, HPIEC of, 5

Circular dichroism of immunoglobulin G peptides, 10

Column length, effects on HPIEC, 4

Competitive labeling, HPLC studies of, 103–109

Conalbumin, as protein probe, physical parameters of, 158

Corticotropin-like intermediary lobe peptides, HPLC of, 67–71

Corticotropin-releasing factor, see Amunine

Creative kinase
  HPIEC of, 5
  isoenzymes, HPIEC of, 5

Cytochrome c, HPIEC of, 5

D

Desacetylthymosin α_1, preparation of, 174

E

Echis carinatus venom enzyme, prothrombin activation by, HPLC studies on, 181–188

Edman degradation, protein microsequencing by solid-phase type of, 195–209

Elongation factor Tu of E. coli, HPLC in microsequencing of, 199

Enkephalins, HPLC studies on, 211–219

Enterotoxin from porcine E. coli, 243–251

Epidermal growth factor (EGF), transforming growth factor dependent on, HPLC of, 111–118

Escherichia coli, porcine, isolation of enterotoxin from, 243–251

F

Field-desorption mass spectrometry, leucine-enkephalin studies using, 211–219

Fluorescent techniques, in selective detection of peptides, 221–233

1-Fluoro-2,4-dinitrobenzene, competitive labeling by, HPLC studies of, 103–109

N^a-Formyldesacetylthymosin, amino acid sequence of, 174

G

Gel permeation high-performance liquid chromatography, polypeptide purification by, 141–149

Globin fragments, in studies of silicas for HPLC, 82

Glucagon, chemical properties of functional groups of, HPLC of, 103–109

Glycopeptides
  of ceruloplasmin and immunoglobulin D, HPLC of, 47–54

Glycoproteins, HPLC of, solid-phase Edman degradation in, 198–209

Glycylglycylglycine, fluorescent detection of, 227

H

Hemoglobin(s)
  abnormal, cord blood screening for, 17–22
  HPIEC of, 5
  as protein probe, physical parameters of, 158
  detection of, 17
    by HPLC, 20, 22

Hemoglobin D Punjab, HPLC separation of, 22

Hemoglobin F
  detection of, 17
    by HPLC, 20–22

Hemoglobin G Philadelphia
  HPLC separation of, 19, 20
  neonatal screening for, 20–22

Hemoglobin S
  detection of, 17
    by HPLC, 22

Heptafluorobutyric acid (HFBA), peptide isolation using, 65–71

Hexokinase isoenzymes, HPIEC of, 5

High-performance ion-exchange chromatography (HPIEC)
  of adenylsuccinate synthesase isozyme, 189–193

packing materials for, 2–3
pore-diameter effects in, 3–4
of proteins, 1–7
list of, 5
High-performance liquid chromatography (HPLC)
of abnormal hemoglobins, 17–22
of α-N-acetyl-β-endorphin$_{1-26}$, 253–261
of adenovirus type 2-encoded proteins, 73–80
of alcohol dehydrogenase isoenzymes, 39–46
of amunine, 233–241
of ceruloplasmin glycopeptides, 47–54
competitive labeling studies using, 103–109
of enkephalins, 211–219
gel permeation type, see Gel permeation, HPLC
of immunoglobulin D glycopeptides, 47–54
of immunoglobulin G peptides, 9–16
in peptide purification by protein microsequencing by solid-phase Edman degradation, 195–209
of proteins, 189–197
variables in, 95–101
reverse phase
of amyloid P peptides, 29–39
of chymotrypsin and trypsin, 23–27
in peptide isolation, 65–72
High-performance size-exclusion chromatography
calibration standards for, 61
of hydrolyzed plant proteins, 55–64
HPIEC, see High-performance ion-exchange chromatography
HPLC, see High-performance liquid chromatography
Human growth hormone peptides, fluorescent detection of, 230–231
Hypothalami, leucine enkephalin in, HPLC studies on, 211–219

I

Immunoglobulin D (IgD)
glycopeptides
amino acid sequence of, 53
HPLC of, 47–54
Immunoglobulin G (IgG)
HPIEC of, 5
peptic digestion of, 11–16
peptides, HPLC of, 9–16

Insulin
HPIEC of, 5
HPLC of, 97–100
Insulin B chain, fluorescent detection of, 227
Interferon, HPIEC of, 5

K

Kentsin, fluorescent detection of, 227

L

Lactate dehydrogenase
HPIEC of, 5
isoenzymes, HPIEC of, 5
β-Lactoglobulin
HPIEC of, 5
HPLC of, 97–100
Leucine-enkephalin
fluorescent detection of, 227
measurement using HPLC, 211–219
β-Lipotropin peptides, HPLC of, 67–71
Lipoxygenase, HPIEC of, 5
Loading capacity, effects on HPIEC, 4
Lysozyme
HPIEC of, 5
as protein probe, physical parameters of, 158

M

Meizothrombin, formation of, 183
α-Melanotropin (α-MSH) peptides, HPLC of, 67–71
Met-enkephalin, fluorescent detection of, 227
Mobile phases, effects on HPIEC, 6
Mobile-phase velocity, effects on HPIEC, 4–6
Monoamine oxidase, HPIEC of, 5
Myoglobin
HPIEC of, 5
as protein probe, physical parameters of, 158
size exclusion behavior of, 154, 155, 156
Myosin, HPLC of, reverse-phase, on large-pore silicas, 162–172

N

Neoglycopeptides, HPLC of, 173–180
Neurophysins
amino acid sequences of, 134
reverse-phase HPLC of, 129–139

Neurotensin, fluorescent detection of, 227

O

Organic resins, as HPIEC ion exchanger, 2
Ovalbumin
    HPIEC of, 5
    HPLC of, reverse-phase, on large-pore
        silica, 162–172
    as protein probe, physical parameters of,
        158
    size-exclusion behavior of, 154, 155
    in studies of silicas for HPLC, 86–93
Oxytocin, fluorescent detection of, 227

P

Peptides
    isolation by HPLC, 65–72
P-component
    primary sequence of, 34
    reverse-phase HLPC of peptides of, 29–39
Packing materials, for HPIEC, 2–3
Pepsin, as protein probe, physical parameters
        of, 158
Peptides
    fluorescent detection of, 221–233
    of immunoglobulin G, HPLC of, 9–16
pH, effects on size-exclusion chromatography
        of proteins, 151–159
Phosphorylase b, HPLC of, reverse-phase, on
        large-pore silicas, 162–172
Pigs, enterotoxin from E. coli of, 243–251
Pituitary polypeptides, purification by gel
        permeation HPLC, 142–148
Polistes mastoparan, fluorescent detection of,
        227
Polyethylenimine (PEI), as HPIEC ion
        exchanger, 3
Polypeptides, purification by gel permeation
        HPLC, 141–149
Pore diameter, effects on HPIEC, 3–4
Pressinoic acid, fluorescent detection of, 227
Proctolin, fluorescent detection of, 227
Pro-opiomelanocortin (POMC), purification by
        gel permeation HPLC, 146–148
Proteins
    HPIEC of, 1–7
    HPLC, 189–193
    variables in, 95–101

reverse-phase HPLC of, on large pore
    silicas, 161–172
size-exclusion chromatography of, pH
    effects on, 151–159
Protein microsequencing, by solid-phase
    Edman degradation, 195–209
Prothrombin
    activation of, size-exclusion HPLC of,
        181–188
    nomenclature of, 182

R

Ranatensin, fluorescent detection of, 227
Reverse-phase high-performance liquid
        chromatography
    of amyloid P peptides, 29–39
    of chymotrypsin and trypsin, 23–27
    on large pore silicas, 161–172
    of neurophysins, 119–127
    peptide isolation by, 65–72
    silicas used in, evaluation, 81–94
    of transforming growth factor, 111–118

S

Serum apolipoproteins, preparative size-
        exclusion chromatography of, 119–127
Sickling syndromes, neonatal screening for, 17
Silicas
    large pore, protein separation on, 161–172
    used in HPLC, evaluation of, 81–94
Size-exclusion chromatography (SEC)
    high-performance, of hydrolyzed plant
        proteins, 55–64
    preparative, of human serum
        apolipoproteins, 119–127
    of proteins, pH effects on, 151–159
Somatostatin, fluorescent detection of, 227
Soy protein hydrolyzed, HPSEC of, 57–63
Soybean trypsin inhibitor, HPIEC of, 5
Spheron, as HPIEC ion exchanger, 2
Surface-modified ion exchangers, for
        HPIEC, 3

T

α-Thalassemia, neonatal screening for, 17, 22
β-Thalassemia, neonatal screening for, 21
Tonin, purification by gel permeation HPLC,
        145–146

Transforming growth factor (TGF)
  HPLC of, 95, 100–101
  purification of, 96–97
    by HPLC, 111–118
Trifluoroacetic acid (TFA), peptide isolation
    using, 65–71
Trypanosoma congolense, variable surface
    glycoproteins of, HPLC in
    microsequencing of, 199–209
Trypsin
  HPIEC of, 5
  purification by reverse-phase HPLC, 23–27
Tryptylglycine, fluorescent detection of, 227
Tuftsin, fluorescent detection of, 227
Tyrosylglycine, fluorescent detection of, 227

U

Umbilical cord, blood screening for
    hemoglobin disorders by, 17–22

V

Variable surface glycoproteins (VSG), of
    Trypanosoma congolense, HPLC in
    microsequencing of, 199–209

X

Xenopsin, fluorescent detection of, 227